Physical Methods
for Inorganic
Biochemistry

BIOCHEMISTRY OF THE ELEMENTS

Series Editor: Earl Frieden
Florida State University
Tallahassee, Florida

Volume 1   BIOCHEMISTRY OF NONHEME IRON
Anatoly Bezkorovainy

Volume 2   BIOCHEMISTRY OF SELENIUM
Raymond J. Shamberger

Volume 3   BIOCHEMISTRY OF THE ESSENTIAL
ULTRATRACE ELEMENTS
Edited by Earl Frieden

Volume 4   BIOCHEMISTRY OF DIOXYGEN
Lloyd L. Ingraham and Damon L. Meyer

Volume 5   PHYSICAL METHODS FOR
INORGANIC BIOCHEMISTRY
John R. Wright, Wayne A. Hendrickson,
Shigemasa Osaki, and Gordon T. James

A Continuation Order Plan is available for this series. A continuation order will bring delivery of each new volume immediately upon publication. Volumes are billed only upon actual shipment. For further information please contact the publisher.

Physical Methods for Inorganic Biochemistry

John R. Wright
Southeastern Oklahoma State University
Durant, Oklahoma

Wayne A. Hendrickson
Columbia University
New York, New York

Shigemasa Osaki
Hybritech, Inc.
San Diego, California

and

Gordon T. James
University of Colorado Health Sciences Center
Denver, Colorado

PLENUM PRESS • NEW YORK AND LONDON

375081

Library of Congress Cataloging in Publication Data

Main entry under title:

Physical methods for inorganic biochemistry.

(Biochemistry of the elements; v. 5)
Bibliography: p.
Includes index.
1. Spectrum analysis. 2.Bioinorganic chemistry—Technique. 3. Biomolecules—
Analysis. I. Wright, John R. (John Ricken), 1939–   . II. Series.
QP519.9.S6P48   1986   574.19′214   85-28191
ISBN 0-306-42049-X

© 1986 Plenum Press, New York
A Division of Plenum Publishing Corporation
233 Spring Street, New York, N.Y. 10013

Printed in the United States of America

IN MEMORIAM

The year 1983 marked the passing of Jon Adam Wright, father of John R. Wright, and Shizuko Osaki, beloved wife of Shigemasa Osaki. This volume is dedicated to them.

Preface

This volume is intended for students and professionals in diverse areas of the biological and biochemical sciences. It is oriented to those who are unfamiliar with the use of physical methods in studies of the biological elements. We hope the reader will find the material a helpful reference for other volumes of this series as well as the general literature, and some may see ways to adopt these techniques in their own pursuits. Every effort has been made to avoid an abstruse presentation.

It should be clear that one individual cannot be expert in all the disciplines considered here (and the authors recognize that fact with sincere humility). As may be expected of an introductory reference, most of our attention was focused on the commonly used methods. To balance this, we have included a few examples of approaches which are promising but relatively undeveloped at this time. Also, an emphasis has been placed on element selectivity. It is impossible to envision the course of future events, and a volume which deals with instrumentation is especially prone to become outdated. Nevertheless, any valid approach to a scientific question should be applicable indefinitely.

Since there are four authors, the portions prepared by each should be identified. Chapter 6 (X-ray diffraction) was written by Wayne Hendrickson, Ph.D. Gordon James, Ph.D., prepared Chapters 11 (microprobe methods) and 12 (neutron activation analysis), while Shigemasa Osaki, Ph.D., contributed Chapter 10 (kinetic methods). John Wright, Ph.D., is responsible for the remaining material. We especially wish to recognize our wives for their support throughout this effort.

Shigemasa Osaki wishes to express his gratitude to Miss J. Blackburn, Mr. R.C. DeLeon, and Mrs. L.W. Kates for their assistance in the preparation of Chapter 10. John Wright is indebted to Barbara Wright, Cathi Craig, and Pat James for their help in typing the bulk of the manuscript, and the effort was aided by facilities made possible through National Institutes of Health Grant RR08003 (an institutional grant). Wayne Hendrickson thanks Steven Sheriff, Janet Smith, and Richard Honzatko for critical reading of Chapter 6. Steven Sheriff made especially valuable contributions to Table 6-2. He is also grateful to Janet Smith for Figure 6-6, to Jane Richardson for Figure 6-7, and to Alwyn Jones for Figure 6-8. Wayne Hendrickson's portion of the manuscript was supported in part by a grant, GM29548, from the National Institutes of Health, and it was written while he was a member of the Laboratory for the Structure of Matter at the Naval Research Laboratory in Washington, D.C.

Contents

Introduction

In attempting to write a book treating the physical methods and their application to the special problems associated with the biological elements (Raymond, 1977), several questions were raised about what should be included. All of the known techniques are applicable to some degree, but if one conducts even a cursory inspection of the literature it will be apparent that there has been no lack of imagination in applying both old and new tools to the study of biomolecules and their interactions. A respectable literature accumulates rapidly in all areas, and the total effect is a deluge of information. Broad treatment of all these subjects goes well beyond the space and purpose of a single book.

Careful consideration of the types of evidence one may obtain from the various approaches will cause one to realize that all methods are not equal in their ability to yield unambiguous information about the behavior of a given type of atom or a group of atoms within a biomolecule; that is to say, some methods are more specific than others. (Some techniques, e.g., infrared spectroscopy, are virtually prohibited in the aqueous environment.) For example, light scattering experiments will provide estimates of the sizes and shapes of protein molecules. We might use these techniques to characterize a conformational change which occurs when a certain enzyme encounters and binds an activating metal ion. However, any observed changes will reflect geometrical alterations on the scale of the macromolecule, and may involve rapid equilibrations between two or more discrete conformational states.

This sort of evidence is by no means impertinent, but it does not provide much insight concerning the metal ion and its immediate environment. Contrast this situation with a ^{39}K-nuclear magnetic resonance experiment in which the potassium ion is observed to interact with a biomolecule. One is now receiving information from the ion itself in the form of chemical shifts and relaxation times (i.e., linewidth information),

1

and from such data a picture of the ion and its coordination environment may be formed.

Clearly, some methods are better suited than others for answering the questions of the bioinorganic chemist. We have thus chosen to emphasize techniques which discriminate between the different elements. We might ask, for example, why does the enzyme glutathione peroxidase contain selenium? What is so unique about a selenium atom that the evolutionary process has chosen it over all other atoms for this particular role? To answer our inquiry we will need to probe the selenium part of this enzyme in detail, e.g., to determine its valence and charge density, bonding environment and ligands, etc. It will be particularly informative to observe how these properties change when the enzyme interacts with a substrate. The enquiry is thus one into fine structure and dynamics, and it will require resolution on the scale of individual atoms. This is chiefly the domain of molecular spectroscopy and diffraction methods.

The design of the book has been to include in each chapter a discussion of the type of information one may obtain from the method in question. Limitations have been given special consideration. Much misinformation has been permitted to enter the literature out of inattention to the restrictions inherent in the method in use *and* the biological system under investigation. Each chapter contains an introduction to the method and its associated instrumentation, and the potential applications are illustrated by citing examples from the primary literature. The latter will hopefully convey to the reader the idea that physical instrumentation—however sophisticated or expensive—is fairly inert; as in all areas of science, there is no substitute for ingenuity and care in the design of an experiment.

A cardinal rule unifying these techniques is to avoid placing overreliance upon any one method. Evidence from two or more sources is often complementary, and this has been brought out in many places. Also, on the biological side of the question we should be aware of the highly coherent or synergetic quality of the interactions taking place within living entities. The nonequilibrium thermodynamic models of ordering phenomena (Haken, 1977) provide a quantitative basis for the more intuitive holistic views expressed by biologists for many years. When we reduce living cells to their component parts many significant characteristics will be lost. At the protein level there is always the possibility of artifactual behavior. In keeping with this concept, the illustrative examples range from "simple" one-protein cases through supramolecular aggregates to *in vivo* conditions. Fortunately, some of the techniques allow an examination of the living state. Less emphasis has been given to model systems even though these are indispensable in providing points of reference to more established chemical knowledge.

1.1 Physical Methods

In Figure 1-1 the major physical methods are grouped from left to right as their emphasis progresses from individual nuclei, through more complex structures, to the molecular entity. One will observe that the unifying principles of symmetry and group theory (Bunker, 1979; Cotton, 1963) bridge between these experimental tools and the chemical bonding concepts. It is at this level of interaction between data and theory that chemical insight is gained. All three areas (chemical theory, symmetry, and experimental methodology) are equally important, and the individual who hopes to use the physical methods must be aware of this fact.

Each physical method is, in the correct sense, a communication channel (Shannon and Weaver, 1949; Brillouin, 1964) between an observer and an atomic or molecular structure under investigation. In every case a source or *information carrier* (e.g., a light beam, electron beam, or radio frequency field) is *modulated* when it interacts with a sample of matter. Chemical information is thus coded upon the carrier.

Unfortunately, the information emerging from currently available in-

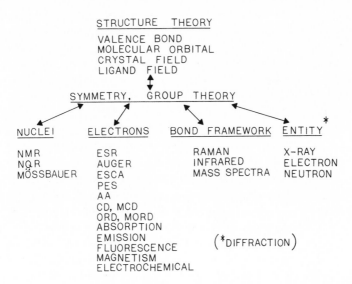

Figure 1-1. Relationships between the various physical methods and chemical theories. Abbreviations: ORD, optical rotatory dispersion; CD, circular dichroism; MORD, magnetic optical rotatory dispersion; MCD, magnetic circular dichroism; ESCA, electron spectroscopy for chemical applications; PES, photo-electron spectroscopy; AA, atomic absorption; ESR, electron spin resonance; NMR, nuclear magnetic resonance; NQR, nuclear quadrupole resonance.

strumentation is not of the direct variety and requires further *processing* in the perspective of chemical theory before one is able to visualize molecular features (e.g., geometry, electron densities, bond polarities, etc.). Also, some of the methods produce relatively more information in binary bits per data set than others, and it is possible to classify them (roughly) as channels of high, medium, or low content. On this scale the technique of X-ray diffraction would rate high while electrochemical methods would be relatively low.

The orientation of this volume introduces an additional criterion of comparison: element selectivity. Table 1-1 and the following discussion attempt to evaluate this important property.

1.1.1 Nuclei

The nuclear resonance methods fall into the medium to high range of channel capacities. Moreover, these methods are element selective, which is a simple consequence of the fact that it is the nucleus which determines an atom's identity. Mössbauer (Gonser, 1975) and nmr spectroscopy (Berliner and Reuben, 1978; James, 1975) have virtually absolute selectivity since natural linewidths and molecular effects upon line positions are quite small in relation to the spectral differences between the various isotopes. Nqr spectroscopy (Semin *et al.*, 1975) is more ambiguous since the chemical environment has a much larger effect upon line positions, although uncertainties may be removed by isotope replacement experiments.

1.1.2 Electrons

The electron-related methods are mostly medium-capacity information channels. They are also less selective than the nuclear resonances, a result of the many pairwise interactions experienced by molecular bonding and nonbonding electrons. However, under some circumstances one may be certain that the electronic property depends upon a particular type of atom. For example, if chemical assays have established that a protein contains a transition element it is reasonably safe to associate any observed paramagnetism with that element. If the protein yields an esr spectrum, one would also attribute it to the transition element; furthermore, the esr spectrum may contain information in the form of electron–nuclear coupling interactions which reveals the identity of the nu-

Table 1-1. A Comparison of Spectrometric Methods

Spectroscopic method	State	Region probed in the biomolecule
Nmr	in vivo, in vitro	Depends on the nucleus chosen. Abundant nuclei yield evidence from broad regions of the molecule, e.g., ^1H, ^{14}N, while a lone nucleus, e.g., a metal ion, may probe only a localized region.
Nqr	Crystal	Similar to nmr, but crystal-induced asymmetry may create more than one environment for the same nucleus, leading to a greater number of lines than equivalent nuclei.
Esr	in vivo, crystal, glassy	The region probed depends on the extent of delocalization of the unpaired electron(s), usually small (one atom) to moderate (several atoms).
Mössbauer	Crystal, glass	Small, usually the immediate environment of the ^{57}Fe atom.
Diffraction methods	Crystal	Whole molecule. These methods are difficult, but they are presently the only means for deriving an unequivocal space-filling structure.
ESCA	Crystal, amorphous	Whole molecule. Energy levels are grouped so that it may be possible to identify atom types. Works best when the atom of interest is present as a minor component of the macromolecule, e.g., a transition metal ion in a metalloprotein.

cleus or nuclei associated with the unpaired electron (Knowles *et al.,* 1976). Note again the desirability of observing nuclear-based phenomena. In other cases an unpaired spin may be more delocalized, as in the example of the free radical derivative of vitamin E (Kohl *et al.,* 1969) shown in Figure 1-2.

Selectivity is greatly increased when the electron involved in the spectral effect belongs to one nucleus in an isolated atom (e.g., atomic absorbance) or is one occupying an inner shell and therefore is influenced by a particular nucleus (e.g., X-ray fluorescence). Of the electron spec-

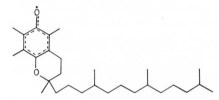

Figure 1-2. A free radical derivative of vitamin E (α-tocopherol). The molecular orbital containing an unpaired electron is delocalized over the region shown by the dotted line. As a result, the esr spectrum contains information in the form of hyperfine coupling constants originating in nuclei in the immediate environment of the ring (e.g., the methyl substituent protons). More distant groups, such as those in the side chain, are not observed in the esr spectrum. Adapted from Kohl *et al.* (1969).

troscopies, ESCA and Auger are more selective since inner shell atomic electrons are characterized (Brundle and Baker, 1979; Carlson, 1975) while probes of the outer (valence) energy levels (as in PES) become difficult or intractable for high molecular weight structures. Conventional ultraviolet and visible absorption spectra are also based on valence electron energy levels and tend to be nonselective unless it has been established by other means that a chromophore involves a particular type of atom (or a group of atoms). Examples are metal–ligand charge transfer bands and the metalloporphyrins. Chromophore electrons responsible for visible light absorptions are never delocalized more than a small fraction of the volume of a protein molecule but it should be evident that a more localized structure, e.g., as is common among charge transfer bands, provides better selectivity.

This concept applies to methods which detect chromophore asymmetry and geometry (Djerassi, 1960), e.g., ORD and CD. In spectral regions well away from asymmetric chromophores, ORD is nonspecific while CD does not produce a response; however, in the region of an optically active absorption, ORD and CD may convey information from a rather localized portion of a macromolecule. This volume places more emphasis upon MCD since the phenomena involved are linked to the paramagnetism of transition metal ions. MCD is applicable even if the chromophore is not intrinsically asymmetric.

1.1.3 Bonding Framework

The methods which characterize molecular framework components are infrared, Raman, and mass spectroscopy. All of these depend on bond strengths (i.e., bond orders) and nuclear masses, and they are information channels of medium to high capacity. Also, the elements associated with the spectral lines may be identified unequivocally through isotope replacements.

Infrared and ordinary Raman are quite limited in a macromolecular environment due to excessive spectral cluttering. Information is received from all portions of the structure, and the vibrations of individual bonds are more or less lost among the others. In contrast, resonance Raman depends on an accentuation of the vibrational modes in the vicinity of a light-absorbing group (Carey, 1978) and thus narrows the region under examination. This simplifies the spectrum while introducing a considerable increase in sensitivity. The method is particularly applicable to metalloproteins and metalloenzymes since protein-bound metals often have near-ultraviolet or visible absorption bands.

Mass spectroscopy produces large amounts of information since its high channel capacity ranks it with X-ray diffraction and nuclear magnetic resonance. It is a powerful tool for covalent structures ranging up to about 10^3 Daltons, and fragmentation of macromolecules at a surface is possible (e.g., SIMS). The detection and identification of selenocysteine in glutathione peroxidase is a fairly straightforward mass spectroscopic procedure (Kraus *et al.*, 1983). However, cluttering is inevitable in the case of the larger molecules as the ionizing bond cleavage may occur at many different weak points. As a result of these limitations, the method has been excluded from this volume. The reader is referred to a fairly recent reference for mass spectroscopic methods (Middleditch, 1979). Biological applications are indirectly related to function, although applications range from metabolic studies to peptide sequence determinations (Zaretskii and Dan, 1979; Horning *et al.*, 1978; Jellum, 1977).

1.1.4 Whole Molecule Methods

Of the methods which observe the molecule as an entity, X-ray and neutron diffraction are by far the most applicable to problems of bioinorganic chemistry since their resolution approaches the scale of atoms. X-ray diffractometry of small molecules (Stout and Jensen, 1968) leads to a three-dimensional map of electron density, permitting one to distinguish, for example, between N and S, but not between Cl^- and S^{2-} since the latter are isoelectronic (Schugar *et al.*, 1976; Birker and Freeman, 1976). In the cited X-ray structure determinations, the central atom of a cluster could be S^{2-} or Cl^-. Assignment as Cl^- was based on supporting chemical evidence that the reaction required Cl^-.

In practical X-ray crystallography of proteins the resolution does not quite reach the atomic level, and hydrogen atoms are "invisible." Nevertheless, the density map may be compared with chemical data (e.g., the primary sequence), allowing one to locate the amino acids. Also, the heavier transition metals may stand out separately. X-ray diffraction has been especially valuable in characterizing the structural details of metalloproteins but is less informative where the lighter elements are concerned.

Neutron diffraction is based on interactions with nuclei rather than electrons. It has the advantage of being sensitive to lighter nuclei, but in order to use this method one must have access to a nuclear reactor for a source of neutrons. In spite of equipment expenses, neutron diffraction has promising applications in biological problems.

The electrochemical methods also characterize molecular entities. With some qualifications the "specific ion" electrodes have a degree of element selectivity but are only applicable to relatively loosely bound ions, e.g., Cl^-, Na^+, Ca^{2+}, Mg^{2+}, etc., and certainly not the covalently attached metal ions of many metalloproteins. Redox potential and conductance measurements are low-information approaches which generally lack element selectivity. Sensitive assays for specific elements may make use of atomic absorption spectroscopy.

1.1.5 Kinetic Methods

In principle, any physical property, spectral feature, etc., may be recorded as a function of time to obtain dynamic information. In practice, to date, only a few of the possibilities have been explored. Biochemical systems are inherently dynamic and it is this property which is most closely associated with function. Thus, a whole chapter (Kinetic Methods, Chapter 10) has been devoted to that subject.

1.2 Probable Future Trends in Physical Instrumentation

1.2.1 Double Resonance Methods

It was noted earlier that by comparing the results from more than one physical approach we are usually able to obtain a better picture of molecular properties. This is a simple expectation from information theory, which holds that each received increment of nonredundant information is a removal of uncertainty. As a general rule the evidence produced by the different methods is fairly nonredundant, though not completely so.

The mere comparison of data from differing methods is a *passive* process. Double resonance is the *active* version of this concept. In double resonance the sample is irradiated in one part of its spectrum while effects are observed in another. The term "manipulation" has been applied to these processes (Mehring, 1976), but one could just as well view them as means for increasing information channel capacity (e.g., by providing a second communication channel), which is in fact what they are.

Double resonance methods are not new. Electron *n*uclear *d*ouble *r*esonance (ENDOR) was an early development in esr spectroscopy (Feher, 1959) while spin decoupling and *i*nter *n*uclear *d*ouble *r*esonance (INDOR) are well-known tools of nmr spectroscopy. Our reviews of the physical literature uncovered a considerable amount of work with new techniques involving wide-ranging regions of the spectrum. Also a great many potential combinations simply have not been explored. Several novel examples of double resonance have therefore been included in the following chapters, and it is expected that similar developments will continue to improve the capacity and selectivity of the various methods for handling complex biological substances.

1.2.2 Dedicated Computers

The recent microelectronic revolution has rapidly found its way into the laboratory with the arrival of a host of "microprocessor"-operated instruments. While these developments have somewhat extended the utility of preexisting designs, economics is the underlying issue. One may now build a sophisticated instrument around a relatively inexpensive processor/memory/analog–digital converter unit which has been programmed to simulate the effect of cams, gears, discrete electronic components, and the like. The instrument is cheaper to build because physical parts have actually been replaced by software equivalents, i.e., they now reside in a memory. This is the so-called software decision.

Such instruments are "on-line" in the sense that data goes directly from detector to computer. Some prefer "in-line" (Ratzlaff, 1978) if the user cannot change the programming. Obviously, with enough memory, further reduction of the data is possible, and to a small extent this is already being realized. However, a more idealized instrument would be one which could acquire a data set *and* compare it against elegant chemical theories. Then it would decide how to pursue further testing (e.g., perhaps double resonance) and report its conclusions, all of this being automatic. A large processor with a throughput at or above 1×10^6 single-precision floating-point calculations per second would be adequate for such a task. In contrast, an existing 8080A or Z-80 microcomputer/arithmetic processor combination (Beavers *et al.*, 1980) with supporting memory and converter cost less than \$2000 (in 1980) but had a throughput of only 5 ×

10^3 floating points per second (roughly, a 1024-point Fourier transform in three seconds). Ultrafast, microelectronic computers are possible as a result of the Josephson junction (Rosenberg *et al.*, 1980), although the economy gained in microcircuit fabrication is offset by the need for liquid helium-based cooling systems. It is more probable that very high speed integrated circuits will be developed from gallium arsenide technology, which has already reached the medium scale level of integration (Bond, 1984). A more immediately practical approach is based on the grouping of currently available, inexpensive eight- and sixteen-bit microprocessors into throughput-oriented "multiprocessors" (Nadir and McCormick, 1980) which can effect order of magnitude improvements in the data handling rate. These developments are leading toward extremely powerful small computers. At this writing (1984), a microcomputer-oriented coprocessor which slightly exceeds the one million per second floating-point rate stated above has become commercially available for under $10,000.

All of this means that computers will not only become more attainable (and therefore abundant) but also quite adept. The major cost factors now appear to be shifting toward programming rather than the equipment itself. As in the case of double resonance, the long-range significance of computer/instrument integration lies in a substantial improvement of the information channel, specifically in the rate at which it delivers information in a usable form. Examples will be seen in the Fourier transform applications of Chapter 2. Perhaps it is correct to believe that biological applications of the physical methods are coming of age. Automation certainly has the potential for reducing much of the abstruse qualities surrounding the subject.

References

Beavers, C. R., George, S. E., Robinson, J. L., and Wright, J. R., 1980. Linking several moderate usage laboratory instruments to one microcomputer, in *Personal Computers in Chemistry (1979) ACS Symposium,* P. Lykos (ed.), Wiley Interscience, New York.

Berliner, L. J., and Reuben, J. (eds.), 1978–onward. *Biological Magnetic Resonance,* all volumes, Plenum Publishing Corp., New York.

Birker, P. J. M. W. L., and Freeman, H. C., 1976. Metal binding in chelation therapy: X-ray crystal structure of a copper (I)–copper (II) complex of D-penicillamine, *J. Chem. Soc., Chem. Commun.* 1976:312.

Bond, J., 1984. Circuit density and speed boost tomorrow's hardware, *Computer Design* 23:210.

Brillouin, L., 1964. *Scientific Uncertainty and Information,* Academic Press, New York.

Brundle, C. R., and Baker, A. D. (eds.), 1979. *Electron Spectroscopy: Theory, Techniques and Applications,* Vols. 1–3, Academic Press, New York.

Bunker, P. R., 1979. *Molecular Symmetry and Spectroscopy,* Academic Press, New York.

Carey, P. R., 1978. Resonance Raman spectroscopy in biochemistry and biology, *Quart. Rev. Biophys.* 11:309.

Carlson, T. A., 1975. *Photoelectron and Auger Spectroscopy,* Plenum Press, New York.

Cotton, F. A., 1963. *Chemical Applications of Group Theory,* Wiley-Interscience, New York.

Djerassi, C. (ed.), 1960. *Optical Rotatory Dispersion: Applications to Organic Chemistry,* McGraw-Hill, New York.

Feher, G., 1959. Electron spin resonance experiments on donors in silicon. I. Electronic structure of donors by the electron nuclear double resonance technique, *Phys. Rev.* 114:1219.

Gonser, U. (ed.), 1975. *Topics in Applied Physics Series,* Vol. 5, *Mössbauer Spectroscopy,* Springer-Verlag, Berlin.

Haken, H., 1977. *Synergetics. An Introduction,* Springer-Verlag, New York.

Horning, E. C., Carroll, D. I., Dzidic, I., Nowlin, J. G., Stillwell, R. N., and Thenot, J. P., 1978. Quantitative bioanalytical mass spectrometry, *Acta Pharm. Suec.* 15:477.

James, T. L., 1975. *Nuclear Magnetic Resonance in Biochemistry: Principles and Applications,* Academic Press, New York.

Jellum, E., 1977. GC-MS in the study of inborn errors of metabolism. An overview, in *Mass. Spectrom. Comb. Tech. Med. Clin. Chem. Clin. Biochem., Symp.,* (M. Eggstein and H. M. Liebich, eds.), University of Tuebingen Press, Tuebingen, Germany, p. 146.

Knowles, P. F., Marsh, D., and Rattle, H. W. E., 1976. *Magnetic Resonance of Biomolecules: An Introduction to the Theory and Practice of Nmr and Esr in Biological Systems,* Wiley-Interscience, New York.

Kohl, D. H., Wright, J. R., and Weissman, M., 1969. Electron spin resonance studies of free radicals derived from plastoquinone, α- and γ-tocopherol and their relation to free radicals observed in photosynthetic materials, *Biochim. Biophys. Acta.* 180:536.

Kraus, R. J., Foster, S. J., and Ganther, H. E., 1983. Identification of selenocysteine in glutathione peroxidase by mass spectroscopy, *J. Am. Chem. Soc.* 22:5853.

Mehring, M., 1976. High resolution Nmr spectroscopy in solids, in *Nmr, Basic Principles and Progress,* (P. Diehl, E. Fluck, and R. Kosfeld, eds.), Springer-Verlag, New York, p. 3.

Middleditch, B. S. (ed.), 1979. *Practical Mass Spectrometry, A Contemporary Introduction,* Plenum Press, New York.

Nadir, J., and McCormick, B., 1980. Bus arbiter streamlines multiprocessor design, *Computer Design* 19:103.

Ratzlaff, K. L., 1978. Microprocessors in minicomputer applications, *Am. Lab.* 10:17.

Raymond, K. N. (ed.), 1977. *Advances in Chemistry Series,* No. 162, *Bioinorganic Chemistry—II,* American Chemical Society Publications, Washington, D.C.

Rosenberg, R., Kuan, T., and Hovel, H. J., 1980. Thin films: Applications in energy, optics and electronics, *Physics Today* 33:40.

Schugar, H. J., Ou, C., Thich, J. A., Potenza, J. A., Lalancette, R. A., and Furey, W., 1976. Molecular structure and copper (II)–mercaptide charge-transfer spectra of a novel copper cluster ion, *J. Am. Chem. Soc.* 98:3047.

Semin, G. K., Babushkina, T. A., and Yakobson, G. G., 1975 (English Translation.) *Nuclear Quadrupole Resonance in Chemistry,* Halsted Press, New York.

Shannon, C. E., and Weaver, W., 1949. *The Mathematical Theory of Communication,* University of Illinois Press, Urbana, Illinois.

Stout, G. H., and Jensen, L. H., 1968. *X-Ray Structure Determination, A Practical Guide,* Macmillan, New York.

Zaretskii, Z. V. I., and Dan, P., 1979. Energy and metastable characteristics in peptides. I. Metastable/daughter ion ratios as an aid in peptide sequencing, *Biomed. Mass Spectrom.* 6:45.

Nuclear Magnetic Resonance (Nmr)

The extent of this chapter parallels the widespread acceptance of nmr as a tool for biochemical investigation (see the series *Biological Magnetic Resonance* by Berliner and Reuben, 1978; Wasson, 1984; James, 1975; Dwek, 1973). For example, if one examines current issues of a journal such as *Biochemistry* it is scarcely possible to find one which does not contain at least one biological nmr paper. The spectroscopic method described here is known as high-resolution nmr, and as such is a powerful tool for the elucidation of chemical structures in solution. More recently it was discovered that high-resolution spectra may also be obtained from solids and liquid crystals, something that was not anticipated in the early period. Thus, the method is particularly applicable to cellular components, which range from dissolved entities such as proteins and intermediary metabolites to more condensed phases of the type made up of phospholipids and sterols in biological membranes.

Structural methods such as nmr are used by biochemists primarily to establish the key relationships between chemical structure and function (Berliner and Reuben, 1978) (the term function is not applicable to most nonbiological molecules). In many cases an atom or group of atoms located at a site of specific binding is used to *probe* biological function. The term "probe" appears many times in this chapter and elsewhere, and since authors vary in their usage of this word, some definitions are necessary. *Extrinsic* probe will refer to those which are foreign to the system under investigation, examples being fluorescent dyes, nitroxide spin labels, or any other molecular entity which introduces a nonbiological structure. By contrast, *intrinsic* probe will mean a reporter group occurring naturally within the biological molecule. Natural abundance ^{15}N can be used as an intrinsic probe of protein function. One of the outstanding advantages of nmr is its ability to use the intrinsic probes contained within biomolecules.

2.1 The Phenomenon

Every atomic nucleus with a non-zero ground state spin quantum number, I, possesses a magnetic dipole moment. If the nucleus is placed within a uniform magnetic field, H_0, the dipole will assume one of $2I+1$ possible quantized orientations with respect to the field. For example, if $I = 1/2$ (as in the case of ordinary hydrogen) there will be two such orientations, and the energy difference between them will depend on the intensity of the applied field. As shown in Figure 2-1, the spin state in which the nuclear moment, μ, opposes the applied field is a high energy level while the other is one of low energy. Relatively more nuclei will be found in the lower energy state.

At any applied field, H_0, there will be a unique photon energy, $h\nu_0$, which will match the energy difference between the upper and lower levels. Light with a frequency ν_0 is thus resonant with the nuclear spin system, and by its absorption the relative populations of the levels will tend to be equalized. Opposing thermal mechanisms allow the system to relax toward equilibrium. As a result, radiation of frequency ν_0 will be continuously absorbed by the sample. In contrast, frequencies above and below ν_0 will not be absorbed. The condition of resonance is given by Eqs. (2-1) and (2-2) where β_N is the nuclear magneton:

$$h\nu_0 = (\mu\beta_N H_0)/I \qquad (2\text{-}1)$$
$$\nu_0 = \gamma_N H_0/(2\pi) \qquad (2\text{-}2)$$

In a practical nmr spectrometer, ν_0 is in the radio frequency portion of the spectrum while H_0 falls in the range of tens of kilogauss. γ_N is the magnetogyric ratio.

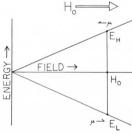

Figure 2-1. Coordinates of spin energy versus applied field. E_H is the energy of a nucleus of spin 1/2 which has its magnetic dipole opposed to the field while in the case of E_L, the field and dipole vectors are pointing in the same direction. At H_0, $E_H - E_L = h\nu_0$, where ν_0 is the resonant frequency corresponding to H_0. From this diagram it is easily seen that a stronger instrument magnet results in a higher resonant frequency. The higher energy level is the $+1/2$ nuclear spin state while the lower one is $-1/2$. In the case of an unpaired electron in a magnetic field (as in esr spectra), the signs are reversed, i.e., the higher level is the $-1/2$ electron spin state.

2.2 Multinuclear Nmr

In Eq. (2-1) the factor $\mu\beta_N/I$ varies from isotope to isotope. As a result, the resonant frequencies of the different elements range from a few megahertz to hundreds of megahertz in a constant applied field. Table 2-1 lists the characteristics of the biologically interesting isotopes standardized at a field strength of 10.00000 kG (Weast, 1972).

An outstanding property of nmr should be obvious: even in modest applied fields the resonances are so far apart that each isotope may be observed without overlapping interference from any others. The method thus has absolute specificity for isotopes in the typical high-resolution instrument. As noted in Chapter 1, this is a most desirable property in a method intended for the study of individual elements. Unfortunately, the property also creates an engineering problem since the radio frequency source must be optimized at definite frequencies. As a result, the detector, usually a module for a given element, tends to be far more expensive than the lamps, detectors, and even dispersive elements of other forms of spectroscopy. Instruments with high-resolution, multiple isotope capabilities are now becoming widely available.

Multinuclear capabilities are not inherent in the designs of older spectrometers (and many newer ones), most of which are optimized for ^{1}H or ^{13}C. The reason for this lies in the widely divergent frequencies presented in Table 2-1. Much greater versatility is realized through the use of a wide-ranging frequency synthesizer and broad-band amplifiers in the transmitter/receiver unit (Santini and Grutzner, 1976a; Ackerman and Maciel, 1976), leaving for modification only four to six tuning components in the sample impedance matching network. Nevertheless, a conventional ^{1}H-nmr spectrometer may be adapted for lower-frequency operation by changing its RF oscillator crystal to an appropriate lower frequency while increasing all tuned circuit capacitances and inductances in the RF transmitter and the receiver RF amplifier to maintain a *constant reactance* value, X, for each capacitor (C) and inductor (L).

$$X_C = 1/(2\pi\nu C) \tag{2-3}$$
$$C_2 = C_1(\nu_1/\nu_2) \tag{2-4}$$
$$X_L = 2\pi\nu L \tag{2-5}$$
$$L_2 = L_1(\nu_1/\nu_2) \tag{2-6}$$

Subscripts 1 and 2 denote original and new values, respectively. Using this method, a 60-MHz proton spectrometer was successfully modified for ^{23}Na operation (Keeton and Wright, 1980) using the scaling constant

Table 2-1. Nmr Data for Biological Isotopes

Isotope	Spin $(I)^a$	Resonant frequency[b] (MHz)	Intrinsic sensitivity[c]	Natural abundance (%)	γ_N (Mrad s^{-1}G^{-1})
^{1}H	1/2	42.5759	1.00	99.985	2.6751×10^{-2}
^{2}H	1	6.53566	9.65×10^{-3}	1.5×10^{-2}	4.1065×10^{-3}
^{3}H	1/2	45.4129	1.21	(radioactive)	2.8534×10^{-2}
^{10}B	3	4.5754	1.99×10^{-2}	19.58	2.8748×10^{-3}
^{11}B	3/2	13.660	0.165	80.42	8.5828×10^{-3}
^{13}C	1/2	10.7054	1.59×10^{-2}	1.108	6.7264×10^{-3}
^{14}N	1	3.0756	1.01×10^{-3}	99.63	1.9325×10^{-3}
^{15}N	1/2	4.3142	1.04×10^{-3}	0.37	-2.7107×10^{-3}
^{17}O	5/2	5.772	2.91×10^{-2}	3.7×10^{-2}	-3.6267×10^{-3}
^{19}F	1/2	40.0541	0.833	100	2.5167×10^{-2}
^{23}Na	3/2	11.262	9.25×10^{-2}	100	7.0761×10^{-3}
^{25}Mg	5/2	2.6054	2.67×10^{-3}	10.13	-1.6370×10^{-3}
^{29}Si	1/2	8.4578	7.84×10^{-3}	4.70	-5.3142×10^{-3}
^{31}P	1/2	17.235	6.63×10^{-2}	100	1.0829×10^{-2}
^{33}S	3/2	3.2654	2.26×10^{-3}	0.76	2.0517×10^{-3}
^{35}Cl	3/2	4.1717	4.70×10^{-3}	75.53	2.6211×10^{-3}
^{37}Cl	3/2	3.472	2.71×10^{-3}	24.47	2.182×10^{-3}
^{39}K	3/2	1.9868	5.08×10^{-4}	93.10	1.2483×10^{-3}
^{41}K	3/2	1.0905	8.40×10^{-5}	6.88	6.8518×10^{-4}
^{43}Ca	7/2	2.8646	6.40×10^{-3}	0.145	-1.7999×10^{-3}
^{50}V	6	4.2450	5.55×10^{-2}	0.24	2.6672×10^{-3}
^{51}V	7/2	11.19	0.382	99.76	7.031×10^{-3}
^{53}Cr	3/2	2.4065	9.03×10^{-4}	9.55	-1.5121×10^{-3}
^{55}Mn	5/2	10.501	0.175	100	$\pm 6.5980 \times 10^{-3}$
^{57}Fe	1/2	1.3758	3.37×10^{-5}	2.19	8.6444×10^{-4}
^{59}Co	7/2	3.8047	3.57×10^{-3}	1.19	2.3906×10^{-3}
^{63}Cu	3/2	11.285	9.31×10^{-2}	69.09	7.0906×10^{-3}
^{65}Cu	3/2	12.089	0.114	30.91	7.5957×10^{-3}
^{67}Zn	5/2	2.663	2.85×10^{-3}	4.11	1.673×10^{-3}
^{77}Se	1/2	8.118	6.93×10^{-3}	7.58	5.101×10^{-3}
^{79}Br	3/2	10.667	7.86×10^{-2}	50.54	6.7023×10^{-3}
^{81}Br	3/2	11.498	9.85×10^{-2}	49.46	7.2244×10^{-3}
^{87}Sr	9/2	1.8452	2.69×10^{-3}	7.02	-1.1594×10^{-3}
^{95}Mo	5/2	2.774	3.23×10^{-3}	15.72	-1.743×10^{-3}
^{97}Mo	5/2	2.832	3.43×10^{-3}	9.46	-1.779×10^{-3}
^{111}Cd	1/2	9.028	9.54×10^{-3}	12.75	-5.672×10^{-3}
^{113}Cd	1/2	9.445	1.09×10^{-2}	12.26	-5.934×10^{-3}
^{115}Sn	1/2	13.922	3.50×10^{-2}	0.35	-8.7474×10^{-3}
^{117}Sn	1/2	15.168	4.52×10^{-2}	7.61	-9.5303×10^{-3}
^{119}Sn	1/2	15.869	5.18×10^{-2}	8.58	-9.9708×10^{-3}
^{127}I	5/2	8.5183	9.34×10^{-2}	100	5.3522×10^{-3}
Nonbiological but useful					
^{135}Ba	3/2	4.2296	4.90×10^{-3}	6.59	2.6575×10^{-3}
^{137}Ba	3/2	4.7315	6.86×10^{-3}	11.32	2.9729×10^{-3}

continued

Table 2-1. (continued)

Isotope	Spin $(I)^a$	Resonant frequency[b] (MHz)	Intrinsic sensitivity[c]	Natural abundance (%)	γ_N (Mrad s^{-1}G^{-1})
^{199}Hg	1/2	7.59012	5.67×10^{-3}	16.84	4.76901×10^{-3}
^{87}Rb	3/2	13.931	0.175	27.85	8.7531×10^{-3}
^{205}Ti	1/2	24.570	0.192	70.50	1.5438×10^{-2}

[a]Nuclear spin in units of $h/2\pi$.
[b]All resonant frequencies are those which would occur in a field of 1.00000 tesla. 1 tesla = 10^4 gauss.
[c]This scale is based on resonances which have the *same linewidth*. Comparison also requires equimolar concentration of the particular isotopes.

$\nu_0(^1\text{H})/\nu_0(^{23}\text{Na}) = 3.8$. The method only works when the modification is to a lower frequency.

Isotopes which are receiving attention and which have obvious applicability to biological problems include ^{1}H, ^{2}H, ^{13}C, ^{15}N, ^{17}O, ^{19}F, ^{23}Na, ^{29}Si, ^{31}P, ^{33}S, ^{35}Cl, ^{39}K, ^{63}Cu, ^{67}Zn, ^{77}Se, ^{79}Br, ^{81}Br, ^{111}Cd, ^{113}Cd, ^{117}Sn, ^{119}Sn, and ^{127}I. The difficulty in applying these specific probes usually relates to instrument sensitivity, often pressed to the limit when one is attempting to detect a single atom within a biopolymer. The isotope ^{139}La and other lanthanides may be used as isomorphous replacements for Ca(II) (Colman *et al.*, 1972). Also, toxicological investigations may make use of ^{75}As, ^{113}Cd, ^{119}Hg, and ^{207}Pb. These applications have not exhausted the possibilities of Table 2-1. The newer nmr nuclei have been examined in a two-volume review by Laszlo (1983 and 1984).

In the perspective of information theory (Shannon and Weaver, 1949), nmr opens a channel of communication between the observer and a specific element within a biomolecule. Under some conditions, e.g., double resonance, the communication may be two-way. At this point it will be appropriate to consider the types of information provided by nmr, and the limitations of the method. There are basically three types of *structure-related* phenomena observed in nmr spectra: shifts, linewidth effects, and nuclear spin–spin splitting patterns. The first two are more important in studies of biopolymers, at least currently.

2.3 Nmr Phenomena Related to Molecular Structure

2.3.1 Chemical Shifts

Nmr would not have chemical applications if ν_0 were not somewhat dependent on the electronic environment at a nucleus. The actual effects are small, amounting to no more than a few hundred ppm of the resonant frequency. Fortunately, linewidths are usually narrow enough to permit the resolution of chemically-shifted absorptions. Chemical shifts derive from several magnetic properties of the atomic and bonding electrons which either augment or oppose the applied field, creating, respectively, deshielding or shielding at the nuclei in question. A diamagnetic effect is one which opposes the field while paramagnetic effects have the opposite quality.

Diamagnetic shielding occurs when the applied field induces a circulating current of atomic electrons. The circulating current, in the sense of a small electromagnet, causes a weak magnetic field which *opposes* the external one. Consequently, resonance will occur at a slightly higher applied field than would be observed for the "naked" nucleus. The diamagnetic effect is due to the immediate atomic electrons and not those of neighboring atoms. However, inductive effects which remove electron charge modulate the degree of diamagnetic shielding, e.g., the proton resonances of —CH$_2$— occur at a lower field than those of —CH$_3$ in Figure 2-2.

Interatomic diamagnetic effects are observed when a molecular orbital's symmetry permits circulating currents which encompass a polyatomic structure. An example is the π-electron system of the benzene ring, shown in Figure 2-3. In this case the induced field acts over a greater distance, causing an apparent paramagnetic (or deshielding) effect at the

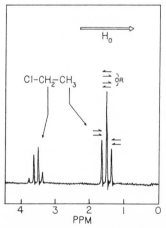

Figure 2-2. The ^{1}H-nmr spectrum of ethyl chloride. Protons on the —CH$_2$— group are deshielded due to proximity to the Cl atom, thus resonating at a lower applied field. The origin of the triplet pattern associated with the —CH$_3$ resonance is explained by the pairs of nuclear dipole vectors above each line. These represent the small magnetic fields produced by the two —CH$_2$— protons adjacent to the —CH$_3$ group. The low-field line appears at a lower applied field (H_0, indicated by the larger vector at the top) because both of the —CH$_2$— spins augment the external field. The central line of the —CH$_3$ triplet is twice as tall since there are two equivalent ways to arrange the —CH$_2$— spins. Finally, the high-field line occurs at a higher field because the —CH$_2$— spins oppose the applied field. The quartet associated with the —CH$_2$— resonance may be explained in similar fashion. One will note that the patterns are skewed. This is a second-order effect due to quantum mixing, and it becomes more pronounced when the resonances lie closer together. In general, a nucleus coupled equally to n adjacent spins will produce an $n+1$ multiplet.

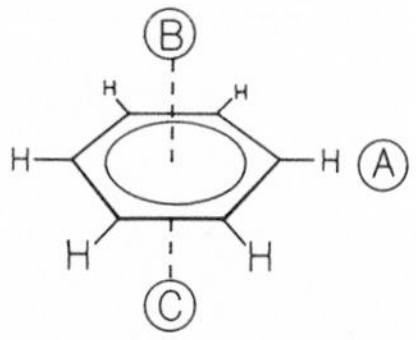

Figure 2-3. Chemical shift effects near aromatic structures. Nuclei in the ring plane (A) will be deshielded, while those above and below it (B and C) will be shielded. As a result of deshielding, protons and aliphatic substituents which are attached to a benzene ring or similar aromatic structure resonate at lower applied fields.

ring protons (A) and the opposite effect at positions above and below the ring (B and C).

Other paramagnetic effects occur when the field causes mixing of a ground state electron with a low-lying excited orbital. Depending on geometry, the effect in the vicinity of the responsible group may be either diamagnetic or paramagnetic. Examples are shifts observed near π-bonded atom pairs (Dwek, 1973).

Contact shifts are relatively large resonance displacements which occur when a molecule contains an unpaired electron (McConnell and Robertson, 1958). Figure 2-4 presents an example of broadened, contact-shifted hydrogen resonances in a metal atom cluster which contains six paramagnetic Cu(II) ions. One should note that one of the resonances is found well beyond the spectral range expected of diamagnetic molecules, i.e., >12 ppm. Contact shifts are so named because they are due to a finite electron spin density at the nucleus in question. Pseudocontact shifts (McConnell and Robertson, 1958) are through-space interactions which occur when the g-tensor of the unpaired electron is anisotropic. More will be said of electron paramagnetism and its effects on the nmr spectra of transition metal-containing molecules (see Section 2.7.2.4.3).

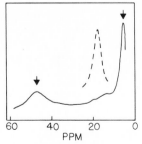

Figure 2-4. Contact-shifted ^{1}H-nmr spectra of clusters of the type $Cu(I)_8Cu(II)_6L_{12}Cl^n$, where n is the net charge and L the coordinating ligand. Legend: (----), L = $NH_2CH_2C(CH_3)_2SH$; (—), L = $HO_2CCHNH_2C(CH_3)_2SH$ (penicillamine). The references for the chemical shift scale (shown in ppm) are the methyl resonances of the uncoordinated ligands. In the case of penicillamine, an asymmetric carbon atom causes nonequivalence between the two methyl groups, resulting in the shifted resonances at -5.56 and -46.3 ppm (arrows). The second ligand contains two equivalent methyl groups, which produce a single resonance at -17.9 ppm (dashed line). An undetermined amount of pseudocontact shift is present in these spectra. The usual chemical shift range of protons in organic compounds is about 12 ppm. Adapted from Eggleton et al. (1978).

2.3.2 Linewidths

Nmr absorption lines are usually broader than the instrument limit of resolution, especially in biological studies. Natural linewidth is a result of the Heisenberg principle, which requires that the uncertainty in energy, ΔE, must be inversely proportional to the lifetime, Δt, of the nuclear spin state, i.e.:

$$\Delta E \Delta t \approx h \tag{2-7}$$

In a more usable form, the relationship between lifetime and linewidth is simply:

$$\Delta \nu_{1/2} = 1/(\pi T_2) \text{ (in Hz)} \tag{2-8}$$

2.3.2.1 Spin–Lattice (T_1) and Spin–Spin (T_2) Relaxation Times

One class of mechanisms determining spin lifetime results in spin–lattice relaxation, T_1. Fluctuating electric and magnetic fields within the molecule create forces which cause the spin populations to relax toward equilibrium, i.e., any excess energy of excitation will be dissipated in this mode. Another mechanism involves energy exchanges between nuclei of like kind whereby the transition of one spin from high to low energy causes another to undergo a transition from low to high energy. Spin populations are unaffected in the latter process, T_2.

Linewidths are generally broader in large molecules as a result of relatively slower rotational motion and longer correlation time, τ_c. In a small, rapidly rotating molecule the magnetic dipoles which induce relaxation are more nearly averaged to spherical symmetry; consequently, relaxation events are infrequent and the spin state lifetime is longer. The pertinent relationships between τ_c and relaxation times are shown in Eqs. 2-9–2-11.

$$\frac{1}{T_1} = \frac{3\gamma_N^4 \hbar^2}{10 r^6} \left(\frac{\tau_c}{1 + 4\pi^2 \nu_0^2 \tau_c^2} + \frac{4\tau_c}{1 + 16\pi^2 \nu_0^2 \tau_c^2} \right) \tag{2-9}$$

$$\frac{1}{T_2} = \frac{3\gamma_N^4 \hbar^2}{10 r^6} \left(\frac{3\tau_c}{2} + \frac{2.5\tau_c}{1 + 4\pi^2 \nu_0^2 \tau_c^2} + \frac{\tau_c}{1 + 16\pi^2 \nu_0^2 \tau_c^2} \right) \tag{2-10}$$

For quadrupolar nuclei:

$$\frac{1}{T_1} = \frac{1}{T_2} = \frac{3}{40} \left(\frac{2I + 3}{I^2(2I - 1)} \right) \left(1 + \frac{\eta^2}{3} \right) \left(\frac{e^2 Qq}{\hbar} \right)^2 \tau_c \tag{2-11}$$

Effects of motion upon linewidth are also seen in the broad lines characteristic of solids and mesomorphic states (Kittel, 1968; Chapman, 1967; Borsa and Rigamonti, 1979). T_2 is the dominant relaxation time in these cases. Molecular groups which rotate about an axis, e.g., CH_3—, are less affected by molecular size.

Relaxation is induced when fluctuating magnetic fields appear in the nuclear environment. Equations (2-9) and (2-10) describe relaxation in terms of a *dipolar mechanism* in which the nuclei are interacting with the magnetic dipoles of nearby magnetic nuclei. The dependence of relaxation upon internuclear distance is of the form $1/T_1$ (dipole–dipole) $\propto 1/r^6$. Thus, for atoms with large radii the dipolar mechanism may not be efficient, and other mechanisms may predominate.

Spin rotation-induced relaxation is characterized by an inverse temperature dependence. This mechanism is important for a number of elements (see Section 2.7.1.7).

Chemical shift anisotropy also introduces a relaxation mechanism. In this case a fluctuating magnetic field is produced when molecular rotation causes the chemical shielding at the nucleus to vary. Anisotropic effects are often quite significant when the nucleus in question is located adjacent to a π-electron system.

2.3.2.2 Chemical Exchange Equilibria, Linewidths, and Lineshapes

In cases where the resonating nucleus exchanges between two environments but does not experience a chemical shift, spectral lineshapes such as those shown in Figure 2-5 may be observed (Binsch, 1969). In curve A the nucleus in question belongs to a small molecule which is exchanging *slowly* between the solvent and a binding site on a macromolecule. The lineshape, which may be described as a sharp peak on a fattened dome, is a result of superimposed resonances due to solvated

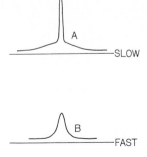

Figure 2-5. Slow and fast exchange between two sites which differ in relaxation rate but not chemical shift. In case *A* the areas under the broad and narrow components are approximately equal, reflecting equal populations of the two environments. Often, this is not the case, and one of the components will be exaggerated. Case *B* displays a single line with a width intermediate between those of the two components shown in *A*.

(narrow line) and macromolecule-bound (broad line) states. If the macromolecule is a typical protein, the broadline component may escape direct detection, and one will only witness a reduced intensity when the protein is brought into solution with the small molecule.

Curve B of Figure 2-5 occurs when chemical exchange is faster than the time scale of the nmr detection process (determined by ν_0). In this case the addition of the macromolecule leads to a concentration-dependent line broadening since the nucleus is now averaging two environments. Figure 2-6 should be compared with Figure 2-5 since the former shows the analogous effects which occur when the two sites differ in chemical shift but not relaxation rate. Obviously, these are limiting cases and a real equilibrium may involve *both* types of effects.

In a two-site binding model the substrate, S, combines with protein, P, in an equilibrium with a small molecule/protein complex, C:

$$S + P \rightleftharpoons C \qquad (2\text{-}12)$$

In the limit of rapid equilibrium between the solvated and complexed states any nmr characteristic. X, e.g., a shift, linewidth, or a derivative of these, may be related to $[S]$ and $[C]$:

$$[C] = \frac{X_{\text{obs}} - X_s}{X_c - X_s} [S] \qquad (2\text{-}13)$$

X_s and X_c are the values of the nmr characteristic in the solvated and bound states, respectively, while X_{obs} is the observed (average) value under fast exchange conditions. The equilibrium expressions of Eq. (2-12) and Eq. (2-13) may be combined with little difficulty (Gutowsky and Saika, 1953; Nakano *et al.*, 1967):

$$\text{slope}_0^{-1} = \frac{([S] + [P] - [C])}{X_c - X_s} + \frac{1}{K(X_c - X_s)} \qquad (2\text{-}14)$$

Here slope is defined as the rate of change of the nmr characteristic with respect to the concentration of added protein, e.g., $\Delta(X_{\text{obs}} - X_s)/\Delta[P]$.

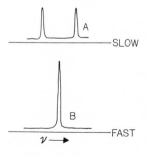

Figure 2-6. Slow and fast exchange between two sites which differ in chemical shift but not relaxation rate. The two sites in A are shown equally populated, but in real systems this is usually not the case. Under fast exchange conditions, *B*, one observes a single line which averages the chemical shifts of the two sites indicated in *A*. If the two sites are unequally populated and in rapid exchange, the single resonance will be located closer to the chemical shift of the more populated site. Broadline conditions will exist when the exchange rate is intermediate between fast and slow.

The subscript 0 indicates use of the limiting value as $[P]$ approaches zero. This is necessary since viscosity effects accelerate with increasing concentration and will affect linewidths significantly.

The experimental values of $slope_0$ and X_s are readily measured. X_c may be estimated through the Stokes–Einstein equation (Abragam, 1961), or determined experimentally by setting $[S]{\cdot}K \gg 1$:

$$X_c \cong \frac{[S]}{[P]} (X_{\mathrm{obs}} - X_s) + X_s \qquad (2\text{-}15)$$

A recursive method may then be used to estimate $[C]$ and K (Nakano et al., 1967). As the equilibrium constant K increases, the second (intercept) term of Eq. (2-14) becomes quite small under practical nmr conditions. At still larger values of K, the exchange rate is no longer rapid and Eq. (2-13) is not valid. The treatment of lineshapes is further complicated when spin–spin splitting must be taken into account (Suzuki and Kubo, 1964).

One should recognize that the effects produced in biological systems may not be accurately described by Eq. (2-14) or Figures 2-5 and 2-6 since (1) *multiple site* binding may occur and (2) the sites may involve a combination of both shift and linewidth effects. Swift and Connick (1962) have derived equations which describe four limiting cases where *both* linewidth and shift may be affected:

Case 1 (shift > linewidth; slow exchange):

$$\Delta\nu \approx f_c/\tau_c' \qquad (2\text{-}16)$$

where $\Delta\nu$ = linewidth increase, f_c = fraction complexed, and τ_c' = lifetime of the complexed state.

Case 2 (shift > linewidth; fast exchange):

$$\Delta\nu \approx f_c\tau_c'(\nu_c - \nu_s)^2 \qquad (2\text{-}17)$$

where ν_s = shift in the solvent and ν_c = shift in the complex.

Case 3 (shift << linewidth; slow exchange):

$$\Delta\nu \approx f_c/\tau_c' \qquad (2\text{-}18)$$

Case 4 (shift << linewidth; fast exchange):

$$\Delta\nu \cong f_c/T_{2c} \qquad (2\text{-}19)$$

where T_{2c} = transverse relaxation in complex.

2.3.2.3 Quadrupolar Nuclei and Linewidths

If the nucleus under scrutiny possesses an electric quadrupole moment there will be an additional mechanism for line broadening since the nucleus will now experience forces originating in fluctuating electric field gradients. The linewidths of quadrupolar nuclei are generally much broader than those which do not have a quadrupole moment, and the dominant effect is described by Eq. (2-11). As a result, the nmr spectra of those isotopes contain less information, examples being ^{25}Mg ($I = 5/2$) and ^{35}Cl ($I = 3/2$) (see Sections 2.7.1.8.2 and 2.7.2.3). On the other hand, quadrupole relaxation may effectively decouple one nucleus from another of interest (Bushweller et al., 1971), leading to a spectrum which is easier to interpret. The latter effect is often observed in boron–hydrogen and nitrogen–hydrogen spin coupling interactions, typically as a simplification of the spectrum of the nonquadrupolar species at lowered temperatures.

2.3.2.4 Paramagnetic Broadening

Paramagnetic broadening is due to the presence of an unpaired electron (e.g., a transition metal ion or a free radical) in the vicinity of the nucleus under observation. It is a strong effect simply because the magnetic moment of the electron is approximately three orders of magnitude larger than the nuclear moment. However, it is not necessarily true that the presence of a paramagnetic center will obliterate the nmr spectrum. A very rapidly relaxing *electron* spin system is effectively uncoupled from nuclear relaxation mechanisms. Figure 2-4 is an example of paramagnetic broadening, which is usually bad in Cu(II) complexes. Other transition metal ions produce much milder broadening effects (see Section 2.7.2.4.3).

2.3.3 Spin–Spin Coupling Patterns and Coupling Constants (J)

The phenomenon of nuclear spin–spin coupling (Roberts, 1961) is shown in the very simple case of Figure 2-2. Insets above one group depict the environment leading to the observed effects. For example, the methyl protons are subjected to the external field plus or minus the weak fields of the two nearby methylene protons, leading to the observed three line multiple pattern, i.e., there are three ways to arrange the lesser fields with respect to the strong field. The central line is coincident with the actual chemical shift of the methyl protons (methylene spins cancel) and is more intense for statistical reasons, i.e., there are *two* ways to realize it (shown in the figure). The spin–spin coupling pattern only involves nuclei which differ in chemical shift. Spin–spin coupling may be as large as 50 mG for nuclei spaced one bond apart and too small to resolve beyond four or five bonds. The coupling effect is conducted through the valence electrons, not space, and J values do not depend on the applied field. Heteronuclear coupling (e.g., ^1H and ^{13}C) also occurs (Dwek, 1973).

Of the three types of phenomena, spin–spin coupling is the one least likely to be detected in the spectra of biological macromolecules simply because line broadening usually obscures it (see Section 2.2.2). Where it does exist as heteronuclear coupling, and the molecule is large, decoupling is warranted.

Although the natural nmr linewidths seen in polymers often prohibit a direct detection of spin–spin splitting, it should not be inferred that coupling information is completely inaccessible. For example, accurate J values may be obtained from poorly resolved doublets if the spin system is first manipulated with a $90°$-τ_{delay}-$180°$-τ_{delay} pulse sequence (see Section 2.4.4.4) and a gated (i.e., time-share) mode of detection and decoupling is then programmed (see Section 2.4.4.4; Campbell et al., 1978). Decoupling reduces the doublet to a singlet, which is nulled to zero amplitude if τ_{delay} satisfies the following equation:

$$\tau_{\text{delay}} = 1/(4J) \equiv \tau_{\text{null}} \tag{2-20}$$

A careful measurement of the null point may be obtained by plotting τ_{delay} versus amplitude and interpolating for the value of τ when amplitude is zero. J may then be calculated using the above equation. As an example of this method, the leucine-17 methyl resonances of the lysozyme peptide were decoupled by irradiating the methine proton. The observed τ_{null} correspond to J (vicinal) = 5.55 Hz (Campbell et al., 1978).

The simple spin–spin splitting pattern of Figure 2-2 occurs with rapid rotation about the C—C bond, but in some structures this is not possible. It is well-established that coupling depends on the dihedral angle between interacting nuclei (Dwek, 1973; Roberts, 1961), and the method may thus be used to obtain stereochemical information about hindered structures. The Karplus function (Karplus, 1963) is employed to determine conformation from spin–spin coupling constants. The original equation is $J(\phi) = A \cos^2\phi + B \cos\phi$, where $J(\phi)$ is the spin–spin coupling constant and ϕ the dihedral angle formed between vicinal hydrogens (H—C—C—H). This equation is limited to tetrahedral carbon structures, but modified versions have been applied to other elements (Hruska et al., 1970; Quin et al., 1980).

2.4 Instrument Characteristics

Modern nmr makes use of a wide variety of signal acquisition and processing methods. These are often confusing to those who are unfamiliar with the literature, and the present section describes the more important features of nmr instrumentation. One should realize that the methodology is continually improving (Wasson and Corvan, 1978). The reader is referred to ANSI/ASTM E 386-78 for standard definitions of the terms and symbols associated with nmr spectroscopy.

2.4.1 Dispersion

The dispersion of a spectrometer is a measure of its ability to spread out the different wavelengths. In an optical instrument using visible light the different wavelengths are dispersed through an angle by refraction (with a prism) or diffraction (using a grating). The distance between the dispersive element and the point at which measurements will be made determines the degree of separation of the spectral lines (or bands), and a longer path is said to be more dispersive.

There is no comparable divergence of wavelengths over a distance or angle in a magnetic resonance spectrometer. Instead, it is the magnetic field strength which isolates the lines. Also, as one might expect, dispersion increases in direct proportion to field strength [see Eqs. (2-1) and (2-2)]. It should be observed that this quality of the instrument applies only to chemical shifts. Natural linewidths and spin–spin coupling are field-independent phenomena.

The term $(\nu_0 - \nu_i)$ may be taken as a measure of the dispersion between a reference line at its resonance frequency ν_0 and a line of interest at ν_i. Shift differences amount at most to only a few hundred ppm when defined *relative* to the irradiating frequency, e.g., δ (relative shift) $= (\nu_0 - \nu_i)10^6/\nu_0$. This form is field independent and will be used throughout the chapter. The reader will observe that our coordinate system places negative values to the *right* of the reference line, which is the opposite of the usual manner of constructing a graph. This peculiar practice originated in the early years of nmr.

The importance of high disperison in biological applications cannot be overstated. This fact is readily appreciated when one considers the numbers of atoms even in small proteins and polynucleotides. With numerous atoms there are more opportunities for chemical shift similarities, and the result will be a partial or complete superpositioning of the resonant lines. The problem is illustrated in Figure 2-7, which shows partially resolved ^{13}C absorptions of a fairly small protein. Superpositioning in ^1H spectra is even worse. On the other hand, resonances from atoms present in small numbers, e.g., activating metal ions, may be much more amenable to study, though often at the expense of long detection times (see Section

Figure 2-7. The ^{13}C-nmr spectrum of ribo-
nuclease A at ν_0 = 15.1 MHz. Although ri-
bonuclease A is a relatively small molecule
at 14,000 daltons, only a few of the resonances are resolved. Most resonances merge
into complex envelopes.

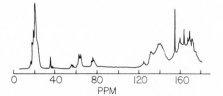

2.7.2.2). In any case one is more likely to resolve single atoms at a higher
field strength.

Analytical nmr instruments which operate at 20 kG or lower are of
quite limited use in direct studies of biopolymers. These may, however,
be used to observe small molecule/biopolymer binding interactions (see
Section 2.7.3). Field strengths on the order of 150 kG (Oldfield et al.,
1978a) are better matched to the biological problem. The state of the art
of driven superconducting magnets is about 300 kG (Levi, 1977), suggesting that nmr has much room for future growth.

2.4.2 Resolution

In an optical spectrometer the diffraction-limited resolution (or blur
limit) is determined by the clear aperture of the system, e.g., larger mirrors
and dispersive elements produce higher resolution. The equivalent property in a magnetic resonance spectrometer is based on field homogeneity
and is illustrated in Figure 2-8. The exaggerated left-hand example shows
an inhomogeneous field in which the strength increases from top to bottom. Nuclei at B will thus come into resonance sooner than nuclei at A,
assuming the field is being swept from low to high intensity. The result
is the illustrated smearing of resonances over a range which is large relative to the actual linewidth. The example to the right presents the ho-

Figure 2-8. Field homogeneity and linewidth. The left-hand
example shows the sample in an inhomogeneous magnetic field
and the instrument-broadened resonance below. In the right-
hand case, the field is homogeneous throughout the sample
volume. As a result, the resonance is very narrow. In a well-
tuned spectrometer, the observed linewidth is often the natural
linewidth. Since the area under the resonance is constant, an
improvement in resolution increases signal amplitude, e.g., the
example at the lower right would be much taller than the figure
indicates. Large pole faces and small air gaps minimize the
edge effects shown in the figure.

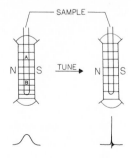

mogeneous field case in which resonance occurs simultaneously throughout the cuvette.

High-resolution nmr is made possible by including several "shim"
coils around the sample cell. These create field gradients along or around
several axes, and by adjusting the shim currents it is possible to markedly
improve field homogeneity. In addition, cell imperfections are averaged
by spinning (the spin axis may be perpendicular, parallel, or tilted to the
field). A rather high resolution, e.g., <0.3 Hz, is feasible.

Field homogeneity must be as high as possible during *any* magnetic
resonance measurements. However, it should be noted that the longer
correlation time of a biopolymer leads to inherently broader lines. In-
strument-limited resolution is thus unlikely in most biologically oriented
investigations.

2.4.3 Sensitivity

The most limiting quality of nuclear magnetic resonance is found in
its sensitivity. This will be appreciated when one considers the small
differences between the energy levels produced even by strong fields.
Equation 2-21 is a version of the Boltzmann factor, an expression proportional to the probability that a given thermal molecular collision will
supply enough energy for a transition (Moore, 1962).

$$f_B = \exp(-h\nu_0/kT) \tag{2-21}$$

The value of f_B may be examined for a 100-MHz instrument operating at
25°C, i.e., $\nu_0 = 1.00 \times 10^8 \text{ s}^{-1}$, $T = 298 \text{ K}$, $k = 1.38 \times 10^{-6} \text{ erg K}^{-1}$,
$h = 6.63 \times 10^{-27} \text{ erg s}$. It is found that the exponent is 1.61×10^{-5} and
the value of the factor virtually one. This means that essentially all collisions may cause the transition, and as a result, the higher energy levels
will be only slightly less populated than the lower ones. It also creates a
situation where the signal may be comparable to background noise originating in the thermal collision process (which causes random fluctuations
in the receiver output).

The signal-to-noise power ratio, S/N_p, defines instrument sensitivity.
This ratio will be improved by (1) going to a higher frequency or (2)
operating at a lower temperature. As stated earlier, a high frequency/high
field condition is also desirable from the standpoint of creating good dispersion. Proteins and other biopolymers are rarely soluble beyond 30 mM.
Even if the spectral dispersion is great enough to resolve a single atom
in a protein, the resonance may have a poor S/N_p ratio. Large signal-to-

noise ratios are experienced only in concentrated solutions of small molecules, and such conditions are of slight interest to the biological scientist.

There is little room for improvement by cooling since the freezing points of water and heavy water set a limit to biological conditions. However, in CW detection (see Section 2.4.4.1) the value of S/N_p will increase with applied RF power up to a point and then diminish. This phenomenon of signal loss at high input power is known as saturation, and it is the result of causing the populations of the upper and lower energy states to equalize, whereby further net absorption of RF energy becomes impossible. Signal detection requires a finite (though small) difference between the upper and lower states, and the spin–lattice relaxation mechanism (T_1) maintains this difference. Thus, slowly relaxing nuclei (long T_1; usually narrow lines) will be easily saturated. Conversely, the input power may be greater for short-lived nuclei. There is an optimum irradiation for any given situation. In the case of pulsed mode spectrometers there is also a limit to input power, e.g., that which achieves a 90° (or perhaps 180°) rotation of the spin vector. Exceeding these values achieves nothing (see Section 2.4.4.4).

In view of the above sensitivity limitations it is not surprising that biologically oriented nmr has been preoccupied with the signal-to-noise ratio problem. One outcome has been a trend to superconductor-based, very high field spectrometers. Biological nmr has also stimulated innovative electronic designs and signal processing methods (to be discussed in the sections immediately following). One area of no small importance concerns the sample-holding cuvette itself, with the trend being to larger cells.

Large bore cells are not new, but interest has intensified in the past three years (Oldfield *et al.*, 1978a). For theoretical reasons (Abragam, 1961) the larger cells should increase S/N_p. Early efforts failed to confirm this anticipated improvement, but more recent changes in probe design and the use of 20–30-mm cuvettes in a "sideways spinning tube" (i.e., field parallel) arrangement have led to promising results (Oldfield *et al.*, 1978b). The trend will probably continue in this direction.

2.4.4 Data Acquisition

2.4.4.1 Continuous Wave (CW) Detection Methods

The simplest and oldest method for obtaining nmr spectra involves slowly and steadily changing the field strength so as to scan (or sweep) a region of interest. Due to the relatively narrow range of chemical shifts for any given isotope, this is easily accomplished by increasing or decreasing the current through a pair of Helmholtz coils located in the magnet gap. Alternatively, the field may be kept constant and the frequency swept. These arrangements are illustrated in Figure 2-9. The transmitter/sample/receiver section usually forms a type of bridge circuit in which resonances are detected as impedance-based voltage changes (e.g., at resonance the cell absorbs radiation and the bridge is unbalanced). Such changes may then be amplified and sent to a recorder. The term "continuous wave" refers to the constant amplitude quality of the transmitter's radio frequency (RF) output, and it is used here to distinguish these techniques from pulse detection modes, which will be discussed in Section 2.4.4.4.

Line distortion results if the sweep rate is excessive. Proper slow scan conditions will be met if (1) T_1 and T_2 are significantly shorter than the time required to scan the narrowest line's width, i.e., $\Delta\nu_{1/2}$ (see Section 2.3.2) and (2) the receiver's response to a change is fast relative to the line scanning interval. Spectral scanning requires on the order of five to ten minutes in these simple methods.

2.4.4.2 Signal Improvement (Averaging)

It is a common misconception to believe that S/N_p is a fixed instrument characteristic. In reality, the instrument noise component—appearing as irregular voltage swings in the receiver output—averages to zero. As a result, signal averaging or filtering techniques may be used to

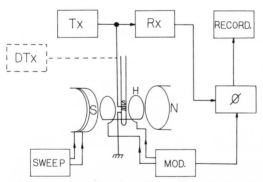

Figure 2-9. Block diagram of a simple continuous wave (CW) nmr spectrometer. Legend: Tx, radio frequency source or transmitter; Rx, receiver; ϕ, phase-sensitive detector; Record., stripchart recording unit; Mod., field audio frequency modulator; Sweep, field sweep circuit which scans the spectrum in step with the X-T recorder; DTx, optional heteronuclear decoupling transmitter. At resonance the sample (small s) absorbs energy from the coil, which reduces the impedance of its tuned circuit and changes the RF level at the receiver input. Also, this unbalance is modulated at an audio frequency by means of the Helmholtz coils (H) since the relatively weak modulator field is alternating toward and away from H_0. The resulting weak signal is converted into a direct current signal by auto-correlating the receiver output with a reference signal from the field modulator. The latter process takes place in the phase-sensitive detector.

increase S/N_p. A straightforward means for accomplishing this is the very simple circuit shown in Figure 2-10(A). The more sophisticated circuit of Figure 2-10(B) is an active filter (Berlin, 1977). These are known as low-pass filters since their output follows signal and noise components with frequencies from DC up to about $1/T_c$, where the time constant $T_c = RC$ (R = resistance in ohms and C = capacitance in farads). Frequencies much higher than $1/T_c$ are strongly attenuated in the output. As shown in Figure 2-10(A) the output has a better S/N_p since the large-amplitude, higher-frequency swings have been rejected in the averaging process. The advantage of the active filter lies in its ability to amplify the signal while producing a more abrupt cutoff of high frequencies.

The method just described is applicable to *any* electronic instrument and indeed is used to enhance the S/N_p of CW nmr spectrometers. However, it will be noted that this advantage is gained at the penalty of having to further slow the scanning rate (which was already quite low). The time to scan a given feature must be long relative to T_c in order to avoid a distorted signal. The inverse relationship between S/N_p and the data acquisition rate leads to an important principle which applies to all of the methods of this volume, i.e., the relative sensitivity advantage of a given instrument is best judged in terms of *the time interval required to attain a desired S/N_p ratio*. In a biologically oriented experiment one is not always permitted long data acquisition times since the preparations under study tend to denature. Thus a fast data acquisition rate is a high-priority consideration.

2.4.4.3 Computer of Average Transients (CAT)

In the low-pass filter methods [Figures 2-10(A) and 2-10(B)] one could conceive of going to ever slower scans and larger time constants, with the result that the output would be virtually DC. Amplification would then reveal very weak signals. Unfortunately, there is a limit to this because all high-gain DC amplifiers suffer from drift and interference, e.g., ulti-

Figure 2-10. Low-pass noise filters. Circuit *A* is a simple RC filter. The capacitor presents an infinite resistance to direct current but is an effective short to high-frequency alternating currents. This circuit attenuates frequencies above the threshold at which $R \sim X_c$. Circuit *B* is an active second-order Butterworth filter.

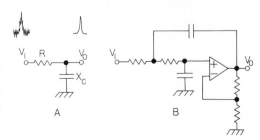

mately the transients from turning on a nearby appliance will be sensed even through a well-regulated power supply.

A better averaging method involves repeatedly scanning the spectrum under normal slow-sweep conditions and summing the digitized receiver output in a computer. The spectrum is divided into M channels spaced at equal intervals, and M is usually in the range 500–16,000, depending on the frequency span and instrument dispersion (more points are needed in high-dispersion instruments). After completing a predetermined number of scans (N) the digitized spectrum may then be scaled for a graphic output or further processed in the computer. One will note that a signal component is additive while noise swings are random about zero. Consequently, in the summing process, the signal to noise ratio (expressed as an amplitude) is enhanced (Cooper, 1977), increasing as $N^{1/2}$.

Amplitude drift will not be very significant during the relatively short scan interval, and it is improbable that power line interference will repeatedly occur in the same channel. Thus, the DC imperfections associated with low-pass filtering are circumvented. This method does require good field reproducibility, since a given resonance peak must occur in the same channel each time. Field lock circuits permit rather high field reproducibility, and the computer may also check the channel of a prominent line after each scan for the purpose of calculating and applying small registration corrections. Finally, most modern spectrometers have provisions for homogeneity optimization (using the lock channel) during a measurement. Nevertheless, hours of averaging are required if one intends to work in the submillimolar concentration range.

2.4.4.4 Pulsed Fourier Transform (FT) Methods

The *rate* of attainment of a desired S/N_p is essentially identical in the two enhancement methods described above even though the computer averaging process has the longer-term advantage. Slow-scan CW detection is limited by having to construct the spectrum sequentially, i.e., line by line. Further improvement in data acquisition rate demands a more parallel input process. In other words, it will be necessary to examine *more of the spectrum per unit time*. With this in mind we might conceive of two separate receiver/transmitter units operating simultaneously (one for the upper half and the other for the lower half of the spectrum). This would double the acquisition rate. Continuing in this line we would arrive at M receiver/transmitter units corresponding to the M channels of a signal averager, a fully parallel (though impractical) arrangement. Fortunately, it is possible to duplicate the characteristics of such an instrument using a single wide band receiver (i.e., one which responds to rapid changes), a correspondingly wide band irradiation, and a combination of signal averaging and Fourier transform procedures (Mullen and Pregosin, 1976).

In Fourier transform nmr the whole spectrum is irradiated simultaneously by a brief but high-power RF pulse. Uniform illumination over the whole region, i.e., a constant power spectral density, is approximated by using an appropriate pulse modulation, chief among the methods being square pulses, triangular envelopes, and band-limited white noise, as shown in Figure 2-11. The populations of the energy levels are thus altered from equilibrium and begin relaxing immediately after the RF input. During the latter period (known as the free induction decay or FID interval) audio frequency signals (corresponding to the differences between the relaxing nuclear resonances and a stable low-power RF reference) appear in the receiver output, as shown in Figure 2-12.

Free induction decay signals are unintelligible but are nevertheless periodic functions related to the absorption spectrum by a Fourier transform, as shown in Eq. (2-22) and Figure 2-12. In a practical FT spectrometer this conversion is based on a discrete transform function [Eq. (2-23)] and is accomplished in a digital computer.

$$F(\nu) = \int_{-\infty}^{+\infty} f(t) \, \exp(-2i\pi\nu t) \mathrm{d}t \qquad (2\text{-}22)$$

$$F(m) = \frac{1}{N} \sum_{n=0}^{N-1} f(n)\{\exp(-2i\pi/N)\}^{mn} \qquad (2\text{-}23)$$

Furthermore, Cooley and Tukey have shown that Eq. (2-23) contains redundant operations, leading to a further simplification of the computational process (Cooley and Tukey, 1965) known as the fast Fourier transform (FFT). There are several variants of the FFT (Rabiner *et al.*, 1969; Singleton, 1969), and all are able to transform large data arrays in seconds. The speed of the transform obviously depends on the throughput of the computer.

The size of the Fourier transform depends on the desired resolution

Figure 2-11. Envelopes used to modulate H_1 in pulsed Fourier transform spectrometry. The rectangular envelope (*A*) is commonly used but does not illuminate the spectral region uniformly, i.e., its Fourier transform has nodes at regular intervals. The noise- or pseudonoise-modulated envelope (*C*) produces a more even illumination of the spectrum.

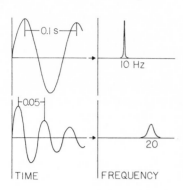

Figure 2-12. Time domain signals and their frequency domain equivalents. FID signals shown at the left will transform to the spectra shown at the right. The lower example presents a rapidly decaying FID and its equivalent broad line. If the latter FID were decaying slowly, as in the top example, the line at 20 Hz would also be narrow.

and the extent of the spectrum ($\Delta\nu$) to be recorded. Resolution is simply the reciprocal of the time interval through which the FID is observed, e.g., if the FID is recorded for 5.0 s, the resolution will be 1/5.0 s or 0.20 s^{-1}. To avoid "aliasing" in the FFT the Nyquist criterion must be observed. This merely states that the sampling rate should be twice the maximum anticipated frequency, in this case $\Delta\nu$ for a single phase detector (or $\Delta\nu/2$ for quadrature detection). Thus, if the highest frequency is 10,000 s^{-1} (e.g., a ^{13}C spectrum) and the resolution is 1 s^{-1}, the spectrum will have to be digitized in 20,000 data storage points. The computer memory and computational requirement clearly increases with improvements in resolution and dispersion.

Measurements of T_1 and T_2 are best done with pulsed Fourier transform spectrometers, although T_2 may also be obtained from linewidth in any high-resolution instrument. Since the pulse techniques depend upon "phasing" between the individual nuclear spins, field homogeneity *must* be high. If it is not high, spins will precess about the main field vector at different rates in different parts of the sample, leading to a rapid loss of phase. These effects are shown in Figure 2-13(A) where the main field, H_0, is along the z-axis and the RF field (i.e., the axis of the detector coil) along y. In this typical arrangement the RF detector is sensitive to the component of magnetization in the xy-plane.

At equilibrium the net nuclear magnetization, M_z, lies along the z-axis, and one can envision this as the vector component of all spins precessing (as in a tilted, spinning top) about H_0. Application of a 180° pulse, which is twice the pulse time calculated in Eq. (2-26), rotates M_z to oppose the field, as in Figure 2-13(B), i.e., the spin system is now inverted and will begin relaxing toward its original equilibrium at a rate determined by T_1, passing through a null [Figure 2-13(C)] at time τ_{null}. If a 90° pulse is applied at the end of the τ_{null} period, the z-component of

Figure 2-13. Instrument coordinates and spin manipulations in pulsed FT nmr. Axis z is always parallel to the main field vector while y is along the long axis of the sample cell. H_1 is shown being applied along x in a double-coil instrument. In crossed two-coil designs, H_1 disturbs equilibrium and the exciting pulse will rotate the nuclear magnetization into y rather than x, as shown in D. A 180° RF pulse inverts the spin system, creating a negative temperature, as shown in B. The spin vector

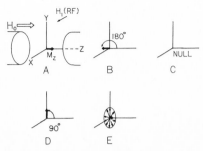

then relaxes via T_1 through a null (C), ultimately arriving at equilibrium as in A. A 90° pulse rotates nuclear magnetization into the xy-plane, which immediately begins to decay via T_2 (E) until equilibrium is again established as in case A.

magnetization rotated into the xy-plane will be zero, and no FID from that resonance will be detected.

In practice a 180°-τ-90° pulse sequence is applied and the FID following the last pulse is transformed to obtain a full spectrum. This process is repeated for different intervals, producing a series of spectra (a stacked plot) from which τ_{null} for each line may be easily determined. A simplified example is shown in Figure 2-14. Time T_1 may then be calculated from τ_{null}:

$$T_1 = \tau_{null}/0.6931 \qquad (2\text{-}24)$$

T_2 is determined by the Carr–Purcell 90°-τ-180° sequence (Carr and Purcell, 1954) or its phase-alternating Meiboom–Gill (Meiboom and Gill, 1958) variant. In this sequence the 90° pulse rotates M_z into the detector's xy-plane [Figure 2-13(D)], initiating a FID, which is a loss of phase (literally, entropy) of the individual spins adding up to the M_x component in the xy-plane. The rate of "unphasing" is determined by T_2. Application of a 180° pulse after an appropriate delay τ will "rephase" the disordered spin system, producing an echo which follows the original 90° pulse by 2τ. It is interesting to note that pulses may be used to manipulate and

Figure 2-14. Spectra obtained by inversion–recovery pulse sequences. τ, the interval between 180° and 90° pulses, *decreases* in the direction indicated by the arrow. τ_{null} has occurred between the third and fourth recordings (from the bottom), and its value may be used to estimate T_1 from

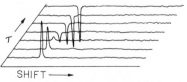

Eq. (2-24). The example spectrum consists of only one line, but the method is obviously applicable to more complex spectra.

organize spin systems. More will be said of this in Sections 2.5.4 and 2.7.1.1.2.

The RF pulse applied in FT nmr must be strong enough to bring all nuclei in the spectral region of interest, $\Delta\nu$, into excitation so that a FID signal may be observed. The strength of the RF field H_1, which must be delivered to the sample may be calculated as follows:

$$H_1 \geq 2\pi\Delta\nu/\gamma_N \text{ (gauss)} \qquad (2\text{-}25)$$

Also, the pulse must be significantly shorter than the nuclear spin relaxation times (T_1 and T_2). A pulse time which will rotate the nuclear magnetization vector 90° may also be calculated (a 180° pulse inverts the spin system):

$$\tau_{90°} = \pi/2\gamma_N H_1 \text{ (seconds)} \qquad (2\text{-}26)$$

Typically, high-wattage pulses of a few microseconds duration are necessary.

If the receiver has only one phase detector, the transmitter carrier frequently must be above or below the resonances of interest, and to avoid fold-over and phase effects it is necessary to reject half of the spectrum with a crystal filter. Quadrature detection improves this situation by using *two* phase detectors with a 90° shift between references (Samuelson *et al.*, 1977), which distinguishes between resonances above and below the carrier. As in Figure 2-15, the RF carrier may then be centered on the region of interest without crystal filtering. This method obviously decreases the pulse energy requirement while improving signal fidelity, since crystal filters have less than perfect responses. Arrangements must be made to avoid overwhelming the receiver, e.g., by using crossed coils or directional couplers (Mullen and Pregosin, 1976).

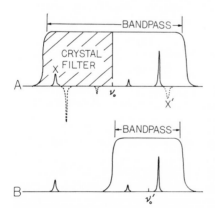

Figure 2-15. A comparison of conventional and quadrature nmr detection. In the conventional method, ν_0 cannot be centered on the region of interest. As shown in *A*, line *x* will fold across, producing the spurious inverted signal *x'*. Similarly, the noise will fold over and degrade S/N_p. For this reason, ν_0 must be located outside the region of interest and a crystal filter is used to prevent noise fold-over. Quadrature detection involves two receivers phased 90° apart, and fold-over is prevented, allowing ν_0 to be centered on the region of interest, as shown in *B*. The smaller bandpass of *B* further rejects high-frequency noise components. Assuming that *x* is actually part of the spectrum, the bandpass of both detectors shown here would have to double. However, the quadrature detector would still have the advantage.

The data handling rate of the FT spectrometer is roughly two orders of magnitude greater than that of a CW instrument. This will be appreciated when one considers the fact that the whole spectrum of a concentrated solution will be obtained by applying a single RF pulse (microseconds), digitizing several thousand points of the FID signal (~ 10 s), and transforming the latter (~ 10 s). This is considerably faster than the typical slow scan method. The FID signal is averaged using the CAT method if the sample concentration is low. After N applied pulses the averaged decay signal is then transformed. Instead of accumulating tens of scans in one hour as in the CW method, FT permits averaging the equivalent of hundreds or thousands. Note that the transform calculation need only be performed once at the end of the FID averaging process.

The periodicity of a FID component transforms the chemical shift while the rate of decay determines linewidth, e.g., a more precipitous drop means a broader line (Figure 2-12). Reflection on the latter feature also suggests that S/N_p progressively worsens in the FID as time passes. It is thus desirable to favor data from the early part of the FID signal. One method known as convolution involves multiplying the accumulated signal with a converging exponential function of the form $f = \exp(-kt)$ prior to transformation. This will introduce a modest line-broadening effect while significantly improving S/N_p. If the sign of the exponent is positive, deconvolution will occur, i.e., line narrowing with an attendant decrease in S/N_p.

Fourier transform spectrometers are generally replacing earlier nmr instruments, and in biological studies they are a virtual necessity. An interesting variant of the Fourier class of instruments is discussed in the following section. Pulsed nmr detection does not always produce an advantage over CW. For example, differences between the two approaches diminish as linewidths broaden beyond about 5 ppm. Spurious ringing may occur in broadline spectra, leading to baseline distortions in the transformed spectrum. These effects may be eliminated by applying appropriate pulsing sequences (Canet et al., 1982). An improvement ratio, FT/CW, may be calculated using a method developed by Ernst and Anderson (1966). In the case of slowly relaxing nuclei, the interpulse interval must be long (see Sections 2.7.2.4.2 and 2.7.1.7). Thus, the advantage of FT nmr is also lost when T_1 is extremely long.

2.4.4.5 Rapid Scan Fourier Transform Methods (Correlation Spectroscopy)

Correlation nmr was developed primarily by Dadok, Sprecher, Bothner-by, Link, Gupta, Ferretti, and Becker (Dadok and Sprecher, 1974; Gupta et al., 1974). This relatively new technique hybridizes the essential concept of pulsed Fourier transform methods, namely that of sampling more of the spectrum per unit time, with the conventional method of scanning a region of interest. In this mode one sweeps through the whole region in a few seconds rather than the usual two to ten minutes. Also, many CW spectrometers may be modified for this application provided the receiver's response is sufficiently rapid.

Figure 2-16(A) shows the outcome of a rapid scan of a ^{1}H-nmr spectrum (Dadok and Sprecher, 1974). The recorded signal has been rendered unintelligible by severe distortions, or so it would appear to the human eye. Figure 2-16(A) is an extreme example of the ringing one observes even in a slowly scanned spectrum of very narrow lines, e.g., as in the case of chloroform's ^{1}H-nmr line. In reality, the distortions are periodic functions of transmitter frequency and line position.

The ringing which follows the passage of a given line is "chirped" in a defined way. (The term "chirp" is used in analogy to a tone which increases in pitch, as in a bird call which goes from low to high frequency.) Ringing frequency in the nmr output has this quality since it is the difference between the relaxing nuclear resonant frequency and the receiver/transmitter frequency at any given instant; this difference will steadily increase after the line is passed. More importantly, the periodic ringing component of the signal averages to zero over a long time interval even though the momentary positive and negative swings might be quite large. In contrast, the spectral component, i.e., the signal which is produced by a slowly scanned spectrum, does not average to zero.

All lines will chirp in like fashion if the sweep rate is constant, the only difference being that narrower lines will ring longer. It should also be noted that frequency differences between a multiplicity of chirp signals are the same as the frequency differences between their originating lines.

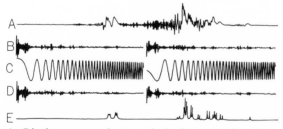

Figure 2-16. Signal processing in correlation nmr. *A* is a fast scan of spectrum *E* and has been rendered unintelligible by ringing and line distortions. *B* shows the real (left) and imaginary (right) Fourier coefficients obtained in the Fourier transform of *A*. *C* is the response due to a single, fast-swept line. In correlation nmr, *D* is obtained as the product of *B* and the complex conjugate of *C*. An inverse transform of *D* yields the undistorted, slow-scan spectrum shown in *E*. In dealing with weak signals, *n* scans of the type shown in *A* are averaged to improve S/N_p. The result is then correlated to obtain *E*. Adapted from Dadok and Sprecher (1974).

The distortion pattern is thus coherent, permitting the use of a coherent signal recovery method (Goldman, 1948) known as cross correlation.

The difficulty in understanding rapid-scan nmr lies in grasping the nature of its correlation process. Equation (2-27) defines a cross correlation function $\phi(T)$ as the integrated product of a signal of interest, $f_1(t + T)$, and a reference signal, $f_2(t)$. In correlation nmr the latter may be the chirped response obtained by scanning a single reference line, as in Figure 2-16(C). In general, the value of $\phi(T)$ will approach or equal zero for periodic components of f_1 and f_2 which have different frequencies (e.g., ringing patterns) but it will be non-zero for the signal component.

$$\phi(T) = \lim_{T \to 0} \frac{1}{2T} \int_{T_1}^{T_2} f_1(t + T)f_2(t)\,dt \qquad (2\text{-}27)$$

It is preferable to express Eq. (2-27) as a discrete summation, as in Eq. (2-28), since in a practical correlator one usually samples the swept spectrum at N points and stores the result in computer memory prior to analysis.

$$\phi(m) = \sum_{n=0}^{N-1} f_1(n + m)f_2(n) \qquad (2\text{-}28)$$

The properties of the last equation may be appreciated by observing one step in the cross correlation of two periodic functions of unequal frequency, as shown in Figure 2-17. Here m is set at zero and the first four products are summed. (The reader is invited to try different amplitudes, frequencies, more discrete points, closer sampling intervals, decaying functions, and different values of m in this fashion). It will be seen that two signals which vary symmetrically about zero and are of *unequal* frequency, as the above, will yield a cross correlation function near or equal to zero, provided one of the functions (the signal in this case) decays

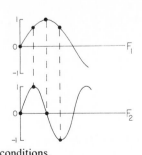

Figure 2-17. Cross correlation. In the interval shown in this figure, $\phi(0) = (0)(0) + (0.0707)(1) + (1)(0) + (0.707)(-1) = 0$. A continuation of this summation will show that unequal frequencies yield $\phi(m) = 0$ over long periods. Obviously this will be insured if one of the components (the signal) decays to zero during the correlation interval. If $F_1 = F_2$, $\phi(m)$ will not be zero. In correlation the latter condition is avoided by having the reference located away from the signals of interest. Correlation is a coherent detection method and thus does not degrade S/N_p, even when S/N_p is near or less than 1. Non-coherent detection methods will diminish S/N_p under such conditions.

to near zero during the correlation interval. The chirp distortions of fast-scan signals are of this type and may be removed by correlation with the chirp from a single resonance obtained under equivalent conditions. Figure 2-16(E) is the cross correlation product of Figure 2-16(A) and a function similar to that of Figure 2-16(C).

Writing a computer algorithm to accomplish Eq. (2-28) is fairly straightforward. However, the calculation will require N^2 complex multiplications and as many additions. The same result may be realized in significantly fewer calculations, using fast Fourier transforms (Cooley and Tukey, 1965). The method involves (1) transforming the fast-scanned spectrum, (2) multiplying with a reference, and (3) calculating the inverse transform of the latter product, a total of $(3N \log_2 N) + N$ calculations, e.g., a spectrum sampled at 1024 positions will require about 1/163 as many calculations as in the direct method. While this is a more sophisticated process than pulsed Fourier spectroscopy, the total time required to produce a desired S/N_p ratio is nevertheless much shorter than that of a conventional slow-scan signal averager. In fact, the access time is typically better than an order of magnitude less for narrow lines. Weak signals are detected by averaging a large number of fast scans to improve S/N_p (the procedure here is no different than in FT or slow-scan averaging). Correlation is then carried out to obtain the undistorted spectrum. Obviously, the method is less efficient when a smaller number of scans are averaged.

One should remember that the slow-scan condition is a relative case, i.e., it depends on the longitudinal relaxation time, T_1, of the spectral lines under scrutiny. In biologically oriented studies one usually deals with polymers. Consequently, both T_1 and T_2 are prone to be shorter and the lines broad under these conditions, permitting one to scan faster and increase RF power input without suffering ringing distortions or line saturation. The relative advantage of correlation thus diminishes, and its application in some situations might be equivalent to processing a slow-scan spectrum to remove insignificant ringing residuals.

It was noted earlier that in pulsed Fourier methods the irradiating power spectrum is uniform and broad, and *all* lines (strong or weak) are brought into resonance. The power spectral density produced by the rapid-scan method is by contrast virtually rectangular over the region scanned, as shown in Figure 2-18. There is a considerable advantage in the latter situation when one wishes to recover very weak signals from a spectrum containing strong resonances. As shown in Figure 2-18, the strong resonance at B is excluded since it falls outside the swept region. However, in a pulsed method all the lines will be excited, leading to the possibility that a weak resonance such as A will be lost in a spectrum dominated by

Figure 2-18. The selectivity of correlation nmr. With the exception of small transition regions at the limits of the scanned region (which depend on sweep time), the power spectral density, P_D, approaches a rectangular envelope (dotted lines). The very strong solvent resonance at B and its associated spinning sidebands (SSB) are thus avoided and do not interfere with the detection of the weak resonance at A. In pulsed FT nmr, both A and B will be brought into resonance simultaneously, and A may be lost due to the dynamic range limitations of the receiver.

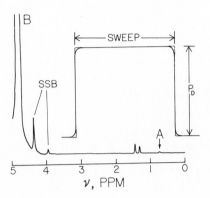

ν, PPM

B. In a pulsed Fourier transform a nulling pulse sequence might be used to null an interfering strong resonance. However, this will also erase weak signals with similar relaxation properties (see Sections 2.4.4.4 and 2.7.1.2.4).

The major advantage of the rapid-scan method lies in its ability to increase the data acquisiton rate of existing CW spectrometers. This is exemplified in Appendix A, which details an inexpensive microcomputer-based correlator for the Varian EM-360 proton spectrometer.

2.4.4.6 Other S/N_p Improvement Methods

Unforeseen future developments may further extend the signal-to-noise ratio of nmr detectors. It is possible to reduce or remove systematic noise (Santini and Grutzner, 1976b), leaving only sample thermal noise to contend with.

2.5 Sample Manipulation (Two-Way Communication)

2.5.1 Multiple Resonance Methods

In simple spin decoupling it is possible to identify the origins of spin–spin splitting patterns by strongly irradiating one absorption feature while scanning through a different region of the spectrum. The nuclei of the irradiated feature are thus forced into frequent transitions between spin states and are decoupled from their normal interactions. Spin–spin splitting patterns which derive from the irradiated nuclei thus vanish, i.e., multiplets will collapse to singlets or at least simplify to the splitting due to remaining unirradiated nuclei, as shown in Figure 2-19 (ascorbate example).

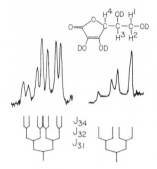

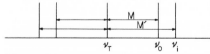

Figure 2-19. Spin decoupling. The complex multiplet shown at the left is the signature due to proton-3 in the ascorbate structure. The interpretation of this splitting pattern (presented as a derivation of doublets below the spectrum) assumes that proton-3 is coupled unequally to protons 1, 2, and 4. Irradiation of proton-4 uncouples the J_{34} component of the splitting pattern, leading to the doublet of doublets shown at the right. This process may be repeated with protons 1 and 2 for confirmation of the assignment.

Homonuclear decoupling (i.e., the irradiated and observed nuclei are of the same type) is easily accomplished in CW detection but does not actually require two separate radio frequency transmitters. Instead, a field is modulated by at least two audio frequency sources to create sidebands (sum and difference frequencies) around the RF carrier, as shown in Figure 2-20. Resonances are observed at ν_0 while ν_i, the irradiating frequency, is varied to track a spectral feature as the magnetic field is swept. In pulsed FT nmr, irradiating frequency is within the receiver's broad passband and would create dynamic range problems (Mullen and Pregosin, 1976). Homonuclear decoupling is also compatible with correlation nmr if the decoupling frequency can be kept out of the receiver's passband. Multiple resonance is possible, e.g., additional field modulators may be added to create more than one irradiation sideband.

Heteronuclear decoupling of ^{1}H is usually required in ^{15}N and ^{13}C pulsed FT nmr. In this case it is necessary to provide a separate, stable transmitter for the irradiation frequency, and both RF units may share a single, double-tuned sample coil (Oldfield *et al.*, 1978b). The method is compatible with all swept and pulsed versions of nmr.

Figure 2-20. Using two modulation frequencies in homonuclear double resonance. The transmitter frequency, ν_T, is modulated by an audio frequency, M ($\sim$10 kHz; see Figure 2-9), which creates sidebands above and below ν_T. As shown in this diagram, the right-hand sideband satisfies the resonance condition. One may thus vary RF power input to the sample by varying the amplitude of signal M. To achieve double resonance the modulator must produce a second, strong signal at audio frequency M'. The resulting sideband at ν_i may be used to irradiate one feature of the NMR spectrum while the spectrum is recorded at ν_0. To do this M' must be variable in order to track the selected resonance while the field varies. In pulsed FT nmr, the field is not varied, and the signal at ν_i must be switched off during the FID period.

INDOR (internuclear double resonance) circumvents the trial and error in the process of spin decoupling used in the identification of structurally related resonances. In this mode ν_0 and field are "parked" at a resonance of interest while the irradiating frequency is swept, the *reverse* of decoupling. The intensity of the observed resonance will vary when the irradiating signal excites coupled nuclei and the recorded amplitude changes provide an identifying fingerprint of the involved spectral feature (this will be apparent when the INDOR signal is plotted alongside the spectrum). Additionally, INDOR detects nuclear Overhauser effects (NOE), i.e., signal enhancements or reductions which occur when nuclei close to those under observation are irradiated (see Section 2.5.2). Double resonance methods are obviously powerful structural tools, but they are more difficult to apply at low concentration.

Broad-band noise irradiation is often used in heteroatom decoupling to improve S/N_p by coalescing multiplets into single resonances, and in many cases (e.g., ^{13}C detection with ^1H decoupling) the method creates further enhancement by the nuclear Overhauser effects (to be discussed in Section 2.5.2).

2.5.2 Nuclear Overhauser Effect

Nuclear Overhauser enhancement (or diminution) occurs when a significant contribution to spin–lattice relaxation comes from dipole interaction of the nucleus in question (x) with a neighboring magnetic nucleus (y) (y can be the same isotope or a different one). If y is irradiated, the line intensity of x will be modified according to Eq. (2-29), where ϵ has the maximum value

$$I_{\text{irradiated}} = (1 + \epsilon)I_{\text{unirradiated}} \qquad (2\text{-}29)$$

$\epsilon_{\max} = \gamma_x/2\gamma_y$. Thus, homonuclear coupling is limited to a 1.5-fold enhancement. One should note that since the magnetogyric ratio is negative in some cases (e.g., ^{15}N), ϵ may also be negative, and by irradiating y one may attenuate the x-resonance or even reduce it to zero (Hawkes *et al.*, 1975; Khaled *et al.*, 1978). The effect diminishes as $1/r^6$, where r is the internuclear distance; hence, atoms with large bond radii do not experience strong nuclear Overhauser effects. NOE is often beneficial in increasing line intensities, e.g., as in the proton decoupling used in ^{13}C-nmr spectroscopy. It also produces structural evidence concerning atom proximities (Noggle and Schirmer, 1971), even in complex biological materials (see Section 2.7.1.5.2 and Figure 2-45). NOE investigations may use difference spectroscopic techniques (Chapman *et al.*, 1978).

2.5.3 High-Resolution Nmr Measurements in Crystalline and Liquid Crystalline Phases

Line-broadening effects in motionally hindered phases, i.e., the crystalline and mesomorphic states of matter and polymers, may be extreme (see Section 2.3.2.1). During the beginning years of high-resolution nmr, interest centered on liquid-state studies, and it was generally assumed that the inherent broadening observed in the more condensed phases of matter posed a barrier which could not be breached. More recently, however, this picture has changed drastically with demonstration of marked line-narrowing effects in the solid state (Mehring, 1976). The techniques fall into the realm of manipulation, in some cases involving a rather high degree of computerized sophistication.

Potential biological applications of these developments are easily appreciated. For example, the element selectivity of nmr may be applied to advantage in the liquid crystalline environment of biological membranes. Conventional methods, i.e., broadline studies, are of the low-information variety. Biopolymers themselves possess quasi-solid character. It is clear that methods developed in this area will help to compensate for at least some of the line-broadening mechanisms which degrade protein spectra.

2.5.4 Spin System Manipulation with Mutiple RF Pulses and Cross Polarization

In the more condensed phases the anisotropic interactions are *not* averaged to zero, which contrasts sharply with the situation in liquids. The anisotropic factors include chemical shift, dipole–dipole (spin–spin) interaction, and quadrupole interaction. The latter two are usually the *dominant* factors in line broadening, and by their removal one is able to observe anisotropic shifts and even spin–spin splitting patterns.

Dipole–dipole interaction is synonymous with the T_2 mechanism discussed earlier (see Section 2.3.2.1), and it may be circumvented by spin dilution, since dipole forces obey an inverse cube law with respect to separating distance. In a practical sense, ^1H is the only naturally occurring isotope which is *not* dilute. Dilution only partially alleviates the problem as the quadrupolar broadening mechanism is not affected thereby.

Multiple pulse methods actively suppress the dipolar and quadrupolar relaxation mechanisms. By applying a four- or eight-pulse sequence of RF bursts to the sample (in which pulse width and interval are carefully controlled) it is possible to reduce the entropy of the spin system (Waugh *et al.*, 1968). Spins are thus brought into phase and T_2 is considerably increased. The free induction decay signal is then recorded and transformed to obtain a much-narrowed spectrum. Figure 2-21 shows spectra of polycrystalline C_6F_{12} with and without the manipulating pulse sequence.

In cross polarization (Pines *et al.*, 1973), dilute spins are first polarized through interaction with an abundant heteronuclear species (which is irradiated), e.g., ^1H. Then the FID of the dilute species is measured while the heteronuclear spins are decoupled by noise irradiation at their resonant frequency. Both methods yield similar results and have the common goal of introducing order into the spin system of the nucleus in question, i.e., they are *coherence* methods.

2.5.5 Spinning at the "Magic Angle"

The average dipole interaction becomes zero when a sample is rotated *very rapidly* about an axis tilted at an angle θ with respect to the external magnetic field, where θ satisfies the relationship tan $\theta = 2^{1/2}$. In this method, chemical shifts are also averaged to their isotropic values (Lowe, 1959). Alternatively, the sample may stand still while the magnetic field is rotated electronically about it at θ by imposing an appropriate video modulation upon H_1 (Lee and Goldberg, 1965). These relatively simple methods may produce substantial line narrowing. It should be noted that magic angle spinning (referred to as MAS in many references) may also lead to line narrowing in the spectra of quadrupolar nuclei (Samoson *et al.*, 1982).

2.5.6 Other Line-narrowing Methods

In a more modest way linewidths may be increased by deconvolution, which is simply the multiplication of a FID signal with an exponentially increasing function (see Section 2.4.4.4). This also has the effect of markedly degrading the S/N_p ratio, and thus introduces a serious disadvantage in a situation where signals are already weak.

Figure 2-21. ^{19}F-nmr spectra of polycrystalline C_6F_{12} at 200°K. The normal spectrum (lower curve) is dominated by dipolar coupling with resulting line broadening. In the upper curve the dipolar coupling has been removed by applying a multiple pulse sequence, and fine structure is revealed. Adapted from Ellett *et al.*, (1970).

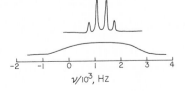

$\nu/10^3$, Hz

2.6 Biological Considerations

2.6.1 Sample Preparation

Nuclear magnetic resonance is almost ideally suited for biological investigation as it permits working in an aqueous solution under conditions (e.g., pH, temperature, ionic strength, etc.) which resemble the cellular environment. The isotope ^1H poses a special problem since water must be replaced with heavy water to circumvent severe interference. Freeze-drying followed by reconstitution with D_2O may be acceptable if the preparation is not easily denatured. Alternatively, the exchange to D_2O may be carried out under more gentle conditions in a stepwise process using ultrafiltration. Inactive isotopes which are loosely bound, e.g., as ions, may be removed by chelation or resin exchange and replaced.

Isotope enrichment or replacement may significantly improve S/N_p, especially for nuclides of low abundance (Table 2-1). Simple exchange techniques are possible in the case of some anions or cations, but biosynthesis, by organisms grown in an isotope-defined medium (Moses *et al.*, 1958; Lutz *et al.*, 1976a; Schaefer *et al.*, 1979), may be the only recourse if the atom of interest is covalently bound to the biomolecule. These methods are inherently very expensive.

In studies of biomolecules with nmr it is almost always necessary to have the biopolymer concentration near the solubility limit, exceptions being in studies of fairly small structures and small molecule–biopolymer interactions (Navon and Lanir, 1972). The experiment is incomplete unless it has been shown that biological function persists under the conditions of the measurement.

2.6.2 Strong Field Effects

It has been often stated that nmr is a nondestructive method. A 300-MHz quantum (2.0×10^{-18} erg), for example, contains *far* less than the amount of energy needed to break even a hydrogen bond (about 3.5×10^{-13} erg). Nevertheless, magnetic field-induced changes in biological functions have been reported. Intense electromagnetic fields apparently change enzyme reaction rates. This property is reminiscent of the kinetic isotope effect (Wayland and Brannen, 1977) and has been attributed to molecular alignment in a strong electromagnetic field since enzymes with paramagnetic centers are the most affected. Other investigators have reported effects of strong magnetic fields on protein solutions and whole, live organisms (Gerasimova and Nakhil'nitskaya, 1977; Aristarkhov *et al.*, 1977; Tenforde, 1979). The subject of magnetic field-induced biological effects should be regarded as controversial, but as the trend continues toward very high-dispersion spectrometers one should be wary of possible artifactual behavior, particularly in kinetic studies.

2.7 Applications of Nmr to the Biological Elements

The major advantage of nmr lies in its ability to receive information from many of the elements present in biological molecules (Harris and Mann, 1979; Axenrod and Webb, 1974; Berliner and Reuben, 1978; Eichhorn and Marzilli, 1979). In some cases the detection process is direct, i.e., the resonance observed is that of the element in question. When this is impossible due to quadrupole relaxation or paramagnetism, often a neighboring atom of a differing kind may show modulating effects caused by the element of interest. In other cases an isomorphorous replacement with a more manageable isotope is possible, sometimes with retention of biological function.

The examples cited hereafter are intended to show both the advantages and disadvantages of nmr in studies of the biological elements. It is impossible to exhaust the subject in so brief a space, and the discussion must be confined to useful isotopes and a few examples of their applications.

Something should be said about the relative usefulness of the different isotopes as nmr probes. There are several factors which mitigate against the detection of many of the isotopes collected in Table 2-1, and the most significant of these factors are strong nuclear quadrupole moments and paramagnetism. In each case the result is severe line broadening.

Quadrupole moments are included in Table 3-1, and it is seen that the value of Q varies considerably. Isotopes which have Q less than 0.1, e.g., ^2H and ^{11}B, experience relatively weak electric quadrupole interactions; consequently the line-broadening effects will be fairly small. Isotopes with Q in the range of 0.1 to 1.0 will experience more pronounced line-broadening characteristics, particularly if they are located in bonding environments of low symmetry. Isotopes with Q greater than 1.0 have strong quadrupole moments. The latter are generally broadened beyond detection in high-resolution nmr instruments and are relatively low-information structural probes.

Paramagnetism alone usually precludes nmr detection of the isotope, and specific examples are found in the transition metal ions. In general, if the product of an isotope's sensitivity and abundance is small (Table 2-1), the element will be difficult, even if the nucleus lacks a quadrupole moment or is not subjected to paramagnetism. Figure 2-22 presents a periodic table which emphasizes these characteristics.

2.7.1 Nonmetal and Main Group Elements

By analogy to studies of simpler inorganic compounds, the properties of metalloenzymes cannot be understood without knowledge of the surrounding ligand groups and their framework structure, i.e., the protein. The latter organic entity is composed of main group elements and hydrogen, and, as will be shown, the nmr spectra of these elements (especially those of carbon, hydrogen, and nitrogen) contain information which can be used to determine the protein's space-filling structure. In practice, this possibility has not been fully realized, but the reader should be aware of the current state of rapid progress toward this goal. The most encouraging results have come from two-dimensional data presentations which reveal either atom-to-atom bonding connectivities or atom-to-atom internuclear distances. These methods deal with structure in the dissolved state (as opposed to crystallographic structure) and are well-suited for observing the molecular properties associated with biological function.

2.7.1.1 Carbon

The relatively low natural abundance of ^{13}C degrades S/N_p but at the same time $^{13}C-^{13}C$ spin–spin splitting is uncoupled due to the low probability of two ^{13}C nuclei being within coupling distance. $^1H-^{13}C$ splitting may also be removed by broad-band decoupling, which enhances S/N_p via the nuclear Overhauser effect (see Section 2.5.2). The chemical shift range of carbon is much larger than ^1H, and ^{13}C resonances are potentially easier to resolve. Spectral cluttering is a problem, even in low-molecular weight proteins, and becomes progressively worse with increasing mo-

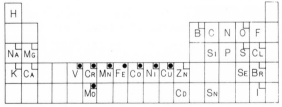

Figure 2-22. Biological elements which have nmr-active isotopes. A square in the upper right-hand corner indicates that the element has no nonquadrupolar isotopes and is thus compromised by quadrupolar broadening effects. Solid circles denote elements which are usually paramagnetic in their biological compounds. Both conditions generally prevent applications of the isotope in biological systems in which the element is covalently bound to a high-molecular weight structure.

lecular weight. Nevertheless, ^{13}C is, in some ways, a more tractable probe than ^{1}H in dealing with large organic structures (Levy and Nelson, 1972; but see Section 2.7.1.2.2).

The soluble silk fibroin fraction isolated from *Bombyx mori* has a relatively monotonous structure consisting mostly of glycine, alanine, and serine residues, but yields a spectrum which is only partly resolved (Figure 2-23). Based on anticipated chemical shifts and the integrated area under each line or band, the ^{13}C-nmr spectrum yields an amino acid composition which agrees closely with the values from an amino acid analysis (Asakura *et al.*, 1984). In other words, nmr offers an alternative, nondestructive approach to amino acid analysis.

2.7.1.1.1 Two-dimensional (2D) ^{13}C-Nmr. Methods which create distinctive dispersions along two (or more) axes, rather than one, lead to a better resolution of the spectral features, much as two-dimensional chromatography and immuno-electrophoresis produce better chemical separations. A crude type of 2D nmr is the "stacked plot" used in the inversion–recovery determination of spin–lattice relaxation times (see Section 2.4.4.4 and Figure 2-14), which creates a display of chemical shift on one axis and interpulse delay time on the other.

Other displays are based on two-dimensional Fourier transform (2DFT) spectroscopy (Terpstra, 1979). In the latter instances the data acquisition proceeds in three distinct steps, as shown in Figure 2-24. In the first, or preparatory step, the spin system is perturbed from equilibrium. Following this, the spin system evolves under the influence of a time-ordered condition, $f(t_1)$ (e.g., coupling or decoupling, spin-echo refocusing, etc.). Then

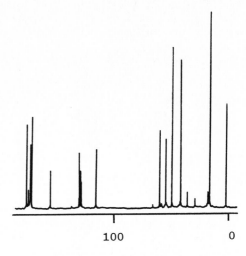

Figure 2-23. The ^{13}C-nmr spectral features of aqueous *Bombyx mori* silk fibroin. The chemical shift scale is in parts per million relative to tetramethylsilane (external standard). This spectrum is uncluttered as a result of the peptide's relatively simple, repetitious structure. Also, the narrow lines are indicative of rapid segmental motion in the peptide's random coil structure. This spectrum has a low degree of line coincidence, and it should be compared with the more typical peptide nmr spectra of Figure 2-33.

100   0

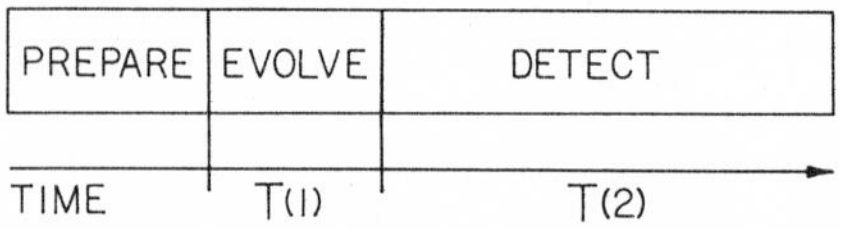

PREPARE	EVOLVE	DETECT
TIME	T(1)	T(2)

Figure 2-24. Two-dimensional nmr. A perturbing effect (decoupling, etc.) is varied during $T(1)$. Period $T(2)$ is equivalent to the FID of ordinary pulsed FT nmr.

in the third period, the spin system is allowed to continue relaxing under different conditions and the FID is recorded as $f(t_2)$. Data are accumulated in the array $f(t_1, t_2)$, which may be transformed to obtain the frequency domain signal, $f(F_1, F_2)$. Techniques which use the 2DFT concept include broad-band homonuclear decoupling, which is particularly applicable to abundant isotopes such as ^1H (Nagayama *et al.*, 1977) and enriched ^{13}C; separated local field (SLF) methods (Rybaczewski *et al.*, 1977), which are appropriate for solid-state studies; J-spectroscopy (Freeman and Hill, 1971); and δ-sorting (Aue *et al.*, 1976).

A simple example of δ-sorting is shown in the 2DFT spectrum of *n*-hexane in Figure 2-25. Axis F_2 maps the proton-decoupled spectrum of three nonequivalent carbons, which is shown projected to the left as viewed from the vantage point marked beside arrow A. The spectrum without decoupling projected at the top is observed from point B. The response surface created by δ-sorting 2DFT resolves these effects, e.g., the quartet signature of the lower resonance clearly identifies it as being due to terminal (methyl) carbon atoms (3 protons and the $n + 1$ splitting rule; see Figure 2-2). In δ-sorting the periods t_1 and t_2 are defined by a gated heteronuclear (i.e., ^1H) decoupler, which is switched on at the beginning of t_2.

Methods which clearly display atom–atom connectivities are highly desirable. When the isotope under scrutiny is dilute, as in the case of natural abundance ^{13}C, and if heteronuclear decoupling is being carried out, then the spectrum will consist of singlets representing each carbon environment. A closer examination of the spectrum will reveal weak satellite resonances arranged symmetrically on each side of the singlets. These are due to minor populations of molecules with two spin active nuclei arranged one bond apart. The satellite resonances are spin–spin

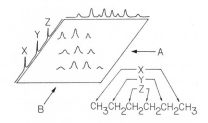

Figure 2-25. Delta-sorting 2DFT ^{13}C-nmr spectrum of *n*-hexane. Adapted from Terpstra (1979).

multiplets which obey the $2I + 1$ rule of multiplicity seen in ordinary nondilute spectra.

A two-dimensional method developed by Bax *et al.* (1981) and known as double quantum coherence (Richarz and Wirthlin, 1981) makes use of satellite resonances so that the nmr information can be presented in an easily interpreted map of atom–atom connectivity. This powerful method is based on a $90°$-τ-$180°$-τ-$90°$-t_1-$90°$-acquisition pulse sequence, where $\tau = (4J_{aa}) - 1$ and J_{aa} is the atom–atom nuclear coupling constant. The variable t_1 is the 2D evolution time, in this case the time required to develop double quantum coherence. This four-pulse sequence nulls the strong signals from isolated nuclei while retaining signals from coupled atom pairs, i.e., the doublets.

A simplified connectivity spectrum is shown in Figure 2-26. The regular chemical shift (based on the FID time, t_2) is on the horizontal axis, while coherence frequency (based on t_1) is the vertical axis. Coupled atoms have the same coherence frequency. Also, the assignment must be consistent with the diagonals, i.e., diagonals cross the midpoint of each line segment connecting a coupled pair. In the hypothetical example of Figure 2-26 the framework is linked C_x—C_y—C_z.

This method can map the framework of molecules up to about molecular weight 500. A major limitation is the requirement for large samples, e.g., approximately one gram. It is clear from this and other examples in this chapter tha pulse programmability is a desirable characteristic in any nmr instrument since it allows flexibility for future developments. Many unique sequences remain to be explored.

Atom connectivity information may also be obtained for nondilute isotopes, e.g., protons (see Section 2.7.1.2.2). The 500-MHz spin-echo correlated ^{1}H-nmr spectrum (SECSY) and J-resolved spectrum of *cyclo-*

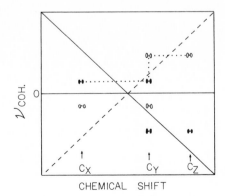

Figure 2-26. Atom connectivity nmr plot (double quantum coherence). The horizontal axis is the chemical shift (^{13}C in this case) while the vertical axis is the double quantum coherence frequency. Satellite doublets shown open are redundant fold-over signals, while those shown solid are real. A non-redundant spectrum is obtained by folding the top and bottom halves together. Since adjacent atoms couple and have the same coherence frequency, horizontal lines connect bonded pairs. Clearly, carbon "x" does not connect to carbon "z." The dotted line shows the correct bonding framework.

CHEMICAL SHIFT

(L-Pro-L-Pro-D-Pro) peptide indicated a structure consistent with that obtained by X-ray crystallography (Kessler *et al.*, 1982). These 2D methods present data in an easily interpreted form and are able to clearly show long-range coupling effects.

The possibilities of 2D nmr have only been partially explored, and these methods apply to *all* isotopes. Copious quantities of information must be stored and processed, and a heavy demand is placed on computer throughput and memory.

2.7.1.1.2 Solid-State ^{13}C-Nmr. High-resolution solid-state nmr has become established with the availability of commercially manufactured instrumentation. Typically, linewidth improvement is based on a combination of spinning at the magic angle and cross polarization (see Section 2.5.3). These are independent processes. Magic angle spinning narrows lines and involves spinning the cell at a field tilt angle $\theta = 54.7°$ and at high speed (~3000 revolutions/s). Further enhancement of sensitivity and linewidth results from cross polarization (Miknis *et al.*, 1979), as illustrated in the four-step process of Figure 2-27; Step 1 is a standard 90° pulse applied along the x'-axis at the ^{1}H frequency, which rotates the magnetization vector into y'. In Step 2 the ^{1}H frequency is given a 90° phase shift, which locks the ^{1}H magnetization along y' with a z frequency component determined by $\gamma_H H_{1H}$. This lock is not for an indefinite period, and Step 3 begins as soon as lock is established. In Step 3 the sample is irradiated along x'' at the ^{13}C frequency. By choosing appropriate amplitudes for H_{1H} and H_{1C} so that $\gamma_H H_{1H} = \gamma_C H_{1C}$ (the Hahn–Hartmann condition; Hartmann and Hahn, 1952), the z components of ^{1}H and ^{13}C are of the same frequency, and energy will be transferred from the abundant ^{1}H spins to the dilute ^{13}C spins (Figure 2-27). After a time the ^{13}C

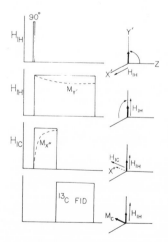

Figure 2-27. Cross polarization nmr spectrometry. The process starts with a 90° pulse applied along x', which rotates the proton magnetization along y'. Immediately after the 90° pulse, a 90° phase shift in the transmitter rotates H_1 along y'. This signal locks the magnetization of ^{1}H in the y'-direction, where decay occurs slowly (dashed line, $M_{y'}$), and irradiation of the ^{1}H nuclei decouples the dipolar interaction responsible for the very broad lines in solids. Simultaneous application of H_{1C} (at the ^{13}C resonance) along x'' under Hahn–Hartmann conditions leads to a buildup of ^{13}C magnetization in the $x'y'$-plane, shown by the dashed line $M_{x''}$. When $M_{x''}$ is established, H_{1C} is removed and the ^{13}C FID is recorded (bottom example). Cross polarization removes the dipolar line-broadening mechanism, increasing T_2 and producing a more acceptable T_1. Magic angle spinning removes chemical shift anisotropic broadening.

nuclei are sufficiently disturbed from equilibrium so that the FID signal may be recorded in Step 4. The Hahn–Hartmann condition decreases T_1.

Line-narrowing methods are particularly applicable in direct probes of biological membranes, where linewidths are intractable by conventional nmr (Urbina and Waugh, 1974). By using an intrinsic probe, perturbing effects such as might be expected of a spin label (or a similar extrinsic probe) are avoided. Figure 2-28 presents spectra of solid *tert*-butyl acetate (Fyfe *et al.*, 1978) both with and without magic angle spinning. Obviously, the solid-state method leads to marked improvements in resolution, permitting measurements of linewidths and shifts under a variety of conditions, e.g., in the presence of metal ions, with artificial or isolated membranes, etc.

Magic angle ^{13}C-nmr spectra of cellulose I in the solid state attain linewidths as small as 5 Hz, permitting assignments of the carbon atoms within the saccharide monomer units (Earl and VanderHart, 1980). It should be noted that the cellulose structure is relatively more monotonous than a protein. Nevertheless, the advent of solid-state nmr has provided a new tool for the study of substances which were hard to probe as a result of insolubility.

2.7.1.1.3 Dissolved Proteins and Nucleic Acids. The determination of structure in aqueous solutions is a major application of ^{13}C-nmr in studies of proteins, nucleic acids, and other biological macromolecules. Various approaches are available to this end (London *et al.*, 1977; Komoroski *et al.*, 1975; Brown *et al.*, 1972), e.g., pH effects, relaxation times, changes on denaturation, ^{13}C enrichment at specific atom sites, the preparation of derivatives, and difference spectroscopy, where the effects of functionally bound entities such as ions and substrates are evaluated. All of these are based on the certainty that both secondary and tertiary structures affect shifts and relaxation times. For example, the geometric arrangement between aromatic amino acids in proteins and nearby groups can lead to upfield or downfield shifts (though generally small) and paramagnetic ions may produce large effects in both linewidths and shifts (see Section 2.7.2.4.3). ^{13}C-Nmr chemical shift data for the amino acids are collected in Table 2-2 (Clerc *et al.*, 1973).

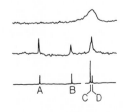

Figure 2-28. The effect of magic angle spinning on ^{13}C solid-state nmr spectra. *Top*: normal spectrum of solid *tert*-butyl acetate ($-100°$C). *Middle*: same sample spinning at the magic angle. *Bottom*: *tert*-butyl acetate dissolved in CD$_3$OD (liquid). Resonance assignments: *A*, carbonyl; *B*, tertiary carbon; *C*, (CH$_3$)$_3$; *D*, acetate CH$_3$. Adapted from Fyfe *et al.*, (1978).

Table 2-2. ^{13}C-Chemical Shifts of Amino Acid Carbon Atoms[a]

Amino acid	Shift, ppm from tetramethylsilane		
	—CO_2H	α-Carbon	Others[b]/type
Glycine	173.5	42.5	— —
Alanine	176.8	51.6	17.3 CH_3
Valine	175.3	61.6	30.2 CH
			17.9 $CH_3{}^c$
			19.1 $CH_3{}^c$
Leucine	176.6	54.7	41.0 CH_2
			25.4 CH
			22.1 $CH_3{}^c$
			23.2 $CH_3{}^c$
Isoleucine	175.2	60.9	39.7 CH
			15.9 CH_3
			25.7 CH_2
			12.5 CH_3
Serine	173.1	57.4	61.3 CH_2OH
Threonine	174.0	61.5	67.1 CH—OH
			20.5 CH_3
Cysteine[d]	172.0	57.5	27.4 CH_2SH
Cystine[d]	175.6	54.7	39.0 CH_2SSCH_2
Methionine	175.3	55.3	31.0 CH_2
			30.1 CH_2S
			15.2 SCH_3
Phenylalanine	175.0	57.3	37.5 CH_2
			~140 Aromatic substitution position
			131.1 Aromatic *ortho*
			130.7 Aromatic *meta*
			129.5 Aromatic *para*
Tyrosine	175.0	57.3	37.5 CH_2
			~133 Aromatic substitution position
			130.5 Aromatic *ortho*
			117.5 Aromatic *meta*
			156.3 Aromatic *para*
Aspartic acid	175.5	53.2	37.6 CH_2
			178.8 Pendant CO_2H
Glutamic acid	175.6	55.7	28.1 CH_2
			34.5 CH_2
			182.3 Pendant CO_2H
Lysine	175.4	55.3	27.2 CH_2
			22.4 CH_2
			30.7 CH_2
			40.0 CH_2NH_2
Arginine	175.2	55.1	28.5 CH_2
			24.9 CH_2
			41.5 CH_2NH
			157.5 Guanidine

continued

Table 2-2. (*continued*)

Amino acid	Shift, ppm from tetramethylsilane			
	—CO₂H	α-Carbon		Others[b]/type
Histidine	174.9	58.8	29.0	CH₂
			~135	—C— imidazole
			118.2	—CH—imidazole
			137.2	N—C—N imidazole

[a]Adapted from Clerc *et al.* (1973).
[b]Order from α-carbon.
[c]Nonequivalent to other methyl group in the molecule due to asymmetry at the α-carbon atom.
[d]As the hydrochloride.

Nuclear magnetic resonance applications involving the abundant biological elements, e.g., ^{13}C, ^{31}P, ^{1}H, and ^{15}N, are qualitatively similar. More examples of the numerous strategies available to the investigator will be treated in the following sections on hydrogen, nitrogen, and phosphorus.

2.7.1.2 Hydrogen

Nearly all of the beginning work with nmr was based on protons. The isotope ^{1}H is both abundant and easily detected. However, abundance is not necessarily an advantage since the one-dimensional spectra of large molecules are not only overly dense from numerous discrete resonances but also further cluttered by spin–spin splitting. This can be somewhat alleviated (at the expense of reduced sensitivity) by diluting ^{1}H with ^{2}H. Serious work with ^{1}H calls for the highest possible dispersion. The state of the art is about 600 MHz (Schwarzschild, 1979) using driven superconductors, and attempts at higher fields will continue. With each increase in field, more detail appears, as may be seen by comparing the two spectra of Figure 2-29. A parallel development, which enhances the resolution of detail at any field, is two-dimensional spectroscopy (Nagayama *et al.*, 1977; Jeener *et al.*, 1979; Kumar *et al.*, 1980; Wagner and Wuthrich, 1982).

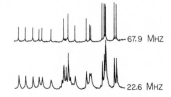

Figure 2-29. ^{13}C-nmr spectra of gramicidin S at low (bottom) and high (top) field strengths. Both instruments might have the same homogeneity-limited resolution in Hz, but the higher-field instrument will produce a larger Δν (in Hz) relative to the resolution limit, resulting in a better separation of the lines. Adapted from Komoroski *et al.* (1975).

67.9 MHZ

22.6 MHZ

Two-dimensional nmr was introduced in Section 2.7.1.1.1 and 2D proton methods which reveal space-filling stucture will be examined in Section 2.7.1.2.2.

   2.7.1.2.1 Nmr Titration Methods. The structure of bovine erythrocyte superoxide dismutase (SD) has been characterized in the crystalline phase using X-ray diffraction (Richardson *et al.*, 1975a,b). This enzyme consists of two 15,600-dalton subunits, each containing one Zn(II) and one Cu(I or II) ion. Crystallographic electron density maps place the copper and zinc ions in close proximity, and histidine residues are implicated in the coordination of these ions.

   Proton spectra of SD obtained at 270 MHz demonstrate that these basic structural features are retained in aqueous solution (Cass *et al.*, 1977a). Convolution difference spectroscopy (Campbell *et al.*, 1973) achieves a considerable resolution enhancement of the histidine imidazole C(2)H protons (~8 ppm) in SD and was used in the cited study (Cass *et al.*, 1977a) to obtain the pH titration curves shown in Figure 2-30. The top family of curves was derived from apo-SD and exhibits the chemical shift dependence of C(2)H resonances on pH, as would be expected of labile histidine groups. In contrast, the lower family, obtained from diamagnetic Cu(I), Zn(II) SD, clearly indicates at least four pH-independent histidine residues, and these are presumed to be participating in metal ion coordination. Two histidines were found to be titratable in Cu(I), Zn(II) SD, consistent with two noncoordinating histidines observed in crystalline SD (Richardson *et al.*, 1975a,b).

   As further evidence, the spectrum of Cu(II), Zn(II) SD lacks resonances from the coordinating histidines. These all lie within 8 Å of the Cu(II) paramagnetic center and are presumed broadened beyond detection in accordance with the Solomon–Bloembergen effect (Solomon, 1955; Bloembergen, 1957). Resonances from protons separated $\geq$ 13 Å from the paramagnetic center should not be significantly broadened. Features

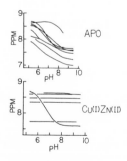

Figure 2-30. ^{1}H-nmr titration curves for superoxide dismutase and its apoprotein. Both ordinates are ^{1}H-nmr shifts relative to TMS. Adapted from Cass *et al.* (1977a).

missing or present in Cu(II), Zn(II) SD compared with Cu(I), Zn(II) SD are consistent with the crystallographic structure.

It should be noted that the nmr-based conclusions about histidine–metal binding in SD do not depend on prior knowledge of the X-ray structure. In the cited nmr study (Cass *et al.*, 1977a), foreknowledge of crystallographic structure was required only in the assignment of non-ligand histidine residues of Cu(I), Zn(II) SD as His-19 and -41, and the assignment of one resonance to an *N*-acetyl group rather than Met-115. Nmr evidence should be viewed as complementary to X-ray diffraction as there is no complete assurance that the crystalline phase conformation will be retained in the aqueous solution phase.

Similar methods were used to establish two metal ion binding sites of unequal affinity in *Bacillus cereus* β-lactamase (Baldwin *et al.*, 1978). This enzyme, which catalyzes the hydrolysis of penicillins and cephalosporins, contains two Zn(II) ions. Three of five histidine residues identified in the apoenzyme shift on the addition of one equivalent of Zn(II) and are no longer titratable. A fourth histidine follows this pattern when the second equivalent of Zn(II) is added.

Histidine residues are relatively accessible through ^{1}H-nmr simply because the associated resonances occur in less cluttered regions of the spectrum. The importance of histidine as a metal binding ligand is established, e.g., as in the cases of carbonic anhydrase (Chlebowski and Coleman, 1976), alkaline phosphatase (Carlson, 1976), and carboxypeptidase (Chlebowski and Coleman, 1976).

The use of ^{1}H-nmr is more compromised in the detection of potentially coordinating residues which contain aliphatic groupings, e.g., aspartic and glutamic acids and their amides; also serine, threonine, cysteine, and methionine. These are difficult to locate with certainty and are better subjects for ^{13}C-nmr (see Section 2.7.1.1), particularly if one uses specific enrichments of ^{13}C (London *et al.*, 1977). Methionine and cysteine are accessible by photoelectron spectroscopy (see Chapter 7). Residues which bear nitrogen-containing pendants such as tryptophan, lysine, histidine, and arginine are good subjects for ^{15}N-nmr (see Section 2.7.1.4). Proton spectra obtained from proteins are generally better-resolved in the aromatic (i.e., low-field) region; thus, one has a better chance of detecting and probing the aromatic amino acids.

2.7.1.2.2 Biopolymer Conformation. The conformation of a compact biopolymer, e.g., a globular protein, should be reflected in modulations of linewidths and shifts, often subtle, in nmr spectra. In the specfic case of ^{1}H resonances, a variety of factors may lead to the observed phenomena. For example, tertiary structure may lead to upfield or downfield shifts of groups positioned near aromatic amino acids. It is well-known from studies of small molecules (Dwek, 1973) that protons located above the ring place are upfield-shifted while the converse is true of those in the ring plane (see Figure 2-3).

Thus, changes in the positions of methyl resonances may provide evidence of conformational alteration associated with protein function. This approach was used in studies of plastocyanin and azurin (Cass *et al.*, 1977b). In azurin the shifts of methyl resonances are upfield, indicative of positioning above aromatic ring planes. The redox status of the copper in azurin could be altered without significant ^{1}H-nmr spectral effects, and similar results were obtained for plastocyanin. These findings were interpreted as evidence of little or no conformational alteration between the Cu^{1+} and Cu^{2+} forms of these proteins.

In the case of gramicidin A, which has the sequence shown in Figure 2-31, the methyl resonances of one valine and three leucines are upfield-shifted by proximity to the tryptophan residues (Glickson *et al.*, 1972). As validation of this postulate, hydrogenation of the latter pendants reverses the conformationally determined methyl shifts. The cited study also made use of another method, deuterium exchange, to distinguish between internal and solvent-exposed protons. All four indole NH protons were found to be exchangeable (i.e., they disappear from the spectrum on adding D_2O) as was also true of the ethanolamine OH, but the peptide hydrogens did not exchange, indicating participation in solvent-inaccessible, intramolecular hydrogen bonding.

Based on infrared and X-ray crystallographic evidence (Popov and Zheltova, 1971), the participation of peptide carbonyls in hydrogen bonding results in a charge relay. This property of the peptide backbone, related to the planarity of the HNCO grouping, also leads to electron depletion at the amide hydrogen and downfield shifting (Linas and Klein, 1975). A

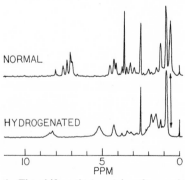

NORMAL

HYDROGENATED

10   5   0
        PPM

Figure 2-31. ^{1}H-NMR spectra of normal and hydrogenated gramicidin A (with some B and C contamination). Normal gramicidin A has the sequence HCO-L-Val-Gly-L-Ala-D-Leu-L-Ala-D-Val-L-Val-D-Val-L-Trp-D-Leu-L-Trp-D-Leu-L-Trp-D-Leu-L-Trp-NHCH$_2$CH$_2$OH. Hydrogenation converts the tryptophan indole rings into saturated structures, which cancels any upfield shifts induced in groups located above or below the indole rings (see Figure 2-3, B and C). The arrow marks a resonance which shifts downfield on hydrogenation and is thus presumed to be located above an indole structure in normal gramicidin A. The shift scale (ppm) is referenced to tetramethylsilane. Adapted from Glickson *et al.* (1972).

comparison of the proton nmr spectrum of vitamin D-induced bovine intestinal calcium binding protein with the established X-ray crystallographic structure revealed similar conformations in the dissolved and crystalline states (Dalgarno *et al.*, 1984).

The intramolecular interactions resulting from a polypeptide's tertiary structure may restrict the mobility of some pendants while others rotate freely. It has been shown that the resonances of tyrosine and phenylalanine may experience temperature-dependent linewidth effects consistent with varying rates of rotation about the C_β—C_γ bond (Campbell *et al.*, 1976). Resonances of a specific tyrosine of horse ferrocytochrome C were found to be markedly temperature dependent. The rate of flips was determined both by computer modeling and independent cross saturation measurements, and the temperature dependence of both sets of data proved to be colinear on an Arrhenius plot (Figure 2-32), corresponding to a rather high activation energy of 17 kJ mol^{-1}. A similar tyrosine was observed in the cytochrome C of several other species.

Studies of the latter type are particularly fruitful since the biopolymer under investigation has been characterized structurally by X-ray crystallography (Dickerson and Timkovich, 1975). One has a better chance of reaching a rational interpretation of solution behavior since the gross features of tertiary structure are known. Conclusions based upon nmr evidence alone is likely to be ambiguous. Very promising measures of success have been realized in theoretical calculations of the favored secondary and tertiary structure of biopolymers, based on known primary structure (Kuntz *et al.*, 1979; Maxfield and Scheraga, 1979; Finkel'shtein and Ptitsyn, 1978; Davies, 1978). Thus, a merging of theory with information originating in primary sequence determinations and physical measurements such as nmr (and others such as Raman; see Section 8.6.1) may offer an alternative to the more established X-ray and neutron diffraction methods.

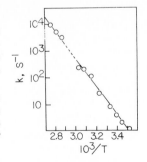

Figure 2-32. Arrhenius plot of the logarithm of rotational rate about the C_β—C_γ bond in tyrosine of ferrocytochrome C (ordinate) versus reciprocal absolute temperature (abscissa). The slope corresponds to an energy barrier of 17 kJ mol^{-1}. The value k (rate) was derived from the relationship $I_a/I_b = (1/T_1)/[(1/T_1) + k]$, where T_1 is the spin–lattice relaxation time and I_a and I_b are, respectively, the intensities after and before irradiation of a resonance which transfers magnetization to the protons under scrutiny. T_1 was obtained by the 180°-τ-90° pulse sequence. Adapted from Campbell *et al.* (1976).

Noncrystallographic conformational studies begin with a knowledge of polymer primary structure and then apply semiempirical energy-minimizing calculations to determine the favored conformational states (Dunfield *et al.*, 1978). Adequately resolved and assigned nmr resonances are then examined for consistency with the probable conformer states. It may be possible to discard many of the structures, thus narrowing the possibilities to a few.

This method was applied to the C-terminus hexapeptide of the growth hormone antagonist, somatostatin (Knappenberg *et al.*, 1979). Energy calculations for amino acid residues 9–14 indicated contributions from several possible compact (folded) structures and semi-extended forms. Some of these cannot explain the nonequivalence found between threonines 10 and 12 while other conformations predict a ring-current shift induced by the proximity of phenylalanine 11 and the threonine 10 pendant (in D_2O). In this particular study the similar results obtained with sub-peptides 10–12 and 10–13 were offered as evidence of the usefulness of peptide fragments in conformational studies. However, in view of probable cooperativity in large peptide (or other polymer) chains, it remains to be seen if fragment studies will lead to a deduction of the space-filling properties of a large, functional peptide polymer.

Biological applications of 2D proton nmr have progressed along two major lines which promise to greatly enhance studies of proteins in the dissolved state. The technique known as correlated spectroscopy (COSY) identifies networks of spin–spin interaction *through* the bonding framework (Nagayama *et al.*, 1980), while two-dimensional nuclear Overhauser spectroscopy (NOESY) yields internuclear distances *through space* (Kumar *et al.*, 1980; Boyd *et al.*, 1984). It is possible to obtain both types of data simultaneously (Gurevich *et al.*, 1984), and a variant of the NOESY method further reduces spectral clutter and therefore ambiguity (Wagner, 1984). The NOESY method detects relatively close range proton–proton interaction (see Section 2.5.2), and with biomolecular concentrations on the order of 10 mM, NOE interactions between protons should be detectable up to 5 Å with an accuracy of about 0.5 Å using a high-field instrument (Keepers and James, 1984).

The NOESY pulse sequence is as follows: $(90°)$-τ_1-$(90°)$-τ_m-$(90°)$-τ_2, where τ_2 is the FID interval for the first Fourier transform and the second transform is over τ_1, the systematically varied interval between the first two pulses. The value of τ_m is the characteristic time of z-magnetization exchange by cross relaxation between interacting spins. Figure 2-33(A) is the 2D NOESY spectrum of a small protein, basic pancreatic trypsin inhibitor. In Figure 2-33(A), the contour along the diagonal corresponds to the conventional 1D spectrum, while off-diagonal signals, referred to as cross peaks, mark pairs of spins which interact by the Overhauser

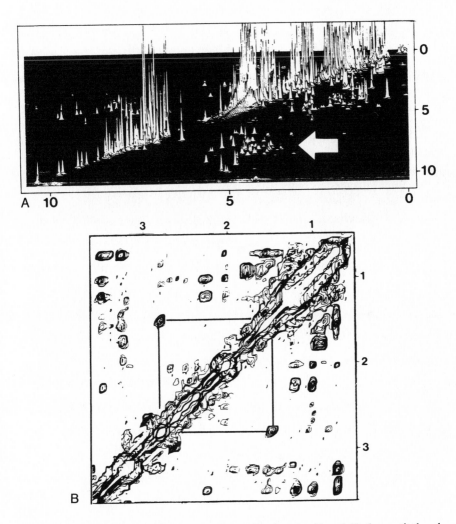

Figure 2-33. 2D-NOESY spectra of peptides. In both cases (A and B) the vertical and horizontal scales are in parts per million relative to tetramethylsilane external standard. In (A) the proton spectrum of basic pancreatic trypsin inhibitor (BPTI) is shown as a stacked plot (adapted from Wagner and Wuthrich, 1982). The contour of the diagonal (lower left to upper right) is the ordinary one-dimensional spectrum, while off-diagonal cross peaks (a group is marked by the arrow) are due to NOE interactions. Cross-peak amplitudes may be related to the proximity of the interacting pair of protons. Data acquisition for a spectrum of this type requires approximately one day. In (B) the 2D NOESY proton spectrum of glucagon interacting with micelles is shown as a contour plot, i.e., essentially a topography map of the kind of spectrum shown in (A). It is seen that the cross peaks are symmetrical about the diagonal (i.e., half of the 2D plot is redundant). Vertical and horizontal lines from equivalent cross peaks (a pair is shown for illustration) intersect with the diagonal, indicating the chemical shifts of the pair of protons responsible for the NOE interaction. A closer inspection of (B) reveals numerous strings of cross peaks along verticals or horizontals, which are potentially due to a specific proton interacting with others in its immediate environment. The contour plot is the preferred form of data presentation in structural work.

61

effect [a group of off-diagonal signals is marked by an arrow in Figure 2-33(A)]. The intensity of these cross peaks is a function of the interproton distance (Keepers and James, 1984), and the chemical shift of each proton is found, respectively, by erecting vertical and horizontal lines to intersect the diagonal [this is best done using a 2D spectrum presented as a contour plot, as in the example of Figure 2-33(B)]. From a more general point of view, all of the homonuclear spins which are near enough to interact with a given proton must show as cross peaks along the same vertical or horizontal line. Collectively, these form a network describing proton proximity relationships, and if the proton spacing information (cross peak intensity) is included, the NOESY spectrum can be translated, in principle, to the three-dimensional structure of the molecule.

With increasing molecular weight, cluttering leads to superimposed cross peaks and ambiguity. A 600-MHz proton spectrometer begins to approach the degree of dispersion needed for small protein structure characterization. The method has been used to assign sequence-specific resonances in the proton spectrum of DNA fragments, including the *Lac* repressor binding domain (Zuiderweg *et al.*, 1983) and to carry out conformational studies on other nucleic acid polymers (Feigon *et al.*, 1983a,b). Basic pancreatic trypsin inhibitor, a small protein of molecular weight 6500, has been subjected to 2D NOESY analysis (Wagner and Wuthrich, 1982). Based only on the known peptide sequence (58 amino acid residues) and the NOESY spectrum obtained by means of a 500-MHz proton spectrometer, all backbone and beta-hydrogen signals were assigned with the exception of those of Arg 1, Pro 2, and Pro 13 and the amide proton of Gly 37. Also, the side chain assignments included all but those of Pro 2, Pro 13, the *N*-delta protons of Asn 44, the peripheral protons of the lysines, and two of the arginine residues. These studies were instrumental in demonstrating the suitability of nmr for sequence-specific assignments, based on atom-to-atom proximities of adjacent residues. Two-dimensional double quantum ^{1}H-nmr measurements on proteins (Wagner and Zuiderweg, 1983) can distinguish amide protons from all others and indicate if amide and α-protons belong to the same spin system, based on *remote* connectivities.

In other studies (Pardi *et al.*, 1983), the chemical shifts of backbone amide protons were found to be conformationally dependent, varying by more than 0.5 ppm in the range of accessible geometries. Current thinking holds that chemical shift and bond connectivity information (e.g., the COSY spectrum) should be used in the sense of structural refinement while NOE-defined proximities should be the primary structural basis. The power of the latter is illustrated in studies of the peptide hormone glucagon (Braun *et al.*, 1983). It was found that micelle-bound glucagon *lacked* long-range NOE effects between peptide segments 5 and 9, 11 and 29, 5 and 16, and 19 and 29, which should have been present if the peptide had adopted a globular tertiary structure. Instead, the nmr evidence defines an association in which the backbone is extended and roughly parallel to the micelle surface. Calculations which treat NOESY spectra may make use of pseudo structures for the 20 common amino acids, i.e., their internal structural constraints lead to signatures in the nmr spectrum (much as an ethyl group can be recognized in the 1D nmr spectrum of a simple compound). This approach to computation was described by Wuthrich *et al.* (1983). The reader is referred to other good examples of NOESY and similar 2D methods in peptide structure determinations (Strop *et al.*, 1983a,b).

 2.7.1.2.3 CIDNP Applications in Conformation Studies. Chemically induced nuclear polarization (CIDNP) occurs when a chemical reaction generates nuclei in excited spin states (Granger, 1974; Lepley and Closs, 1973; Muus *et al.*, 1977). Under these conditions the affected nuclei emit rather than absorb RF radiation and appear as amplitude deflections with inverse polarity. Spin excitation may occur in thermochemical reactions, e.g., as in the classical thermal decomposition of dibenzoyl peroxide (Bargon *et al.*, 1967). Photochemical activation may lead to the same result. For example, irradiation of dyes such as fluorescein creates a triplet excited state. The latter may abstract a phenolic hydrogen atom (Figure 2-34) and then return it to create nuclear spin excitation (Muzskat and Weinstein, 1975; Muzskat, 1977a,b,c).

 This method has been reported to produce spin-polarized nuclei in a variety of biological phenolic structures, including the tyrosyl groups of $ZnCl_2$-modified bovine insulin, which contains four tyrosines among a total of 51 amino acid residues (Muzskat *et al.*, 1978). The latter example is shown in Figure 2-35. An arrow marks a resonance position assigned as *ortho* protons of the tyrosyl residues. In Figure 2-35(B) the sample has received a one-second irradiation from a 2500-W mercury/xenon source

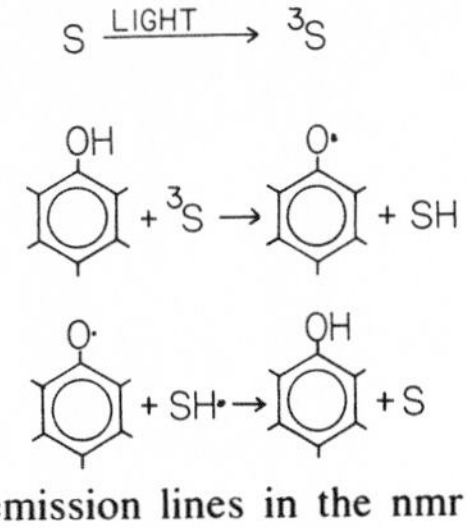

Figure 2-34. Chemical processes which may be responsible for the photo CIDNP effect. In the first step, a dye molecule absorbs a photon and enters an energetic triplet state (^3S). The latter species then encounters a phenolic molecule, and the phenolic O—H bond is homolytically cleaved, as shown in the second step. Finally, the radicals produced in the second step recombine (as shown in the third step), forming the original substances. However, the nuclear spin system becomes inverted during this process even though no apparent reaction has occurred. Nuclei adjacent to the phenolic functional group thus appear as emission lines in the nmr spectrum. Adapted from Muzskat *et al.* (1978).

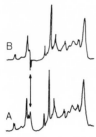

Figure 2-35. Photo ^{1}H-CIDNP effect in a solution of bovine insulin containing fluorescein (9.7 mM insulin; 1.3 mM fluorescein). Spectrum A was obtained without light while spectrum B was produced when light pulses from a 2500-W Hg/Xe lamp were applied before each nmr pulse (FT detection at 90 MHz). Deflections below the baseline are due to lines which *emit* rather than absorb at ν_0, clearly the case for the resonance position marked by the arrow. Adapted from Muzskat *et al.* (1978).

prior to each RF pulse (FFT mode). The *ortho* protons now appear in emission, confirming the energy transfer from the dye to the macromolecule. Both spectra were recorded in the presence of 1.3 mM fluorescein at pD 2.4 and ν_0 = 90 MHz. Polarization diminishes at higher pD, presumably as a result of ionization effects at the phenolic OH, but CIDNP activity is significant at physiologically realistic pD values, as shown in Table 2-3.

CIDNP may be used as a structural tool, e.g., as in the discrimination between groups buried within the macromolecule (and therefore inaccessible to chemical interaction) and those at the solvated periphery (Kaptein *et al.*, 1978). Conformational effects or direct interactions with tyrosyl residues occurring during metal ion binding could be assessed by these methods. Polarization of nuclear species other than ^{1}H is also possible (Pople *et al.*, 1968; Bubnov *et al.*, 1972).

Studies which did not use CIDNP have produced evidene of hydrophobic tyrosine residues. For example, ^{1}H-nmr and thermal stability data for the vitamin D-dependent, calcium-binding protein of intestinal mucosa are consistent with a long tyrosyl residue oriented within a pocket containing phenylalanines and isoleucine (Birdsall *et al.*, 1979). A significant data base for tyrosine CIDNP effects in proteins has not been accumulated

Table 2-3. Influence of pD on the Photo-CIDNP Effect in a Phenolic Compound[a,b]

pD	Light signal/Dark signal[c]
9.6	1
8.9	-1
6.6	-3
2.0	-5

[a]L-Tyrosyl-L-tyrosine, 10 mM; fluorescein, 0.5 mM.
[b]Adapted from Muzskat *et al.* (1978).
[c]The no-signal baseline is the zero reference, e.g., a ratio of -1 means that the downward deflection below the baseline (in light) is equal in amount to the dark signal's upward deflection above the baseline.

at this writing, and some interpretations of data have been challenged (see Kaptein in the Biological Magnetic Resonances series). The negative deflection of Figure 2-35 appears to be a CIDNP effect, but its assignment is in question.

2.7.1.2.4 Solvent Suppression and Solvent Interactions. The special dynamic range problem encountered with ^1H, namely that of the strong solvent resonance in the presence of weak signals, has been remedied by pre-irradiation methods which saturate and thus null the solvent's interference. While this may work for small molecules, saturation is apt to be transferred into and throughout a macromolecule (Stoesz *et al.*, 1978), leading to unacceptable distortions of the macromolecular resonances. Saturation transfer effects generally increase with increasing molecular weight.

The solvent signal may also be suppressed by applying an appropriate $180°$-τ-$90°$ pulse sequence, with τ chosen to null the solvent's nuclear magnetization. Weak resonances which lie under that of water but differ significantly in relaxation properties may be thus detected by this technique. This method has the disadvantage of nulling other resonances for which T_1 is similar to that of water.

Correlation nmr was discussed earlier (see Section 2.4.4.5). This approach avoids dynamic range problems simply by scanning regions which do not include the water resonance. A combination of correlation nmr and pulse methods for nulling nuclear magnetization minimizes the possibility of overlooking an important weak signal, as shown in Figure 2-36.

Water structure is altered in the presence of metal ions, proteins, and other biological components. The spectrum of ^1H_2O in biological materials is often a featureless line, but the relaxation rate of the protons is mod-

Figure 2-36. Acquisition modes which scan across the chemical shift range, such as correlation NMR, cannot detect weak resonances which are near the chemical shift of the solvent (water). The blind region is denoted by S in the figure. In pulsed FT detection an appropriate $180°$-τ-$90°$ pulse sequence might be used to null the solvent resonance, but this method will also null other spectral features which happen to have a T_1 near the value for water (S'). A combination of the two approaches minimizes the possibility of overlooking a weak resonance, e.g., uncertainty is reduced to the rectangular area at the center of the figure. Solvent resonance is usually a serious problem in ^1H-nmr spectrometry.

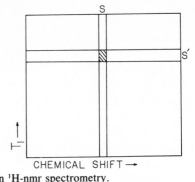

CHEMICAL SHIFT →

ulated by the dissolved entities. In one form of nmr known originally as "Zeugmatography" the nmr image of a whole living organism (based on hydrogen concentration and relaxation properties) may be examined (Lauterbur *et al.*, 1975). Imaging nmr has already reached the stage of commercially available instruments intended for clinical and research applications and may now be compared with X-ray and positron emission tomographic methods (Brownell *et al.*, 1982). Nmr images reveal organ boundaries, lesions, etc., and resolve detail in a human-sized mass on a scale of 2 mm or less.

2.7.1.3 Deuterium and Tritium

Deuterium labeling is highly advantageous in Raman (see Chapter 8) and neutron diffraction studies of biomolecules. Similarly, ^{2}H-nmr may provide information unobtainable by ^{1}H-nmr, particularly in the case of biological membranes (Seelig, 1977; Mantsch *et al.*, 1977). Both isotopes (^{1}H and ^{2}H) are present in natural water, and as a consequence, solvent interference is also present in ^{2}H-nmr. However the latter effect is not as great as in the case of ^{1}H since ^{2}H abundance is only 0.016% and ^{2}H-enriched substances dissolved in ordinary water have a larger ^{2}H(sample)/^{2}H(water) ratio. Also, solvent interference may be further reduced by labeling specific molecular sites with ^{2}H and dissolving the sample in deuterium-depleted water (commercially available).

Unlike ^{1}H, ^{2}H is a quadrupolar nucleus ($I = 1$), and its longitudinal relaxation rate depends entirely on the quadrupolar mechanism, i.e., the presence of electric field gradients and molecular correlation time [see Section 2.3.2.3 and Eq. (2-11)]. This simplicity stands in contrast with ^{1}H, ^{31}P, and ^{13}C, which relax according to a variety of modes.

The quadrupole moment of ^{2}H also proves to be of value in probing molecular orientation and motion in ordered phases such as biological membranes. In most organic compounds the C—D bond has pseudoaxial symmetry, and the electric field gradient vector thus virtually coincides with the bond axis. If the latter is directed at an angle, θ', with respect to H_0, e.g., as a result of restricted molecular motion in a condensed phase, the two nmr transitions due to $+1 \leftrightarrow 0$ and $0 \leftrightarrow -1$ are of *unequal* energy, leading to a quadrupolar splitting of the ^{2}H-nmr resonance into *two* lines. The separation between maxima may be calculated by Eq. (2-30), where $(e^2qQ)/h$ is the quadrupole coupling constant (see Chapter 3).

$$\Delta\nu = \frac{3e^2qQ}{16h} (\overline{3\cos^2\alpha - 1})(\overline{3\cos^2\gamma' - 1})(3\cos^2\theta' - 1) \quad (2\text{-}30)$$

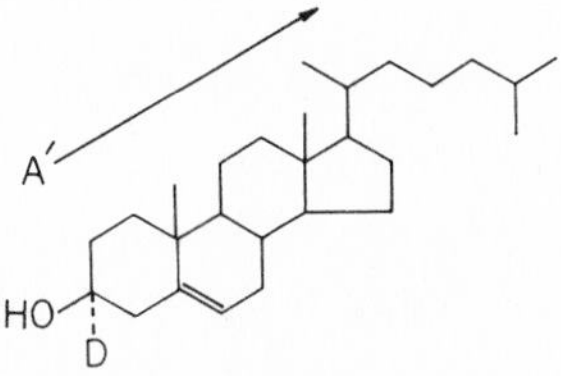

Figure 2-37. Cholesterol deuterated at the 3α-position. The C—D bond is at an axial position on the cyclohexane ring and is oriented approximately perpendicular to the long axis of the steroid molecule, shown as A′.

Oldfield *et al.* (1978b) have examined cholesterol, selectively labeled with ^{2}H at the 3α-position, as a probe of fluidity and orientation within phospholipid bilayers. The cholesterol molecule, shown in Figure 2-37, is rigid, and the major axis of the polycyclic framework (A′) is expected to parallel the hydrocarbon chains of the phospholipid. As a result, the C—D bond is nearly perpendicular to the chains, and under conditions of maximum order it may be concluded from an extension of Eq. (2-30) and empirically determined coupling constants (Burnett and Muller, 1971; Derbyshire *et al.*, 1969) that the maximum $\Delta\nu$ should be 63.75 kHz. Smaller values reflect motional averaging of the quadrupole–bond gradient interaction. Figure 2-38 presents observed quadrupole splitting in the cholesterol–phospholipid system, from which it may be noted that (1) the increased chain motion at higher temperature indeed reduces the value of $\Delta\nu$ but (2) order is introduced at higher temperatures by increasing the cholesterol:phospholipid ratio. The latter result is a consequence of the rigidity of the sterol framework.

This sort of experiment illustrates the uniqueness of ^{2}H in probing the dynamic structure of biomembranes. It should be added that Oldfield *et al.* (1978b) also prepared phospholipids labeled with ^{2}H at specific positions. Their results from ^{2}H-nmr measurements compare closely with neutron diffraction studies.

Tritium is also nmr active and has received some attention (Brevard and Kintzinger, 1978). A consideration of the potential for accidental radiological contamination causes one to raise very serious questions about its suitability in nmr experiments.

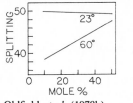

Figure 2-38. ^{2}H-quadrupole splitting of cholesterol-3α-d$_1$ in a lipid bilayer model membrane system. The ordinate is the deuterium quadrupole splitting expressed in units of kilohertz and the abscissa is mole percent cholesterol in the membrane system. An example of the doublet structure seen in the ^{2}H-nmr spectrum is shown at the right. Adapted from Oldfield *et al.* (1978b).

2.7.1.4 Nitrogen

Although virtually all naturally occurring nitrogen is in the form of ^{14}N, quadrupolar broadening in the polar covalent environment limits the use of this isotope to relatively low-molecular weight structures. By contrast, ^{15}N is very difficult to observe at its natural abundance level of 0.36%, but line broadening is not a problem ($I = 1/2$). The S/N_p ratio obtainable for ^{13}C at natural abundance is 30 times that for ^{15}N under comparable conditions (Hawkes *et al.*, 1975). Consequently, work with ^{15}N necessitates high-field spectrometers and large-volume samples (see Section 2.4.3). The 280-fold enhancement possible through isotopic enrichment should not be overlooked. As an example of isotope enhancement, Schaefer *et al.* (1979) were able to detect and distinguish between amine and amide resonances in soybeans grown on ^{15}N-enriched fertilizer using a cross-polarization spectrometer.

The availability of ^{15}N-enriched compounds (such as histidine) is increasing, and these are valuable in studies of active sites and metal–ligand interactions (Kanamori and Roberts, 1983). Studies of more complex biological molecules should benefit from a developing base of nmr data for relatively simpler systems (Mason, 1981), which include examples of nitrogen interacting with paramagnetic centers and ^{15}N bound to putative analogues of nitrogen-fixing enzymes.

T_1 for ^{15}N is quite long in many cases (Lippmaa *et al.*, 1971), but can be reduced to permit a more acceptable pulse repetition rate in FT spectrometers by the addition of paramagnetic agents (Farnell *et al.*, 1972). Nuclear Overhauser enhancement with ^1H decoupling is also significant. However, one should observe that $\gamma(^{15}N)$ is negative while $\gamma(^1H)$ is positive. The NOE may thus range from $1 + \epsilon = 1$ to a maximum value of -3.9 [see Section 2.5.2 and Eq. (2-29)]. If dipole–dipole relaxation is inefficient, $1 + \epsilon$ can be less than one or even zero, defeating the detection process. This problem is to be expected when τ_c is long (Hawkes *et al.*, 1975).

A relatively recent improvement in ^{15}N signal enhancement is based on transfer of polarization from protons to the nitrogens (Morris, 1980). This is basically a liquid-state version of the polarization method used in high-resolution solid-state nmr (see Section 2.7.1.1.2). Synchronized pulse irradiation sequences at both ^1H and ^{15}N resonant frequencies may be followed by decoupling during the FID interval. Alternatively, the decoupling may be omitted. The method is distinct from NOE enhancement, producing a substantial improvement in S/N_p without risking the possibility of signal degradation inherent in the NOE peculiarity of ^{15}N.

At least two advantages of ^{15}N-nmr are easily appreciated. First, since nitrogen is less abundant, spectra are correspondingly less cluttered, and the individual polymer residues are more readily resolved than is the case for either carbon or hydrogen, in peptides and polynucleotides. This can be appreciated from the relatively simple ^{15}N spectrum of gramicidin S (compared with the ^{13}C spectrum) shown in Figure 2-39 (Khaled et al., 1978; Komoroski et al., 1975). Second, nitrogen is more directly involved in metal ion coordination and solvent exchange processes than carbon, and while hydrogen actively participates in solvent exchange, hydrogen, like carbon, is a more indirect reporter of metal coordination. One may expect to observe marked effects in the ^{15}N spectrum. Indeed, the chemical shifts and linewidths of ^{15}N resonances are quite sensitive to paramagnetic ions (Morishima and Inubushi, 1978; Farnell et al., 1972), including impurities, necessitating careful control of sample preparations in spectroscopic measurements. Thus, ^{15}N is probably the method of choice in probing metal ion coordination through nitrogen-containing pendant groups, e.g., especially those of asparagine, glutamine, lysine, histidine, arginine, and tryptophan.

    2.7.1.4.1 Conformation Studies with Peptides Using ^{15}N-Nmr. There are several nmr approaches to the determination of solvent-exposed and solvent-shielded groups. The irradiation of solvent protons has been observed to create an intermolecular nuclear Overhauser effect (NOE) (Pitner et al., 1975). This method is capable of delineating between interior and exterior nuclei, i.e., the NOE will be greatest for those nuclei presented to the solvent layer and therefore exposed to exchange processes. Pitner et al. (1975) applied this method to gramicidin S (GS), a cyclic peptide with the structure shown in Figure 2-40 (see Figure 2-39) in which the two repeating sequences are magnetically equivalent in solution. The intermolecular NOE is consistent with other evidence for a secondary structure involving intramolecular hydrogen bonds at all of the amide

Figure 2-39. A comparison of ^{13}C- and ^{15}N-nmr spectra of gramicidin S. The point group symmetry of this cyclic peptide reduces the number of nonequivalent nitrogens to six. The ^{15}N-nmr spectrum shown here accounts for the five unique amide environments. A pendant group nitrogen of ornithine is not shown. The aliphatic ^{13}C resonance spectrum is conspicuously more cluttered simply because there are more unique carbon atoms in the molecule. Adapted from Komoroski et al. (1975) and Khaled et al. (1978).

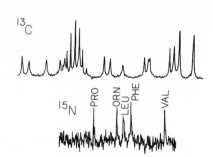

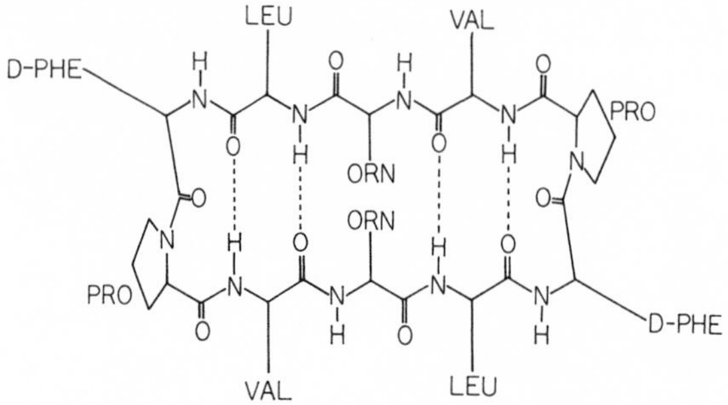

Figure 2-40. Structural features of gramicidin S. Intramolecular hydrogen bonds are indicated by the dotted lines. Adapted from Khaled *et al.* (1978).

nitrogens except those of phenylalanine and ornithine (Stern *et al.*, 1968; Urry and Ohnishi, 1970; Pitner and Urry, 1972).

Studies with ^{15}N-nmr have also produced insight into the structure of GS, which is a model for conformational effects. Khaled *et al.* (1978) obtained a tentative assignment of the five amide nitrogen resonances observed in GS (Figure 2-39) through comparisons with several small peptides containing the appropriate amino acids. The nitrogen spectrum of GS may be enhanced by proton decoupling, as noted above. Under such conditions a selective deuterium exchange should cause a loss of NOE enhancement at solvent-exposed nitrogen atoms. Consistent with this model, the addition of D_2O to a dimethyl sulfoxide solution of GS under conditions of proton decoupling resulted in an attenuation of the ^{15}N resonances assigned as phenylalanine and ornithine amides (Khaled *et al.*, 1978).

In related studies, a combination of ^{1}H, ^{13}C, and ^{15}N-nmr measurements provided evidence of solvent-inaccessible hydrogen bonding in the antibiotic viomycin (Hawkes *et al.*, 1978). In this case, dissolved viomycin could be compared with its X-ray crystallographic structure (Bycroft, 1972). The findings were consistent with closely similar bonding effects and geometries in both crystalline and solution phases.

^{15}N-Nmr would appear to be a particularly good reporter of conformation in view of the proximity of nitrogen to hydrogen atoms involved in hydrogen bonding and solvent exchange. Computerized methods of conformation analysis have been mentioned (see 2.7.1.2.2). It should be noted that these depend on knowledge of the primary sequence of amino acids. Used in conjunction with nmr evidence of intramolecularly hydrogen bonded residues, the number of possible conformational states may be considerably narrowed.

2.7.1.4.2 Nucleic Acids. ^{15}N resonance shifts (relative to external $H^{15}NO_3$) of nitrogen atoms in selected purines and pyridines are collected in Table 2-4 (Levy and Lichter, 1979). The range of shifts is considerable,

Table 2-4. ^{15}N Chemical Shifts of Nucleotides[a,b]

	Uridine	Thymidine	Cytidine	Guanosine	Adenosine
N-1	142.4	142.7	151.8	146.0	234.4
N-3	156.5	154.5	207.9	164.5	221.3
N-7	—	—	—	245.4	239.3
N-9	—	—	—	168.7	168.4
NH$_2$	—	—	93.0	72.0	80.2

[a]In $(CH_3)_2SO$, 0.5–1.0 M. Measured with respect to external HNO_3, conversion constant = 374.0 ppm.
[b]Adapted from Levy and Lichter (1979).

suggesting that ^{15}N-nmr will be useful in structural studies of nucleic acids, particularly the smaller species such as tRNA.

2.7.1.4.3 Alternate Detection Methods. Single bond heteronuclear coupling constants between ^{15}N and ^1H (i.e., J_{NH}^1) in amides are on the order of 90 Hz (Llinas *et al.*, 1970). Heteronuclear INDOR measurements of the type $^1H\{^{15}N\}$ (meaning of the symbolism: ^1H is observed while ^{15}N is irradiated) permit a mapping of the ^{15}N spectrum by means of a suitably modified ^1H-nmr spectrometer, e.g., one in which the probe coil is double-tuned to permit the introduction of an irradiating signal at the ^{15}N frequency (Llinas *et al.*, 1976).

The method is conceptually simple, though time-consuming. ^1H-Nmr spectra are recorded while the narrow ^{15}N decoupling signal is stepped through its frequency range. Whenever the latter coincides with an amide ^{15}N resonance a corresponding ^1H–^{15}N doublet will be observed to collapse in the ^1H-nmr spectrum. The ^{15}N spectrum is thus mapped by noting the ^{15}N irradiation frequencies (determined accurately by a frequency counter) which produce decoupling effects in the ^1H-nmr spectrum.

Measurements of this type were used in conformational studies of alumichrome, an Al^{3+} replacement derivative of naturally occurring ferrichrome (Llinas *et al.*, 1976). The INDOR-mapped ^{15}N-nmr spectrum of alumichrome is shown in Figure 2-41. This technique appears to be limited to relatively small structures since line broadening and clutter impairs ^1H–^{15}N doublet resolution at higher molecular weights. Also such studies are best conducted with ^{15}N-enriched compounds, e.g., the cited authors produced ferrichrome by growing *Ustilago sphaerogena* on a medium containing 99.2% [^{15}N]ammonium acetate.

Llinas *et al.* (1976) also demonstrated that ^{15}N-coupled proton resonances could be identified at natural nitrogen isotope abundance by noise-irradiated internuclear difference spectroscopy. In the latter technique, alternate pulses are gated with ^{15}N noise irradiation. The difference

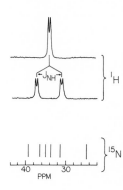

Figure 2-41. Heteronuclear INDOR mapped ^{15}N spectrum of the cyclic peptide/Al^{3+} complex alumichrome. Actual measurements are made in the ^{1}H-nmr spectrum while an irradiating frequency is swept through the ^{15}N spectral region. The top ^{1}H-spectral feature is representative of a measurement at natural abundance ^{15}N. This feature is split by ^{15}N–^{1}H spin–spin coupling when the sample contains 100% ^{15}N, as shown in the lower ^{1}H spectrum. If the enriched sample is irradiated at the appropriate ^{15}N frequency, the splitting collapses and the spectrum will again appear as in the top scan. The line positions marked in the lower ^{15}N spectrum were obtained by stepping the ^{15}N INDOR signal through the nitrogen spectrum and noting positions where ^{1}H-nmr doublets were observed to collapse. The chemical shift reference is [^{15}N]urea in DMSO-d_6. Adapted from Llinas et al. (1976).

FID between irradiated and unirradiated transients is then Fourier transformed to obtain the ^{15}N-coupled proton spectrum at natural abundance.

2.7.1.5 Phosphorus

Natural phosphorus consists entirely of ^{31}P, an easily detected isotope with spin 1/2 and thus inherently narrow linewidths. The chemical shift range of ^{31}P is more than 300 ppm. Not surprisingly, ^{31}P-nmr has been relatively well-explored, including its biological applications (Mavel, 1973; Harris and Mann, 1979).

The obvious biochemical subjects for ^{31}P-nmr include phospholipids and biomembranes, polynucleotides, phosphorus-containing proteins, and a variety of phosphorylated intermediary metabolites. Some of these will be examined below to illustrate the many potential applications of ^{31}P-nmr.

2.7.1.5.1 ^{31}P-Nmr of Nucleic Acids. Phosphorus signals originating in whole cells have been used to characterize differences between normal and malignant tissues (Zaner and Damadian, 1975). The presence of marked abnormalities in the chromosomal material of neoplastic cells suggested to these investigators that nucleic acid structural modifications might be reflected in the ^{31}P-nmr resonance signal. Measurements of spin–lattice relaxation times confirmed significant differences between normal liver (2.33 ± 0.14 s) and Novikoff hepatoma (5.98 ± 0.57 s). The cited authors acknowledge that other cellular factors could contribute to these differences, e.g., alterations in the membrane phospholipids. The point to be made here is that nmr is applicable to complex systems under favorable conditions, and the measurement process is nondestructive.

Phosphorus nmr has been used to probe helix-to-coil transitions of native DNA (Mariam and Wilson, 1979). A total of four partially resolved resonances were detected in both calf thymus and salmon sperm DNA. Signals from the helical and coil forms were observed simultaneously near T_m, and an upper limit could be placed on the interconversion rates ($<<36$ s^{-1} at 70°C). At still higher temperatures three signals (and possibly a fourth) were detected, suggesting three conformer species. The relatively small chemical shift differences were interpreted as evidence that the gauche–gauche conformation occurs in both helical and coil DNA forms.

Probes of the phosphate ester linkages in high-molecular weight DNA (or RNA) are not likely to encounter particularly large chemical shift differences unless the strands are oriented with respect to the field (see Section 2.7.1.5.2). The processes of binding with metal ions and other entities may act to produce more substantial effects. The much smaller transfer-RNA structures are inherently more amenable to study.

2.7.1.5.2 ³¹P-Nmr of Phospholipids and Membranes. Phospholipids (along with stabilizing proteins) aggregate spontaneously to form the characteristic bilayers of biological membranes. A typical phospholipid molecular structure (distearoyl phosphatidylcholine) is shown in Figure 2-42. From this structure it may be seen that the single, magnetically dilute atom of phosphorous occurs in a diester linkage. Magnetic dilution opens the possibility of high-resolution cross-polarization measurements (Pines *et al.*, 1973) in the mesomorphic environment of lipid bilayers. Chemical shift anisotropy dominates the spectrum of phospholipids organized in bilayer phases (e.g., membranes and artifically prepared liposomes) since one degree of translation and two of rotation are considerably restricted.

It is necessary to first characterize the orientation of the chemical shift tensor with respect to the diester linkage before attempting to interpret spectral effects in bilayer phases. In one approach to this problem, single-crystal cross-polarization measurements were obtained for the simple phosphate diester barium diethyl phosphate (BDEP) (Herzfeld *et al.*, 1978). The latter's crystal structure belongs to space group *I2/a* (Kyogoku and Iitaka, 1966) and as a result of symmetry contains only two magnetically nonequivalent phosphates per unit cell.

In confirmation, the single-crystal ³¹P-nmr spectrum exhibits only

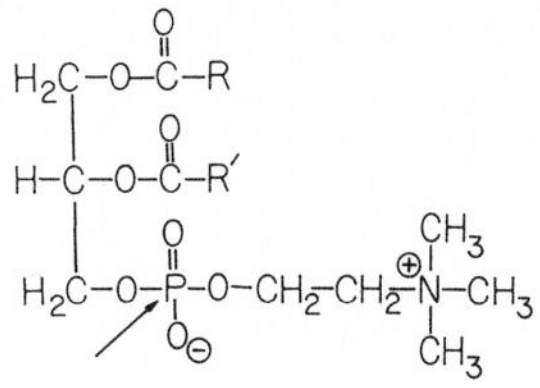

Figure 2-42. A typical phospholipid. The R groups are aliphatic portions of saturated or unsaturated fatty acids. ³¹P-nmr spectra are simplified as a result of the molecule having only one phosphorus atom.

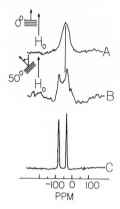

Figure 2-43. Splitting due to chemical shift anisotropy. Recording were obtained with the bilayer sheets oriented (A) perpendicular to and (B) tilted 50° with respect to the magnetic field. The lower recording shows anisotropic shift splitting observed in solid-phase BDEP. Adapted from Herzfeld *et al.* (1978).

two resonances. These depend on the angle between crystal axes and H_0 of the nmr spectrometer (Figures 2-43 and 2-44; Herzfeld *et al.*, 1978). Principal values of the tensor (in ppm) were found to be $\sigma_{11} = -75.9 \pm 1.2$, $\sigma_{22} = -17.5 \pm 0.2$, and $\sigma_{33} = +109.8 \pm 2.5$ relative to external $H_3{}^{31}PO_4$. The values σ_{22} and σ_{33} lie very near the O—P—O plane defined by phosphorus and the nonlinkage oxygens while σ_{11} is directed nearly perpendicular to the plane. Also, σ_{22} is close to the bisector of the O—P—O angle. These results are comparable to independent measurements on phospholipids using other methods (Kohler and Klein, 1976, 1977).

A monodomain of lipid bilayers may be formed by sandwiching a sample of phospholipid between silane surfactant-treated glass plates, annealing at 125°C for several hours, and cooling slowly over a period of hours (Powers and Clark, 1975). The resulting domain consists of bilayers oriented plane-parallel to the glass plates, i.e., in the fashion of a stack of waffles.

Monodomains of hydrated dipalmitoyl phosphatidylcholine (DPPC) have also been examined by ^{31}P-nmr (Griffin *et al.*, 1978). When the bilayer

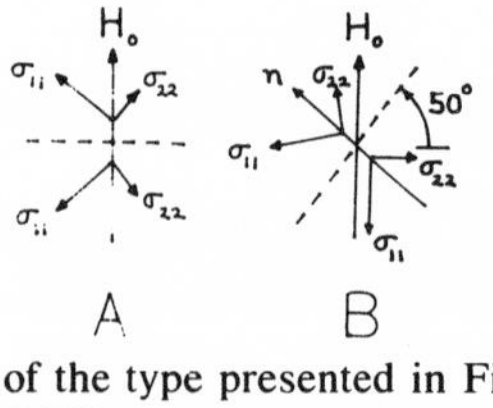

Figure 2-44. Chemical shift tensor components in oriented phospholipid bilayers. In case A the vector resultants of σ_{11} and σ_{22} are virtually identical, and a single resonance is observed. The dashed line denotes the bilayer plane, which is normal to the applied field vector. In B the bilayer normal (n) has been tilted 50° with respect to the applied field. The resultants of σ_{11} and σ_{22} are now unequal, and a two-line spectrum of the type presented in Figure 2-43(B) may be anticipated. Adapted from Herzfeld *et al.* (1978).

sheets are perpendicular to H_0 a single broad resonance is observed, as shown in Figure 2-43. However, on tilting the domain with respect to the magnetic field two broadened resonances appear. (Also shown for comparison in Figure 2-43 are the two resonances of single crystal BDEP.)

Based on the above cited shift tensor orientation within the phosphodiester group the results with DPPC are consistent with an average 50° tilt angle between O—P—O and bilayer planes. Figure 2-44 shows how the doublet structure might arise, i.e., equivalence occurs with H_0 normal to the bilayers but rotation presents differing tensor values above and below the bilayer planes. Layer linewidths observed for mesomorphic DPPC compared with solid BDEP are indicative of dynamic disorder in the monodomain.

Studies of this type are complementary to other methods. For example, low-angle X-ray diffraction measurements or mesomorphic structures yield spacings between layers and planes but not interatomic angles. In this case the evidence for a 50° tilt of the O—P—O plane was unanticipated (Stryer, 1975).

Changes in the conformation of phospholipid head groups may be detected through the nuclear Overhauser effect (NOE) (Yeagle et al., 1977a,b). In phospholipids the dipolar interaction between ^{31}P and ^{1}H involves both $-N(CH_3)_3{}^{+}$ and $P-O-CH_2-CH_2-$ protons of the choline moity. NOE double resonance experiments of the type $^{31}P\{^{1}H\}$ (i.e., those in which the ^{31}P-nmr intensity is observed while the ^{1}H spectrum is swept with an irradiating signal) produce the broad resonance labeled A in Figure 2-45. The abscissa of Figure 2-45 corresponds to the ^{1}H-nmr resonance range. Curve A was obtained from egg phosphatidylcholine vesicles. Spectrum C is the high resolution ^{1}H-nmr recording for egg phosphatidylcholine, and one may note that the maximum of the NOE curve coincides with the position of the narrow resonances due to $-N(CH_3)_3{}^{+}$ protons.

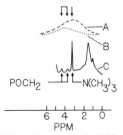

Figure 2-45. Nuclear Overhauser effects (NOE) in phospholipid vesicles. Curves A and B are NOE spectra obtained by recording the intensity of the ^{31}P-nmr signal while sweeping an irradiating signal through the ^{1}H-nmr spectral region. The reference curve (C) is the ^{1}H-nmr spectrum of egg phosphatidylcholine. All three curves are reported in ^{1}H-nmr shifts (ppm) relative to TMS. It is seen that the maximum in the NOE curve for phospholipid along (A) reflects Overhauser effects originating in the $N(CH_3)_3$ protons. On adding cholesterol, the NOE mechanism is altered to involve the $POCH_2$ protons, clearly indicating a conformational change in the polar region of the phospholipid. Adapted from Yeagle et al. (1977 a,b).

On adding cholesterol to the vesicles the NOE curve becomes less intense and shifts downfield, centering over the P—O—CH$_2$—CH$_2$— proton resonances (B). This is indicative of a disruption of head-group interactions due to cholesterol (Yeagle et al., 1977b). Clearly, cholesterol is producing conformation effects in the polar regions of the vesicle even though the steroid's structure suggests more affinity for the hydrocarbon chains. Similar results were obtained with native low-density lipoproteins (Yeagle et al., 1978).

2.7.1.5.3 pH Measurements Using ^{31}P-Nmr. Protonation or deprotonation of a phosphate oxygen would be expected to alter the electron density at phosphorus and, as a result, the chemical shift of the ^{31}P resonance. This expectation is indeed confirmed, and pH-dependent ^{31}P resonance effects are sensitive probes of pH at all levels of complexity, including whole bacterial cells (Navon et al., 1977) and cells, tissues, and organelles (Hollis, 1981). Hydrogen ion shifts are observed for inorganic phosphate and organic forms such as adenosine triphosphate (ATP). A pH calibration curve for the γ-phosphorus resonance of ATP is shown in Figure 2-46.

Using the method cited above, Njus et al. (1978) were able to follow matrix pH changes in chromaffin granules. The quiescent pH was established at 5.65 and was observed to drop by half a unit when external ATP was added, providing evidence for an active proton pump in this organelle's membrane. Chemical shift differences were sufficient to permit a distinction between internal and external ATP. One will note that this method makes use of an intrinsic probe, minimizing disturbances to the system. Lachmann and Schnackerz (1984) describe a method for simultaneously evaluating all pH-dependent resonances.

2.7.1.5.4 Selective Binding Effects Observed with ^{31}P-Nmr. Acid–base titration methods have also been used to observe structural differences among a group of homologous mouse myeloma antibodies (Goetz and Richards, 1978). The immunoglobulins have binding specificity for the haptens phosphorylcholine and L-α-glycerophosphorylcholine. Compared with unbound phosphorylcholine, titration curves of hapten bound to the

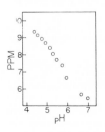

Figure 2-46. Adenosine triphosphate (ATP) as a probe of pH. The ordinate is ^{31}P chemical shift of the gamma phosphate group of ATP expressed in parts per million relative to 88% phosphoric acid. The marked dependence of this resonance upon pH forms the basis of a method for probing the local pH within small vesicles and compartments. Adapted from Nijus et al. (1978).

various antibodies differed not only in pK but also in chemical shift range. The ^{31}P-nmr data demonstrate that while all of the antibodies have the same function, the binding specificities involved differ significantly at the molecular level.

2.7.1.5.5 Kinetic Studies with ^{31}*P-Nmr.* Phospholipase A hydrolysis of the phospholipids of human serum high-density lipoprotein-3 has been followed using ^{31}P-nmr (Brashure *et al.*, 1978). In this reaction, phosphatidylcholine (PC) is converted to lysophosphatidylcholine (LPC). The phosphorus reasonances of PC and LPC differ enough that the reaction's progress may be followed, as shown in Figure 2-47, where deconvoluted peak areas may be related to concentration. PC hydrolysis was found to obey first-order kinetics with $k_{exp} = 3.35 \times 10^{-5}$ s^{-1} ([enzyme]~8.0 × 10^{-9} M; [PC]~9.5 × 10^{-3} M), in agreement with results obtained by pH-stat titrations (Pattnaik *et al.*, 1976). Other workers have used ^{31}P-nmr in studies of enzymic reactions (Cohn and Rao, 1979).

A full treatment of kinetic methods will be found in Chapter 10 of this volume. Again, as in the other cases cited here, kinetic nmr spectroscopy is able to make use of the intrinsic probes contained within biomolecules and thus minimizes disturbances to the system being observed. The more conventional methods for following reaction kinetics (e.g., color-forming reactions) are susceptible to interferences, especially in the complex biological environment. With the availability of multinuclear nmr instruments, kinetic probes have access to most of the biological elements.

Kinetic nmr adaptations may be used to follow the progress of enzyme-controlled reactions in systems of varying levels of sophistication. As an example, the incorporation of glycine-^{1}H$_5$ into *whole erythrocytes* was studied by spin-echo ^{1}H-nmr spectroscopy, which revealed time-dependent resonance changes at 3.54 and 3.76 ppm associated with entry

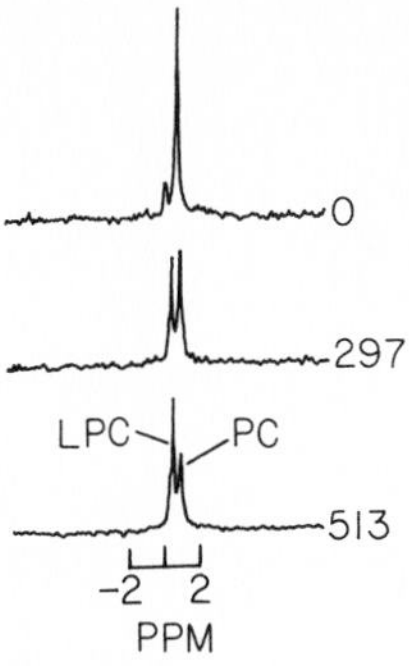

Figure 2-47. The enzymatic hydrolysis of phosphatidylcholine (PC) to lysophosphatidylcholine (LPC) as detected by ^{31}P-nmr. The chemical shifts of the two compounds are different, and the kinetics of the reaction may be followed quantitatively. Kinetic nmr spectrometry is applicable to all types of nmr isotopes. Adapted from Brashure *et al.* (1978).

of the glycine label into the cellular free glycine pool (Anvarhusein and Rabenstein, 1979).

2.7.1.6 Oxygen and Sulfur

Since there are no nonquadrupolar isotopes of sulfur or oxygen and both elements are found in polar covalent environment, studies oriented to biological macromolecules are difficult due to excessive line broadening. The use of ^{17}O ($I = 5/2$) as a valuable structural probe in the low-molecular weight range has been reviewed (Klemperer, 1978). Linewidth effects are minimized by (1) conducting measurements at elevated temperature and (2) using a low-viscosity solvent in which the solute concentration is kept low, which reduces τ_c according to the Stokes–Einstein equation [Eq. (2-31)]. These conditions are not applicable to solutions containing proteins or other biological polymers, although a few biological studies have been attempted (Rodger et al., 1979).

$$\tau_c = 4\pi\eta a^3/(3kT) \tag{2-31}$$

where η is viscosity.

The chemical shift of ^{17}O is large, as shown in Figure 2-48. Isotope enrichment and FT detection methods which compensate for the receiver recovery problems associated with brief interpulse intervals (Canet et al., 1977, 1982) improve the accessibility of this inherently difficult isotope. The ^{17}O linewidth in SO_4^{2-} is on the order of 100 Hz, and in the MoO_4^{2-} ion a relatively sharp line ($\sim$5 Hz) is observed (Lutz et al., 1976b). Interactions between proteins and ions such as SO_4^{2-} (or any other small structure which contains oxygen but lacks a more convenient isotope) can be detected using the ^{17}O resonance.

Studies with sulfur are more difficult, e.g., the ^{33}S resonance of 1 molal aqueous Cs_2SO_4 is a factor of 2×10^{-7} less than that of ^1H (Lutz et al., 1976b). An alternative route to an intrinsic probe of ^{17}O and other quadrupolar species such as ^{33}S is nqr–nmr double resonance (see Section 3.4.1).

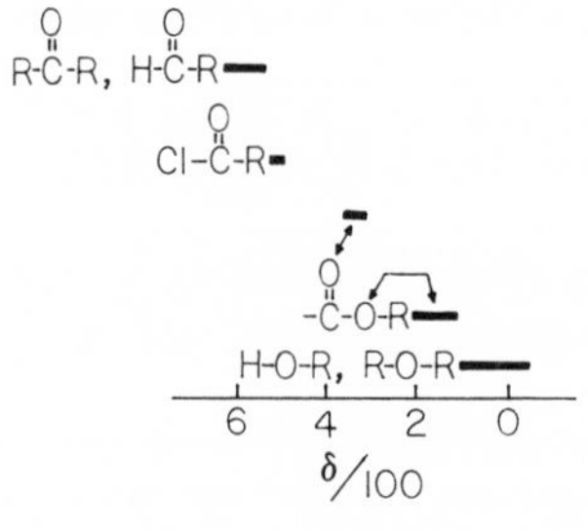

Figure 2-48. Selected chemical shifts of the ^{17}O-nmr resonances in oxygen-containing organic structures. The shift scale in ppm is reported with respect to pure water. It is seen that the ^{17}O-nmr shift range is large, being well over 600 ppm for the types of compounds shown here. Adapted from Klemperer (1978).

2.7.1.7 Selenium

Selenium-77 exists in relatively low natural abundance (7.5%) and has a poor intrinsic sensitivity (Table 2-1), making it a difficult subject for investigation in view of its low concentration in biological forms (e.g., glutathione peroxidase). In compensation, the known chemical shift range of ^{77}Se is ~1500 ppm (Gronowitz et al., 1975), and the absence of a quadrupole moment results in sharp lines and stereoselective spin–spin splitting (Reich and Trend, 1976) in small molecules. ^{77}Se-Nmr parameters of organoselenium forms likely to appear in biomolecules are collected in Table 2-5.

In a preliminary study of ^{77}Se relaxation (Dawson and Odom, 1977) it was found that spin rotation was the predominant T_1 mechanism, i.e., T_1 increases with decreasing temperature or restricted motion when τ_c is long. Since the dipolar mechanism was found to be ineffective in small molecules and will probably be the only T_1 relaxation mechanism in a macromolecular environment, the rate of acquisition of data by FT methods may be compromised in studies of protein-bound Se since the spin system must return to thermal equilibrium between pulses. This could be tens of seconds in the case of ^{77}Se. The large covalent radius of Se mitigates not only against the dipolar relaxation mechanism but also against the nuclear Overhauser effect (see Section 2.5.2), NOE, which is weak

Table 2-5. ^{77}Se-Nmr Parameters of Some Organoselenium Compounds[a,b]

Compound	Temperature (°C)	T_1^{obs}[c] (S)	Chemical shift[d] (ppm)	$\Delta\nu_{1/2}$[e] (Hz)
$(CH_3)_2Se$	60	24.4	0	0.5
	32	7.5		
C_2H_5SeH	−60	9.5	38.7	0.7
	40	1.6		
CH_3SeH	−55	4.3	−130.3	0.5
	40	1.4		
$(C_6H_5CH_2Se)_2$	18	31	411.5	1.4
	55	27		
$(C_6H_5Se)_2$	0	31		
	45	20	480.6	1.1
$(CH_3Se)_2$	0	13	280.6	1.0
	45	9		

[a]Adapted from Dawson and Odom (1977).
[b]Measurements as 19.1 MHz in $CDCl_3$ solvent.
[c]Inversion–recovery pulse sequence. Estimated error <10%.
[d]A positive chemical shift is deshielded with respect to Me_2Se.
[e]Half-height linewidth.

even in the single bond ^{77}Se–^{1}H interaction of selenols (Durig and Bucy, 1977). Maximum possible NOE is 2.62. A less pessimistic view was expressed more recently (Rodger *et al.*, 1979).

Isotope enrichment can produce more than order of magnitude of improvement in the ^{77}Se signal and is advisable in biological studies. In a case like this, a paramagnetic probe, e.g., a Faller–LaMar reagent such as Gd^{3+} (Farnell *et al.*, 1972), might be used to introduce a spin–lattice relaxation mechanism.

2.7.1.8 The Halogens

2.7.1.8.1 Fluorine. All naturally occurring fluorine is in the form of ^{19}F. This isotope is easily detected, has a chemical shift range nearly 100 times that of hydrogen, and does not have a quadrupole moment. Unfortunately, soluble biological molecules containing covalently bound fluorine are uncommon, rare examples being the extremely toxic monofluoroacetate of the *Dichapetalum cymosum* (gifblaar) plant (Chenoweth, 1949) and ω-fluoro fatty acids of related species (Glusker, 1971). Beyond these rare forms, the more general utilization of fluorine is in connection with bone structure (Hoekstra *et al.*, 1974). High-resolution solid-state nmr has been applied to fluorine compounds (Ellett *et al.*, 1970), but biological forms of the element have received little attention.

In contrast, extrinsic probes which contain ^{19}F may be used to circumvent solvent absorptions, which often hamper ^{1}H-nmr measurements, while taking advantage of an easily detected, uncluttered resonance. An example of this type of application is a study of 5-fluorouridine-substituted valine tRNA (Horowitz *et al.*, 1977). Fluorine-substituted substrates and proteins may be used to probe selective binding interactions and conformation (Roberts *et al.*, 1977; Bendall and Lowe, 1976). The label may be a substrate, e.g., a fluorinated carbohydrate, or a modified amino acid residue such as 3-fluorotyrosine or 6-fluorotryptophan (Gerig, 1981). These very effective extrinsic labels retain biological activity. The relative ease of detection of ^{19}F, nearly equal to that of hydrogen, has permitted its use for *in vivo* metabolic studies, e.g., in investigations of the incorporation of 5-fluorouracil into RNA and metabolites (Gochin *et al.*, 1984).

2.7.1.8.2 Chlorine, Bromine, and Iodine. The physicochemical and biological applications of chlorine-, bromine-, and iodine-nmr have been reviewed (Lindman and Forsen, 1976, 1979). All isotopes of these elements are quadrupolar with moments increasing in the order Cl $<$ Br $<$ I. Apart from its utilization by some marine plants (Hoekstra *et al.*, 1974), bromine is the least significant of the three. The more ubiquitous ^{127}I appears in covalent compounds and experiences severe line broadening even in structures of modest molecular weight as a consequence of its large quadrupole moment ($Q \sim -0.7$ barns). Nuclear quadrupole resonance may have far more potential as a direct probe of iodine in its biomolecular forms and will be treated in Chapter 3. Of the three elements, chlorine has widespread biological significance as the simple anion. Consequently, ^{35}Cl-nmr studies of the chloride ion may be conducted under biologically realistic conditions.

Linewidths ~ 10 Hz are observed in aqueous solutions of ^{35}Cl$^-$. The detection of this isotope is comparable to that of ^{13}C in natural abundance. When substances with an affinity for Cl$^-$ are added to NaCl or KCl solutions, marked broadening effects are observed (rather than shifts). ^{35}Cl-Nmr can be used to probe binding interactions by virtue of its enhanced quadrupolar relaxation when τ_c is extended and gradients are introduced and is thus similar to ^{23}Na, ^{25}Mg, and ^{39}K in the manner of data analysis (see Section 2.7.2).

^{35}Cl-Nmr has been used (Shami et al., 1977) to confirm kinetic evidence (Dalmark, 1976; Passow, 1969) of anion binding to the red cell band-3 transport protein (Rothstein et al., 1975) prior to ion translocation. Information obtainable by nmr is unique, i.e., inaccessible by conventional methods, due to the apparent low affinity of the transport site for Cl$^-$ ($K_d = 67$ mM). ^{35}Cl-line broadening was observed for four red cell extracts (ghosts; a band-3-rich Triton X-100 extract; sialoglycoproteins; and phospholipids) and controls (e.g., Triton X-100). The largest broadening effect, 6 Hz/mg protein, was observed for the band-3 fraction. Furthermore, line broadening by this fraction was markedly diminished by the irreversible anion transport inhibitor 4,4'-diisothiocyano-2,2'-stilbene disulfonic acid (DIDS) in the pH range 6.5–8.5.

Many other applications are possible in view of numerous Cl-dependent enzyme reactions and binding affinities. Results obtained in the previous case may be compared with other experiments in which resonances of a low-molecular weight entity are used to probe selective binding interactions (see Sections 2.7.2.1 and 2.7.3).

2.7.1.9 Main Group Elements of Lesser Abundance

2.7.1.9.1 Boron.   The nutritional requirement for boron among some members of the plant kingdom is well-established (Hoekstra et al., 1974). However, relatively little is known of the biochemistry of boron. Nmr studies with the isotope ^{11}B (80.4% natural abundance) are not difficult and a wealth of literature exists (Noth and Wrackmeyer, 1978). Biological applications of ^{11}B-nmr would be restricted to low molecular weights since this isotope has a quadrupole moment. Line broadening is a nuisance even in the spectra of small molecules but not enough to obscure one-bond $^{11}B-^1H$ coupling.

  2.7.1.9.2 Silicon. The resonance of ^{29}Si is relatively difficult to detect, but in compensation the spin is 1/2. Relatively long relaxation times create problems for FT detection modes (Ernst *et al.*, 1974). Silicon–transition metal interactions in nonbiological systems have been studied using ^{29}Si-nmr (Li *et al.*, 1979; Harris and Kimber, 1975).

  The presence of silicon as solid SiO_2 in the exoskeletons of diatoms and its relatively more recent status as an essential growth and development factor in higher animals (Hoekstra *et al.*, 1974) suggests that ^{29}Si-nmr may be applicable to biochemical questions. Silicon has been associated with connective tissue mucopolysaccharides (Schwarz, 1973), possibly functioning as a cross-linkage agent in a type of ether bridge, as shown in Figure 2-49. As such, silicon is associated with relatively immobile structures. Isotope enrichment and high-resolution solid-state methods are probably necessary in biochemical studies of silicon.

  2.7.1.9.3 Tin. The isotopes ^{117}Sn and ^{119}Sn have similar nmr properties, ^{119}Sn being slightly the favored of the two. These isotopes are amenable to pulsed FT nmr, but the nuclear Overhauser effect does not produce a net signal gain, requiring time-gated decoupling where decoupling is desired. Heteronuclear double resonance of the type $^1H\{^{119}Sn\}$ may be used with relatively simple structures. This terminology means that the irradiating frequency sweeps the ^{119}Sn range while a ^{119}Sn satellite line is observed in the ^1H spectrum (Smith and Tupciauskas, 1978). The method detects only those tin nuclei coupled to hydrogen.

  As in the case of silicon, tin is of interest as a result of its role as an essential element in the nutrition of higher animals (Hoekstra *et al.*, 1974), although relatively little is known of its biochemistry.

  2.7.1.9.4. Lead. The isotope ^{207}Pb ($I = 1/2$) is potentially applicable to studies of the toxicity of this element. Lead may be involved in animal nutrition (Hoekstra *et al.*, 1974).

2.7.2 Metal Ions

2.7.2.1 *Sodium*

  Accumulated experience with ^{23}Na-nmr spectroscopy does not begin to approach that with ^1H or ^{13}C, but the existing literature has been

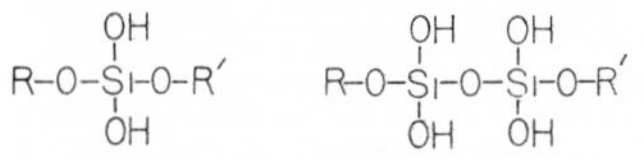

Figure 2-49. Possible biological forms of silicon. The R and R' groups are mucopolysaccharide structures.

reviewed (Civan and Shporer, 1978; Laszlo, 1978). The characteristics of this isotope (Table 2-1) encourage applications in biological studies. First, all naturally occurring sodium *is* ^{23}Na and the intrinsic sensitivity toward the isotope, fully one-tenth that of hydrogen, results in an easily detected resonance. Linewidths for ^{23}Na absorptions generally lie between 5 Hz and 100 Hz. As a result of these properties one may work with 10^{-3} *M* solutions using FT methods (Ernst and Anderson, 1966). In comparison, the concentration of sodium in many *extracellular* fluids is on the order of 10^{-1} *M*. It is fortuitous that soium itself is so readily detected as there are *no* isomorphous cation replacements for this ion (Williams, 1970).

^{23}Na$^+$ chemical shifts have small diamagnetic components of about ± 5 ppm when various solvents are compared, while the paramagnetic contribution may be as large as ± 25 ppm. An extreme case for sodium is the Na$^-$ ion, which is produced on dissolving sodium metal in cryptate-containing solvents (Ceraso and Dye, 1974), fully 63 ppm upfield from NaCl reference. Nevertheless, the range of shifts observed for ^{23}Na$^+$ are small for a heavy element, probably as a result of the relatively *low* polarizability of this ion.

The presence of a paramagnetic component may seem odd as Na$^+$ contains no *p*, *d*, or *f* electrons. This effect has been interpreted as a donation of electrons from solvating or coordinating groups into empty 3*p* orbitals, and the shifts in fact correlate with solvent donicity, as shown in Figure 2-50 (Gutmann and Wychera, 1966). Chemical shifts of ^{23}Na$^+$ are thus expected to correlate with the type of group with which the ion is interacting. There is no justification for describing ^{23}Na line shifts in terms of degrees of coordinate covalency, as theoretical studies strongly favor electrostatic rather than covalent bonding between sodium and its ligands (Dzidic and Kebarle, 1970). Until this question is resolved it will be safer to rely on empirical correlations. ^{23}Na-Nmr shifts will be anticipated when Na$^+$ associates tightly with low-molecular weight coordinating agents. However, in most biological interactions involving macromolecules the observed effect will be line broadening.

The sodium cation exchanges rapidly in most situations, and it is unlikely that shifts or spin–spin splitting patterns will be observed. How-

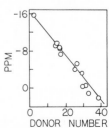

Figure 2-50. The chemical shift of ^{23}Na versus solvent donicity. Shifts in ppm are measured with respect to an aqueous solution of NaCl contained in a separate, coaxial sample cell. Adapted from Laszlo (1978).

ever, linewidths which have been corrected for viscosity (Plaush and Sharp, 1976) yield evidence of the net asymmetry experienced by the coordinated ion. This is a consequence of the model for quadrupole relaxation discussed earlier (see Sections 2.3.2.3 and 2.3.2.1). Even though the isolated sodium ion is spherically symmetric (where q would be zero), solvent fluctuations, temporary coordination vacancies, etc. (Laszlo, 1978) all lead to a nonvanishing electric field gradient in condensed phases and therefore finite linewidths according to Eq. (2-11). Because of the inherent asymmetry at protein/water interfaces, both *selective* and *nonselective* binding interactions with Na^+ are virtually assured of causing linewidth increases.

^{23}Na-Nmr has been used to study monovalent cation association with double-stranded DNA in aqueous solutions (Anderson *et al.*, 1978). Relative affinities of Na^+, tetraethylammonium, and tetrabutylammonium ion toward the nucleic acid were found to be respectively, 20:5:1, and the relaxation rate of bound $^{23}Na^+$ was determined to be $180 \pm 10 \ s^{-1}$. It should be noted that Na^+ is relatively mobile in this sort of environment, in contrast with the more tenacious coordinating qualities of the ionophoric agents mentioned above.

The ^{23}Na-nmr resonance has been used to probe complex systems. Studies of homogenates of halotolerant bacteria (Goldberg and Gilboa, 1978) produced evidence of three sodium environments: (1) an extracellular, aquated form, (2) a distinct "intracellular" form, and (3) a bound form which did not appear to equilibrate rapidly with the other two. The quadrupole coupling constant at site number two could be estimated at 9×10^6 rad/s with a correlation time of 5.5×10^{-7} s.

Coupling constants provide information about the symmetry of a coordinating site as noted above, e.g., small coupling constants reflect high symmetry while large values denote lower symmetry. One cannot make direct observations of tenaciously bound ^{23}Na (and other quadrupole ions) in high-molecular weight proteins since these forms experience line broadening. Rather, strong binding may be inferred through a *decrease* in the intensity of the metal ion resonance when protein is added to the solution. Rapid exchange within (but not between) multiple compartments may be the case if the FID signal of the metal ion proves to be a composite of two or more exponential decays (Goldberg and Gilboa, 1978). An example of computer fitting of relaxation data to decay functions will be given in the Section 2.7.2.2 (potassium nmr).

^{23}Na-Nmr has been used to measure the relative amounts of intracellular and extracellular sodium ion in blood samples (Pike *et al.*, 1984). Normally, the two environments are indistinguishable, i.e., the signals are superimposed, but if a water-soluble lanthanide shift reagent is added to the blood sample, the resonance due to external sodium is relocated considerably upfield or downfield (depending on the shift reagent used), yielding separated resonances due to the two kinds of sodium ion. Relative amounts are determined from the area ratio. This method involves fewer steps than the usual flame photometric technique, since the latter requires washing cells (removing external sodium ion) prior to measurements.

2.7.2.2 Potassium

Although ^{39}K is abundant (93.08%) the intrinsic sensitivity involved in its detection is less than 1/1000 that of ^1H. The situation for ^{41}K is even worse. Consequently, relatively few biological studies (Civan and Shporer, 1978) have involved potassium-nmr.

As in the above examples of ^{23}Na and other quadrupolar nuclei, binding leads to linewidth effects determined by electric field gradients in the bound environment [Eq. (2-11)]. Quadrupolar relaxation in ^{39}K is threefold more pronounced than in ^{23}Na as a result of the Sternheimer antishielding effect (Deverell, 1969). This property depends on the number of atomic electrons, and relaxation rates among the alkali metal ions follow the trend $^{87}Rb^+ > {}^{39}K^+ > {}^{23}Na^+$ in similar environments (Deverell, 1969).

The difficulty involved in the use of ^{39}K-nmr is exemplified by a study of K^+ binding in *whole* striated frog muscle (Civan *et al.*, 1976). Up to 3.8×10^5 FID transients were averaged to obtain usable (but still noisy) time domain signals of the type shown in Figure 2-51. No attempt was made to transform the FID to the frequency domain (to obtain the single, featureless line) but rather the data were trial-fitted by computer to decay functions to obtain relaxation times. T_1 and T_2 were thus determined using the data-fitting method in conjunction with multiple pulse sequences, and due regard was given to field homogeneity, as should be the concern in *all* relaxation studies. Time domain analysis of a *single* resonance is probably the method of choice and is applicable to most studies of "loose" quadrupolar species such as $^{23}Na^+$, $^{39}K^+$, and $^{35}Cl^-$. Fourier transformation to the frequency domain becomes necessary only when the spectrum contains a multiplicity of resonances.

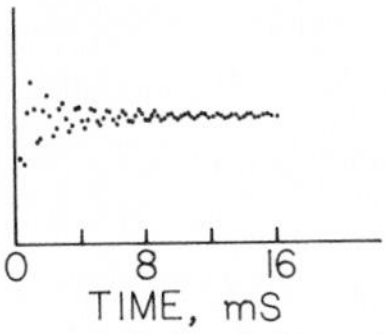

Figure 2-51. Transverse relaxation of ^{39}K nuclear magnetization (M_x) following 90° pulses. Data shown here were obtained in frog muscle at 5.5–7°C and required an accumulation of 383,293 transients in order to attain an acceptable S/N_p. The digitized FID signal gives a good fit with a function of the form $M_x = [7.33 \exp(-1.48t) + 3.32 \exp(-0.222t)] \cos (5.993 + 6.544t) + 15.0$. The early portion of the FID is best seen by viewing the page at a shallow angle. Adapted from Civan *et al.* (1976).

In the cited investigation (Civan *et al.*, 1976) of $^{39}K^+$ in frog muscle it was found that <1% of the intracellular K^+ is immobilized at binding sites (which are in rapid equilibrium with the cytosol). Similar results were obtained in studies of intracellular $^{23}Na^+$ (Berendsen and Edzes, 1973; Shporer and Civan, 1974). It should be noted that much more pronounced relaxation effects are to be expected in simpler systems, e.g., where K^+ is interacting with an isolated species functionally binding for the potassium ion. Nonspecific interaction of ions, substrates, and drugs with complex biological preparations is generally weak.

The isotope ^{87}Rb is an attractive alternative (Reuben *et al.*, 1975) to ^{39}K and ^{41}K in biological studies as a consequence of its natural abundance and an intrinsic sensitivity exceeding that of ^{23}Na (see Table 2-1). Justification for the use of Rb^+ is based on its ability to replace K^+; e.g., rodents survive briefly when K^+ is almost completely displaced by Rb^+, and certain microorganisms survive indefinitely under similar conditions (Suelter, 1974). As noted above (Deverell, 1969), line-broadening effects are more pronounced in ^{87}Rb-nmr (Reuben *et al.*, 1975).

Thallous ion, Tl^+, is chemically similar to K^+ (e.g., charge, ionic radius) and has some potential as a probe for K^+-related studies. However, one should realize that the toxicity of this element (Righetti and Moeschlin, 1971; Gosselin *et al.*, 1976) reflects fundamental differences. ^{205}Tl is particularly suitable for nmr in view of its abundance and intrinsic sensitivity (see Table 2-1). Absence of quadrupolar relaxation ($I = 1/2$) leads to narrow lines and heteronuclear spin coupling with protons (Briggs and Hinton, 1978a) in nonexchanging binding sites (e.g., actins). The chemical shift range of ^{205}Tl is ~700 ppm, and the chemical shift correlates with the Lewis basicity of donor ligands (Hinton and Briggs, 1975), e.g., carboxylate > alcohol > ether > carbonyl.

A "home-built" ^{205}Tl-nmr spectrometer is described in a study of the K^+ ionophores, monensin and nigericin (Briggs and Hinton, 1978b). Chloroform solutions of Tl^+/monensin reveal *two* temperature-dependent ^{205}Tl resonances rather than one, as shown in Figure 2-52, reflecting two species of the 1:1 complex in solution. The study also included complementary evidence from the ^1H-nmr spectrum and made note of crystallographic data for the two distinct forms of Tl^+/monesin$^-$.

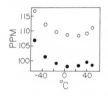

Figure 2-52. The temperature dependence of two ^{205}Tl-nmr signals observed for the Tl^+/monensin complex. The two resonances reflect two distinct forms of the complex in chloroform solutions. The ordinate of the graph presents the ^{205}Tl chemical shift in ppm relative to the shift of $TlNO_3$ in aqueous solution extrapolated to infinite dilution. Adapted from Briggs and Hinton (1978b).

2.7.2.3 Magnesium and Calcium

Although ^{25}Mg and ^{43}Ca are nmr-active isotopes, their "tighter" association with biopolymers can lead to severe line broadening according to Eq. (2-11). Without regard to linewidth effects, the difficulty in detecting ^{25}Mg is comparable to that for ^{13}C (both in natural abundance). ^{43}Ca is more difficult by an order of magnitude. The use of these probes depends on isotope enrichment, and a commercial source exists (Merck and Co., Inc.). Interestingly, ^{25}Mg and ^{43}Ca resonate at similar frequencies, and a single probe can easily adapt to both.

Exploratory studies with ^{25}Mg- and ^{43}Ca-nmr are relatively recent events (Magnusson and Bothner-By, 1971; Bryant, 1969). The linewidth of ^{25}Mg in 2 M aqueous solution is about 7 Hz, but broadens precipitously upon complexation, e.g., for the citrate complex, a relatively small entity, the linewidth was estimated to be $\sim$490 Hz (Magnusson and Bothner-By, 1971). One approach, applicable mostly at low molecular weight, is to estimate bound-state conditions from linewidth titrations, as shown in Figure 2-53, where at low pH the competition with H$^+$ at the ligands results in tractable linewidths (the study cited here appears to have been the first biologically oriented magnesium-nmr study). ^{25}Mg chemical shifts are generally obscured by these marked broadening effects.

Due to a much smaller quadrupole moment, chemical shift predominates over line broadening in the case of ^{43}Ca. Consequently, equillibrium studies may make use of Eq. (2-13) through (2-19) to estimate the fraction of metal ion in the bound state.

Examples of these effects were encountered in a study (Robertson et al., 1978) of ^{43}Ca^{2+} and ^{25}Mg^{2+} binding to γ-carboxylglutamate (D-Gla) peptides which have two such residues in sequence. D-Gla residues occur in the critical NH$_2$ terminal segment of prothrombin zymogen (unless

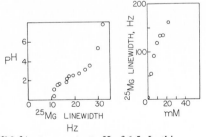

Figure 2-53. The linewidth of the ^{25}Mg-nmr resonance in the presence of chelating agents. The left-hand example is adapted from Magnusson and Bothner-By (1971) and shows the pH dependence of 2 M MgCl$_2$ interacting with 0.041 M adenosine triphosphate. A distinct titration curve is produced, from which equilibrium constants may be derived. The right-hand graph is adapted from Robertson et al. (1978) and shows the gamma carboxyglutamic acid peptide Z-D-Gla-D-Gla-OCH$_3$ interacting with ^{25}Mg^{2+} at a constant pH of 6.5. In this case there is little competition between H$^+$ and Mg^{2+} at the coordinating ligands, and it is seen that the linewidth of the ^{23}Mg resonance increases extremely rapidly. Fast-exchange conditions prevail in both examples.

carboxylation is blocked by warfarin) and are thus presumed to be involved in the Ca^{2+}-dependent thrombin activation of the blood clotting process. Usable spectra (S/N_p = 10/1) were obtained at 6.73 MHz but required the accumulation of as many as 40,000 transients with 79.98 atom% enriched ^{43}Ca when the total calcium ion concentration was 20 mM. Chemical shifts led to an estimated dissociation constant of ~0.6 mM, in good agreement with potentiometric data (Marki et al., 1977). A similar dissociation constant for Mg^{2+} binding was obtained from ^{25}Mg linewidths (Robertson et al., 1978).

An alternative method for probing calcium binding sites involves replacing Ca^{2+} with a paramagnetic lanthanide ion such as Pr^{3+} and observing shift (or line-broadening) effects in the nmr resonances of nearby nuclei. An extrinsic probe study of the latter type is discussed in Section 2.7.2.4.3.

2.7.2.4 Transition Elements

The biologically interesting transition elements (V, Cr, Mn, Fe, Co, Ni, Cu, and Mo) are especially poor nmr probes, at least where high-molecular weight structures are concerned. The potentially useful isotopes are ^{51}V, ^{55}Mn, ^{63}Cu, ^{57}Fe, and ^{59}Co since these are among the most easily detected (see Tables 2-1 and 3-1 and Figure 2-22). With the exception of ^{57}Fe (I = 1/2), all have quadrupole moments, are generally covalently attached to proteins, and will experience extreme line broadening in the macromolecular environment, where τ_c is long [Eq. (2-11)]. Furthermore, most of these exist as paramagnetic entities and are thus subject to an additional, efficient line-broadening mechanism. The latter condition does not apply to Cu(I). The same is also true of V(V), as in the case of the vanadate ion, VO_4^{3-}. Vanadochrome, the vanadium-containing protein of tunicates, contains paramagnetic V(III).

2.7.2.4.1 ^{63}Cu- and ^{65}Cu-Nmr. The linewidth of ^{63}Cu in an acetonitrile solution of $Cu(I)(CH_3CN)_4BF$ was found to be 540 Hz (Lutz et al., 1978). While this is not intractable for an isotope experiencing large chemical shifts, its occurrence in a low-molecular weight structure with virtual tetrahedral symmetry (i.e., one with a vanishing electric field gradient) clearly indicates that linewidths will be broadened beyond detection in biopolymers.

Copper is tightly bound in its various metalloprotein and metalloenzyme forms and in no known case does Cu(I) equilibrate rapidly between the solvent and a weak binding site, reducing the opportunity for two-site line-broadening effects of the type observed in alkali metal nmr (see

Section 2.7.2.1). It should be noted that the resonant frequencies at constant field for ^{23}Na and ^{63}Cu differ by less than 0.3%, and an instrument optimized for the former isotope may be changed to the latter simply by replacing the transmitter crystal (or altering the frequency synthesizer setting). ^{63}Cu is both abundant and easily detected (similar to ^{23}Na) but is limited by the cited linewidth effects, which indeed confine nmr studies to small complexes. Similar results may be anticipated for ^{63}Cu (Lutz *et al.*, 1978).

What is true of diamagnetic copper complexes applies as well to diamagnetic forms of the other transition elements (except ^{57}Fe; see Section 2.7.2.4.2). The example given for copper is intended to illustrate the difficulty involved in using transition metals as intrinsic probes of their metalloproteins. Barring an ingenious development in detection methods, this situation is unlikely to change.

2.7.2.4.2 ^{57}Fe-Nmr. Use of the very hard-to-detect ^{57}Fe nucleus is compromised by the *absence* of an efficient relaxation mechanism, i.e., ^{57}Fe presents the problems associated with extreme *line narrowing*. Slowly relaxing nuclei require long interpulse periods in FT nmr detection methods since the FID signal must decay to the noise level before a new excitation pulse can be applied. This obviously defeats the advantage usually associated with pulsed FT nmr. Similarly, correlation nmr has no advantage over CW methods since the scan rate must be slowed to permit the signal ringing to decay within the selected scan window.

Iron is diamagnetic in a strong octahedral ligand field, but most known biological forms of iron are paramagnetic (exception: the O_2–hemoglobin complex is diamagnetic), and it seems unlikely that functional forms of iron would involve a symmetrically closed coordination field.

^{57}Fe has been used to replace ^{56}Fe in ferredoxins, thus facilitating Mössbauer measurements and creating esr hyperfine splitting patterns (Orme-Johnson, 1973). Electron spin density may be estimated from the magnitude of hyperfine splitting constants (see Chapters 4 and 5). Mössbauer and ESR spectroscopies are thus preferable where iron is concerned.

From this we may conclude that biological nmr studies with transition metal isotopes are currently severely hampered, although work with low-molecular weight entities is possible (Kidd and Goodfellow, 1979). Furthermore, there are no *good* isomorphous replacements for this unique group of elements. *Direct* observations of the transition metals are best achieved through their paramagnetism (esr, static magnetism, and MCD; see Chapters 5, 6, and 9). It remains that *indirect* observation of a paramagnetic ion in a macromolecular environment is possible through effects—often striking—in the nmr spectra of other biological isotopes (such as ^{1}H or ^{13}C). These methods will be discussed in the immediately following sections.

2.7.2.4.3 Contact Shifts, Pseudocontact Shifts, and Paramagnetic Broadening as Applied to the Transition Metals.

Nmr methods which examine the effects of paramagnetic centers in biological systems were treated in an excellent review (Morris and Dwek, 1977). While the cited review is largely oriented to extrinsic probes with lanthanide shift and line-broadening reagents, there is much common ground with the nmr phenomena associated with biological transition elements. In paramagnetic lanthanons, the unpaired electron is relatively localized within an f-orbital, and being thus well-buried within the ion, tends to produce the so-called pseudocontact or through-space effects. In transition elements the unpaired d-electron may participate in bonding, leading to another type of interaction with ligand nuclei (i.e., scalar coupling), which produces the contact nmr shift.

Pseudocontact shifts are described by Eq. (2-32). By inspecting Eq. (2-32) and the spatial coordinates defined in Figure 2-54 one can see that if $g_x = g_y = g_z$ (i.e., isotropic g), then the value of the equation becomes zero. In many cases involving transition metal ions the components of g are unequal, and observed lineshifts may be a composite of contact and pseudocontact effects. One is not able to evaluate the pseudocontact contribution without prior knowledge of the g-tensor and its orientation with respect to the molecular framework. The values of g_x, g_y and g_z may be determined through single-crystal paramagnetic measurements on compounds of known crystal structure (see Chapters 5 and 6).

$$
\begin{aligned}
\left(\frac{\Delta H}{H_0}\right)_p &= \left(\frac{\Delta \nu}{\nu_0}\right)_p \\
&= \frac{-\beta^2 S(S+1)}{9kT} \left\{ \left[g_z^2 - \frac{1}{2}(g_x^2 - g_y^2) \right] \left(\frac{3\cos^2\theta - 1}{r^3}\right) \right. \\
&\quad \left. - \frac{3}{2}(g_y^2 - g_x^2)\frac{\sin^2\theta\cos2\Omega}{r^3} \right\}
\end{aligned} \qquad (2\text{-}32)
$$

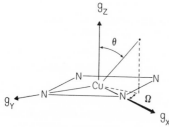

Figure 2-54. A molecular coordinate system for the g-tensor coincident with the d-orbital reference frame. The case shown has axial symmetry and g_x would be equal to g_y. In Cu(II) the unpaired electron is located in the high energy $d_{x^2-y^2}$ orbital, which is equivalent along the x- and y-coordinates but not along z; thus $g_x = g_y \neq g_z$. In other cases, all three tensor components may be unequal. The g-tensor is obviously strongly dependent on symmetry and the types of coordinating ligands present in the complex. In some cases $g_x = g_y = g_z$.

Contact shifts obey Eq. (2-33), where I is the nuclear spin, A_n is the electron–nuclear scalar coupling constant, and γ_e and γ_N are, respectively, the electron and nuclear magnetogyric ratios:

$$\left(\frac{\Delta H}{H_0}\right)_c = \left(\frac{\Delta \nu}{\nu_0}\right)_c = -2A_N \frac{\gamma_e}{\gamma_N} \frac{g\beta I(I+1)S(S+1)}{9kT} \qquad (2\text{-}33)$$

The electron–nuclear spin coupling constant, A_N, is identical to the hyperfine coupling constant of esr spectroscopy (see Chapter 5). A_N is directly proportional to the unpaired electron's spin density, σ_e, at the nucleus in question, as shown in Eq. (2-34), where Q' is the constant of proportionality:

$$A_N \cong Q' \sigma_e \qquad (2\text{-}34)$$

Thus, if the pseudocontact component can be measured or calculated (Golding et al., 1976) and subtracted from the net paramagnetic shift (leaving the so-called isotopic shift), one obtains information which may be related to the unpaired electron's delocalization within the molecule, in turn a reflection of the type of bonding and orbitals involved. Unlike esr, the direction of the shift (relative to the resonance position in the absence of paramagnetism) is determined by the *sign* of the spin density at the nucleus, i.e., downfield shifts denote positive and upfield shifts negative spin densities. The esr spectrum does not provide information of the latter type. Expressed in reciprocal time units, contact shifts are often much larger than the other sources of shift differences.

One may expect contact interactions to be larger among the ligand atoms nearest to the paramagnetic ion. However, there are situations where the pseudocontact term may predominate, e.g., as in the case mentioned earlier of certain lanthanide ions which contain localized, unpaired f-electrons. The latter may be used as probes of ion binding sites, an example being the replacement of Ca^{2+} by Pr^{3+} (Lee et al., 1979).

There are conditions under which contact and pseudocontact shifts may not be observed. In general, when the unpaired electron's spin state is long-lived, magnetic nuclei in proximity to the electron experience an efficient relaxation mechanism and are usually broadened beyond detection. Conversely, when the electron spin state is short-lived, the contact-shifted nmr lines may be quite narrow, while the esr spectrum is broadened. One usually observes the esr spectrum or the nmr spectrum, not both. A near-intermediate case was presented in Figure 2-4 (see also Section 5.2.5.4).

The conditions for observing contact-nmr shifts are $T_s^{-1} \gg A_N$ or $T_e^{-1} \gg A_N$, where T_s and T_e are the electron spin and chemical exchange lifetimes, respectively. Thus, the presence of paramagnetic transition metal ions in biomolecules is expected to be accompanied by marked line broadening or shifting effects in the nmr spectra of nearby nuclei. For example, the ^{13}C-nmr resonance signals of methionine-80 disappear by broadening in Fe(III) cytochrome C. In this case the disturbing entity may be regarded as an intrinsic probe. Similarly, the addition of lanthanide ions to biomembrane vesicles shifts the ^{31}P resonances of the outer phospholipid monolayer but does not perturb the inner, inaccessible monolayer (Yeagle et al., 1978; Hutton et al., 1977).

In another example, the paramagnetic redox inhibitor ion, hexamine chromium(III), causes the residue-25 and -83 signals of a two-iron-two-sulfur ferredoxin to broaden, suggesting that these amino acid units are located near the active site (Chan et al., 1983). Measurements on the pure ferredoxin (without inhibitor) also revealed anomalously long spin–lattice relaxation times for ^{13}C nuclei near the paramagnetic centers. This effect was attributed to rapid electron spin relaxation resulting from antiferromagnetic coupling between the iron atoms (Chan and Markley, 1983).

There are numerous cases in which linewidths are relatively narrow and paramagnetic shifting is the dominant effect (Inubushi et al., 1983; LaMar et al., 1983). Such measurements may reveal the identity of amino acid residues near an active site, and they may also be used to characterize temperature-dependent equilibria between low-spin/high-spin states at a paramagnetic center (LaMar et al., 1983).

One indirect approach to the study of calcium binding sites involves replacing Ca^{2+} with a paramagnetic lanthanide ion such as Pr^{3+} and observing relatively pure pseudocontact shift effects in the ^{1}H- or ^{13}C-nmr spectrum of the biomolecule. This method was used to examine the 50-residue CB-9 Ca^{2+}-binding peptide sequence of troponin-C and parvalbumin using Pr^{3+} as the probe (Lee et al., 1979). The coefficient of $(3\cos^2\theta - 1)/r^3$ was given an assumed value of 1000 ppm/$Å^3$ and the value of $A_2\sin^2\theta\cos2\Omega$ was assumed to be small relative to the $A_1(3\cos^2\theta - 1)$ term (A_1 and A_2 are the coefficients in Eq. 2-32). For $3\cos^2\theta - 1$ values near unity the largest observed ^{1}H-nmr shifts of ~23 ppm (Figure 2-55) correspond to metal–proton distances of 3.5 Å, while at 7-Å spacing the shift diminishes to ~3 ppm. It is thus seen that the shifted resonances must originate in hydrogen atoms on the immediate coordinating ligands.

Lee et al. (1979) observed that the pseudocontact resonances of parvalbumin appear without shifting as the Pr^{3+}/protein equivalence is increased from zero to one. Beyond the 1:1 ratio, no new paramagnetic signals are observed, ruling out additional strong binding sites for Pr^{3+}. The lack of shift changes as a function of the equivalence ratio (below 1:1) is indicative of tight binding and slow-exchange kinetics, in agreement

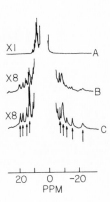

Figure 2-55. Pseudocontact shifts in the ¹H-nmr spectrum of parvalbumin. The shift reagent is Pr^{3+}. The top spectrum is parvalbumin in the absence of Pr^{3+}. Recording B shows several new upfield and downfield lines (see arrows) which appear with Pr^{3+} and protein in a 1:1 molar equivalence. An increase in the molar ratio (curve C) has no appreciable effect on the observed shifts, indicating that slow-exchange conditions prevail, and the Pr^{2+} ions are saturating the available binding sites. The shift scale is in ppm with respect to DSS. Adapted from Lee *et al.* (1979).

with independent measurements (Donato and Martin, 1975). In the fast-exchange limit the observed paramagnetic shift would vary with concentration, as required by Eq. (2-13).

The limitation of the cited study lies in its inability (on the basis of the nmr evidence presented) to identify the specific amino acid residues involved in metal ion coordination. Some insight into this question may be gained through difference spectroscopy, e.g., by subtracting the spectrum of the apoprotein from that of the paramagnetically labeled derivative (Morris and Dwek, 1977). One then observes the originating resonances as negative deflections. Data of the latter type may be confounded by conformationally induced shifts associated with metal ion binding. Fast-exchange conditions are more tractable, e.g., a plot of peak positions versus the concentration of the paramagnetic ion may be extrapolated to zero concentration, revealing the origins of the paramagnetically shifted resonances.

The previous example involved the introduction of an extrinsic probe. Similar effects are found in the nmr spectra of enzymes and proteins which contain transition metal ions. As an example, the ¹H-nmr spectrum of human adult hemoglobin in ordinary water (or D_2O) presents broadened, paramagnetically shifted resonances at -11.9 and -17.6 ppm relative to DSS standard, as shown in Figure 2-56 (Breen *et al.*, 1974). These two resonances, located well beyond the usual shift range, are associated with paramagnetic iron in the α and β subunits, respectively (Lindstrom *et al.*, 1972).

On binding with oxygen the Fe^{2+} paramagnetic moment is lost and the shifted resonances disappear (Figure 2-56, curve C). However, in the partially saturated condition (curve B) it is seen that the resonance at -11.9 is *selectively* diminished, clearly indicating that initial O_2 binding

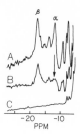

Figure 2-56. Contact shift due to paramagnetic iron in hemoglobin A. In contrast with the convention for ordinary nmr spectra, contact downfield shifts are given negative values, as shown here and in Figure 2-4. The α and β protomer subunits produce the distinct resonances indicated in the top spectrum. Hemoglobin A which has been partially saturated with O_2 shows a preferential loss of the α-subunit resonance (curve B), indicating that O_2 binds more tenaciously with that subunit. The shift scale presented here is with respect to the water resonance. Adapted from Breen *et al.* (1974).

occurs preferentially on the α subunit. This case is another example of the utility of nmr in deciding questions of selective binding and function. It should be pointed out that correlation nmr was used in the investigation to permit the use of ordinary water, thus circumventing possible conformational disturbances in D_2O (Tomita and Riggs, 1970). Contact shifts have also been observed in a ferredoxin (Poe *et al.*, 1971).

Other transition metals produce line broadening rather than shifts. For example, Cu^{2+} is characterized by a long-lived electron spin state (except in some cluster ions), and is usually a paramagnetic relaxation agent. Paramagnetic broadening effects were observed in nmr spectra obtained from a heme-A–copper complex obtained from bovine heart muscle (Bayne *et al.*, 1971). Similar line-broadening effects have been used to probe the interaction of copper with polynucleotides (Berger and Eichhorn, 1971). In studies such as these one may note the resonances which disappear on adding Cu^{2+}, and if the spectrum is correctly assigned, the region of the molecule involved in copper binding may be determined. Furthermore, the linewidths of broadened resonances are functions of the distance between the magnetic nucleus in question and the paramagnetic center (Solomon, 1955; Bloembergen, 1957), which permits a mapping of the geometry involved at the coordination site. Pseudocontact shifts obey an inverse cube dependence on the distance between the paramagnetic center and the nucleus in question. Computational methods for determining the molecular configurations of nonrigid molecules (common among the biopolymers) make use of a time-averaged adaptation of Eq. (2-32) (Armitage *et al.*, 1973) for interpreting the empirical evidence.

Pure contact shifts do not depend on a spatial distance relationship and are not easily related to molecular geometry. Rather, they reflect the delocalization of the unpaired spin. In the case of transition metals it is often extremely difficult to evaluate the relative contributions from contact and pseudocontact interactions, especially in most complex biological molecules.

2.7.2.5 Zinc and Cadmium (The Non-Transition Metals)

2.7.2.5.1 ^{67}Zn-Nmr. The only nmr-active isotope of zinc is ^{67}Zn. This nucleus has a quadrupole moment ($I = 5/2$) and is of limited use in biochemical studies since Zn(II) is bound tenaciously in most of its enzyme and metalloprotein forms. In cases where zinc exchanges rapidly between the aqueous phase and a macromolecular binding site, line-broadening studies of the type described previously for ^{23}Na (see Section 2.7.2.1) may be attempted. The ideal probe for zinc binding sites appears to be ^{113}Cd(II), which will be discussed in the following section.

2.7.2.5.2 ^{113}Cd-Nmr Probes of Zinc Binding Sites. In many cases the Zn(II) binding sites of metalloproteins and enzymes are amenable to Cd(II) replacement. Native metallothionine contains an appreciable amount of Cd(II), and in some cases Cd(II) can replace Zn(II) with retention of catalytic activity (Chlebowski and Coleman, 1976). Since zinc is indeed located at the active site in many of its metalloenzyme forms, probes with the much more tractable isotope ^{113}Cd may not only produce evidence of coordination geometry and strength but also reveal modulations due to *function*.

The chemical shift range of ^{113}Cd is more than 600 ppm, and by virtue of its spin ($I = 1/2$) relatively narrow lines may be expected in both slow- and fast-exchange limits (barring other effects in macromolecular studies). In one investigation the chemical shifts of ^{113}Cd varied by about 50–100 ppm among several enzymes in which ^{113}Cd(II) had been substituted for Zn(II) (Armitage *et al.*, 1978). The results are collected in Table 2-6. Based on this range of chemical shift differences, line broadening may be used to estimate the lifetime, T, of Cd(II) at the binding site when exchange with aqueous Cd(II) occurs within the time limits $10^{-1} \geq T \geq 10^{-5}$ s.

Armitage *et al.* (1978) have probed the zinc-containing enzymes carboxypeptidase A, carbonic anhydrase, alkaline phosphatase, and superoxide dismutase using ^{113}Cd-nmr. Results with superoxide dismutase (SD) may be compared with the ^{1}H-nmr and crystallographic evidence (see Section 2.7.1.2.1). Upon adding two equivalents of ^{113}Cd(II) to the apo-protein, a single resonance, consistent with the two identical subunits, appears 170.2 ppm downfield from 0.1 M Cd(ClO$_4$)$_2$ reference, as shown in Figure 2-57. Then, on adding two equivalents of Cu(II), the cadmium signal vanishes. This result is not surprising in view of the 6-Å spacing between Cu(II) and the zinc site via a bridging histidine ligand. Under these conditions the ^{113}Cd nucleus experiences an efficient paramagnetic relaxation mechanism. Finally, on adding a reductant, which converts Cu(II) to diamagnetic Cu(I), the ^{113}Cd signal reappears *upfield* at 8.6 ppm.

Table 2-6. ^{113}Cd Chemical Shifts for
^{113}Cd(II)-Substituted Zn(II) Metalloenzymes[a]

System[b,c]	pH	Chemical shift[d]
CPD + β-phenylpropionate	6.9	133
BCAB	8.0	214
HCAC	8.1	225.7
HCAC + 4 equiv. Cl$^-$	8.1	239.0
HCAB	9.1	145.5
HCAB + 4 equiv. F$^-$	8.9	—
HCAB + 4 equiv. F$^-$ + 2 equiv. I$^-$	8.9	225.5
HCAB + 4 equiv. Cl$^-$	8.9	241.0
HCAB + 4 equiv. Cl$^-$ + 4 equiv. I$^-$	8.9	225.5
HCAB + 4 equiv. Br$^-$	8.9	236.2
HCAB + 4 equiv. I$^-$	8.9	225.5
AP	6.0	117.2
AP + F$^-$	6.0	123.0
AP + Cl$^-$	6.0	170.0
AP + Br$^-$	6.0	169.3
AP + I$^-$	6.0	169.5
AP + 1 equiv. P	6.5	55; 142
Cd(II)$_2$SOD	6.0	170.2
Cd(II)$_2$Cu(II)$_2$SOD	6.0	—
Cd(II)$_2$Cu(I)$_2$SOD	4.6	8.6
CdCl$_2$ (in standard buffer)	6.0	55.8

[a]Adapted from Armitage *et al.* (1978).
[b]For buffer conditions and sample concentrations see the original reference.
[c]Abbreviations: SOD, superoxide dismutase; AP, alkaline phosphatase; HCAB, human carbonic anhydrase B; BCAB, bovine carbonic anhydrase B; HCAC, human carbonic anhydrase C; CPD, carboxypeptidase.

This effect clearly indicates the interaction between the two subunit metal ions, Cd(II) and Cu(I), and may be due to an increase in electron density at Cd(II) accompanying a suspected release of His-61 in the Cu(I) form of the enzyme (Beem *et al.*, 1977). The broader linewidth could reflect the proposed electron shuttle mechanism of SD action (Fielden *et al.*, 1974), which may occur with residual amounts of Cu(II) present.

Functional binding of phosphate by alkaline phosphatase is readily detected by ^{113}Cd-nmr (Armitage *et al.*, 1978). This enzyme is a 86,000-dalton dimer of two identical subunits. Consistent with this structure, the addition of two equivalents of ^{113}Cd(II) to the apoprotein leads to a single ^{113}Cd-nmr resonance at 117.2 ppm. However, upon adding one equivalent

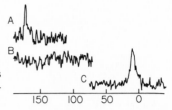

Figure 2-57. ^{113}Cd-nmr spectra of Cd(II) substituted forms of superoxide dismutase. The shift scale in ppm is relative to the resonance position of 0.1 M Cd(ClO$_4$)$_2$.

of phosphate, which binds at serine-99, the ^{113}Cd resonance splits into two at 142 and 55 ppm, i.e., the introduction of a phosphate group on one chain alters the coordination sites of both, presumably by a conformationally propagated effect. Furthermore, addition of the second mole of phosphate does not introduce additional changes. These results reflect negative homotropic interactions between the subunits. Such large shift effects point to the sensitivity of ^{113}Cd as a probe of enzyme–substrate interactions.

It should be noted that ^{113}Cd may be used as an intrinsic reporter of its own toxic interactions. Similarly, the isotope ^{199}Hg ($I = 1/2$) is suitable for toxicological studies involving mercury. The active site of carbonic anhydrase has been probed by ^{199}Hg-nmr in a fashion similar to that described above for ^{113}Cd (Sudmeier and Perkins, 1977).

2.7.3 Nmr Spectra of Small Molecules in Binding Interactions with Biopolymers

The obvious application of nmr in the structural characterization of newly isolated metabolites and hormones (e.g., those in the low-molecular weight range) continues to be an important source of biochemical knowledge but will not be treated in this chapter. The reader is referred to good sources dealing with the structural characterization of small molecules (Abraham and Loftus, 1978; Becker, 1980). Another use of nmr, namely, to probe small molecule/biopolymer binding interactions, is more consistent with the intent of this series as it deals with questions of biological function. These methods have been in use for some time. Studies with small molecules often prove to be less expensive since many older instruments of modest dispersion are applicable. An example is the microcomputer-operated, correlation (fast-scan) spectrometer described in Appendix A.

In most cases the nmr resonances of small molecules and atomic ions are observed to broaden in the presence of a macromolecule (for occa-

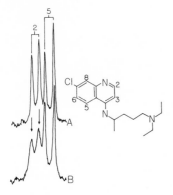

Figure 2-58. Line-broadening effects in fast-exchange interactions of the drug chloroquine with biopolymers. The top recording shows narrow doublets due to protons 2 and 5 in simple solutions of chloroquine. On adding protein to the solution, the resonance due to proton-2 is selectively broadened, as shown in the lower recording.

sional exceptions, see Bushweller *et al.*, 1971; Wright and Klingen, 1972). If the sample under investigation is homogeneous, these effects are due to the relaxation, shift, and chemical exchange phenomena discussed earlier (see Section 2.3.2.2). An example is shown in Figure 2-58, where curve A is the high-resolution ^{1}H-nmr spectrum of the drug chloroquine (aromatic region). In curve B, which was recorded after adding a small portion of bovine serum albumin, all of the lines have been broadened, especially the low-field doublet with the structural assignment shown at the right (Jefferson *et al.*, 1979).

For globular proteins known to bind chloroquine, an nmr parameter defined as the ratio of the relaxation times of two protons $T_2(5)/T_2(2)$ was found to correlate with the molecular weight of the protein, consistent with $[S] \cdot K \gg 1$ [Eq. (2-14)]. Other substances which were not expected to interact with the drug produced negligible limiting slopes. The collected results are shown in Figure 2-59, where the viscosity buildup with polymer concentration (described in the previous section) is pronounced in some cases. Dialyzed lysosomal fractions produced weaker line-broadening effects, and on the basis of their average molecular weights and Eq. (2-14), the observed limit slopes reflected $[S] \cdot K < 1$. Chloroquine is known to accumulate within lysosomes (deDuve *et al.*, 1974), reaching concentrations two orders of magnitude above that of the extracellular medium. On the basis of the cited nmr evidence (obtained at the pH estimated for the lysosomal interior) an uptake mechanism due to direct protein binding appears ruled out. Rather, the evidence is consistent with a proton pump model of weak base trapping (deDuve *et al.*, 1974).

The relaxation times of specific nuclei may be obtained by applying Eq. (2-8) to direct measurements of *linewidth*. Alternatively, if the spectral region is somewhat cluttered, *line amplitudes* (relative to an external

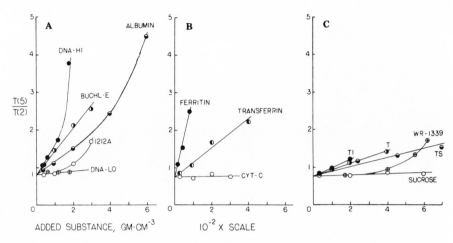

ADDED SUBSTANCE, GM·CM^{-3} 10^{-2} X SCALE

Figure 2-59. Selective broadening versus polymer concentration. Limiting slopes may be related to a binding constant defined as [CQ-bound]/[CQ-unbound]. Method is sensitive only when $[CQ]_T \sim 1/K_{bind}$. Plots A, B, and C have the same scale. The ordinate is the ratio of the linewidth of H-2 to the linewidth of H-5; the unmultiplied abscissa is w/v percent of substances added to the CQ solution. Abbreviations: DNA-HI, high-molecular weight DNA; DNA-LO, low-molecular weight DNA; BUCHL-E, butyrylcholinesterase; 1212A, Guar 1212A polymer; CYT-C, cytochrome-c (III); TI, total insoluble tritosome fraction (membranes); TS, total soluble tritosome fraction (matrix); T, whole tritosomes; WR-1339, tritonizing detergent. Conditions $[CQ] = 4.0 \times 10^{-2}$ M, pH 4.1. Conformationally dynamic polymers (e.g., WR-1339, DNA, Guar) yield poor correlation between molecular weight and limit slope, as expected. Adapted from Jefferson *et al.* (1979).

standard) may be used with Eq. (2-35) to deconvolute the effects of *partially* overlapping resonances (Jefferson *et al.*, 1979)*:

$$T_x = \frac{S_x}{S_R P_x} \left[T_R P_R + \sum \frac{T_i P_i}{1 + 4\pi^2 T_i^{\,2}(\nu_R - \nu_i)^2} \right] \tag{2-35}$$
$$- \frac{1}{P_x} \sum \frac{T_j P_j}{1 + 4\pi^2 T_j^{\,2}(\nu_x - \nu_j)^2}.$$

The latter equation, which must be used iteratively, becomes invalid when the resonances begin to merge. Also, the lineshapes must be Lorentzian.

These methods have been in use for some time. Specific interactions between an antibody and the hapten group involved in its induction were

---

*Definition of symbols in Equation 2-35: T, spin–spin relaxation time; P, relative amplitude of the line at a constant relaxation rate; S, empirical amplitude; ν, chemical shift; subscript x, the resonance in question; subscript R, the reference resonance for which T_2 is known (T_R); subscript i, all resonances except the reference; subscript j, all resonances except the line in question.

clearly evident in early nmr experiments (Metcalfe *et al.*, 1968), e.g., the antibody produced marked broadening in the hapten's resonances. In control experiments with the same hapten, nonspecific γ-globulin fractions produced virtually no line-broadening effects.

A difficulty often encountered in binding interaction studies is strong affinity, leading to modulations of intensity rather than linewidth, i.e., conditions of slow exchange prevail. Misapplication of the fast-exchange approximation may lead to false conclusions. Goldberg and Gilboa (1978) have shown that ions presumed to be "loosely" bound, e.g., Na^+, may not necessarily exchange rapidly in some biological environments. Caution is warranted in all small molecule interaction studies, and the following criteria must be met in order to insure meaningful results (Cohen, 1973):

(1) The exchange must be fast on the nmr time scale;
(2) T_1 must equal T_2;
(3) Controls must be obtained for viscosity alone (by measuring viscosity effects in a medium which has no specific binding affinity for the small molecule or ion);
(4) The possibility of line broadening due to self-association of the small molecule should be considered (by observing the concentration dependence of linewidth in the absence of a biopolymer);
(5) Field homogeneity should be as high as possible;
(6) The system must be free of interfering paramagnetic ions other than those functionally a part of the interacting biomolecule;
(7) Possible relaxation effects in nonbinding nuclei should be considered.

Obviously, pH should be carefully controlled.

2.7.4 Newer Developments in Nmr

2.7.4.1 Advanced Tracer Methods

The use of nmr to detect specific ^2H and ^3H labels has been reviewed (Garson and Staunton, 1979). Nmr labeling with ^{13}C has the advantage of a large chemical shift dispersion, but the required manipulation of the carbon framework of most biomolecules is difficult from the point of view of organic synthesis. The much lower shift dispersion of ^2H may result in the signal of the desired atom not being resolved, even though it is easier and less expensive to place the deuterium label at a specific position. It is possible to combine the best features of both isotopes, using nmr, by transferring polarization from ^2H to ^{13}C using the INEPT pulse sequence. As in all polarization transfer methods, one observes the spin system (I, in this case ^{13}C) while transferring polarization from a coupled heteroatom (S, which is ^{2}H). The pulse sequence (Rinaldi and Baldwin, 1982) is: $90°_x(S)$-τ-$(180°_x(S)$, $180°(I))$-τ-$(90°_y(S)$, $90°(I))$-acquisition, where $\tau = (4J_{IS})^{-1}$. The value of τ can be made selective for the relatively large carbon–deuterium one-bond coupling constant; thus the INEPT enhancement will be localized and will not involve long-range coupling.

In measurements of this type, the ordinary ^{1}H-coupled carbon resonances are greatly attenuated, while the ^{2}H-coupled positions are enhanced. Thus, the labeling information may be compared with *structural information* in the same spectrum. Other isotope combinations are possible, e.g., ^{15}N with ^{2}H.

The isotope ^{99}Tc ($I = 9/2$) is used as an *in vivo* radiolabel in human medicine, e.g., to obtain whole body scans. Since this isotope is also relatively easy to detect by nmr, its use in biological nmr studies has been suggested (Franklin *et al.*, 1982).

2.7.4.2 Miscellaneous Applications

The use of ^{15}N-nmr in biological and related chemical studies has been reviewed (Mason, 1981). Chemical shift sensitivity is usually typical of this isotope, but there are examples in which chemical shift was not altered during functional changes (Timkovich and Cork, 1982).

^{31}P-Nmr has been used to characterize the molecular motions of the intact pBR322 plasmid containing the 2800-base pair human insulin gene (the large quantities necessary for nmr were obtained by cloning in *E. coli*). This plasmid exists as a supercoiled, circular DNA in solution, and its motional characteristics were found to differ substantially from those of the open chain form of pBR322 (Bendel *et al.*, 1982).

Nmr studies at high pressure are possible and have been used to characterize pressure-induced structural changes in heme proteins (Morishima and Hara, 1982). These methods are relevant to studies of deep-sea biological forms.

References

Abragam, A., 1961. *The Principles of Nuclear Magnetism*, Oxford University Press, London.

Abraham, R. J., and Loftus, P., 1978. *Proton and Carbon-13 NMR Spectroscopy: An Integrated Approach*, Heyden & Son, London.

Ackerman, J. J. H., and Maciel, G. E., 1976. Improved broadband NMR spectrometer scheme, *J. Magn. Reson.* 23:67.

Anderson, C. F., Record, M. T., and Hart, P. A., 1978. Sodium-23 NMR studies of cation–DNA interactions, *Biophys. Chem.* 7:301.

Anvarhusein, I., and Rabenstein, D. L., 1979. The Incorporation of ^{2}H-glycine into the glutathione of intact human erythrocytes studied by proton spin-echo Fourier transform NMR, *FEBS Lett.* 106:325.

Aristarkhov, V. M., Piruzyan, L. A., and Tsybyshev, V. P., 1977. Effect of a permanent magnetic field on protein solutions, *Izv. Akad. Nauk. SSSR. Ser. Biol.* 1977:44.

Armitage, I. M., Hall, L., Marshall, A. G., and Werbelow, L. G., 1973. Use of lanthanide nuclear magnetic resonance shift reagents in determination of molecular configuration, *J. Am. Chem. Soc.* 95:1437.

Armitage, I. M., Uiterkamp, A. J. M. S., Chlebowski, J. F., and Coleman, J. E., 1978. ^{113}Cd NMR as a probe of the active sites of metalloenzymes, *J. Magn. Reson.* 29:375.

Asakura, T., Watanabe, Y., Uchida, A., and Minagawa, H., 1984. NMR of silk fibroin. 2. ^{13}C-nmr study of the chain dynamics and solution structure of *Bombyx mori* silk fibroin, *Macromolecules* 17:1075.

Aue, W. P., Bartholdi, E., and Ernst, R. R., 1976. Two-dimensional spectroscopy. Application to nuclear magnetic resonance, *J. Chem. Phys.* 64:2229.

Axenrod, T., and Webb, G. A., 1974. *Nuclear Magnetic Resonance Spectroscopy of Nuclei Other Than Protons.* Wiley-Interscience, New York.

Baldwin, G. S., Galdes, A., Hill, A. O., Smith, B. E., Waley, S. G., and Abraham, E. P., 1978. Histidine residues as zinc ligands in β-lactamase II, *Biochem. J.* 175:441.

Bargon, J., Fischer, H., and Johnsen, U., 1967. Kernresonanz-Emissionslinien wahrend rascher Radikalreaktionen, *Z. Naturforsch.* 22a:1551.

Bax, A., Freeman, R., and Frenkiel, T. A., 1981. An NMR technique for tracing out the carbon skeleton of an organic molecule, *J. Am. Chem. Soc.* 103:2102.

Bayne, R. A., Smythe, G. A., and Caughey, W. S., 1971. Epr and Nmr spectra of a heme A–copper complex isolated from bovine heart muscle, in *Probes of the Structure and Function of Macromolecules in Membranes,* Proceedings (B. Chance, ed.), Academic Press, New York.

Becker, E., 1980. *High Resolution NMR, Theory and Chemical Applications,* 2nd ed., Academic Press, New York.

Beem, K. M., Richardson, D. C., and Rajagopalan, K. V., 1977. Metal sites of copper–zinc superoxide dismutase, *Biochemistry,* 16:1930.

Bendall, M. R., and Lowe, G., 1976. Fluorine-19 nmr probes for proteins: S-trifluoro-methylmercuri-papain, *FEBS Lett.,* 72:231.

Bendel, P., Laub, O., and James, T. L., 1982. Molecular motions of supercoiled and circular DNA. A P-31 nmr study, *J. Am. Chem. Soc.* 104:6748.

Berendsen, H. J. C., and Edzes, H. T., 1973. The observation and general interpretation of sodium magnetic resonance in biological material, *Ann. N.Y. Acad. Sci.* 204:459.

Berger, N. A., and Eichhorn, G. L., 1971. Interaction of metal ions with polynucleotides and related compounds. XV. Nuclear magnetic resonance studies of the binding of Cu(II) to nucleotides and polynucleotides, *Biochemistry* 10:1857.

Berlin, H. M., 1977. *Designs of Active Filters With Experiments,* Howard W. Sams and Co., Inc., Indianapolis, Indiana.

Berliner, L. J., and Reuben, J. (eds.), 1978–. *Biological Magnetic Resonance,* Volume 1 to present, Plenum Publishing Corp., New York.

Binsch, G., 1969. A unified theory of exchange effects on nuclear magnetic resonance line shapes, *J. Am. Chem. Soc.* 91:1304.

Birdsall, W. J., Levine, B. A., Williams, R. J. P., Fulmer, C. S., and Wasserman, R. H., 1979. Proton-magnetic-resonance spectral studies of the intestinal calcium binding protein, *Biochem. Soc. Trans.* 7:702.

Bloembergen, N., 1957. Proton relaxation times in paramagnetic solutions, *J. Chem. Phys.* 27:572.

Borsa, F., and Rigamonti, A., 1979. Nmr and nqr in fluids, paramagnets and crystals, in *Magnetic Resonance of Phase Transitions*, (F. J. Owens, ed.), Academic Press, New York, pp. 79–169.

Boyd, J., Moore, G. R., and Williams, G., 1984. Correlation of proton chemical shifts in proteins using two-dimensional exchange correlated spectroscopy, *J. Magn. Reson.* 58:511.

Brashure, E. B., Henderson, T. O., Glonek, T., Pattnaik, N. M., and Scann, A. M., 1978. Action of α-phospholipase A_2 on human serum high density lipoprotein-3: Kinetic study of the reaction by ^{31}P-nuclear magnetic resonance spectroscopy, *Biochemistry* 17:3934.

Braun, W., Wider, G., Lee, K. H., and Wuthrich, K., 1983. Conformation of glucagon in a lipid–water interphase by ^{1}H nuclear magnetic resonance, *J. Mol. Biol.* 169:921.

Breen, J., Bertoli, D. A., and Dadok, J., 1974. Proton magnetic resonance studies of human adult hemoglobin in water, *Biophys. Chem.* 2:49.

Brevard, C., and Kintzinger, J. P., 1978. Deuterium and tritium in nmr, in *NMR and the Periodic Table* (R. K. Harris and B. E. Mann, eds.), Academic Press, London, p. 107.

Briggs, R. W., and Hinton, J. F., 1978a. A thallium-205 nuclear magnetic resonance investigation of the thallium (I)–valinomycin complex, *J. Magn. Reson.* 32:155.

Briggs, R. W., and Hinton, J. F., 1978b. Thallium-205 and proton nuclear magnetic resonance investigation of the complexation of thallium by the ionophores monesin and nigericin, *Biochemistry* 17:5576.

Brown, C. E., Katz, J. J., and Shemin, D., 1972. The biosynthesis of vitamin B_{12}: A study by ^{13}C magnetic resonance spectroscopy, *Proc. Natl. Acad. Sci. U.S.A.* 69:2585.

Brownell, G. L., Budinger, T. F., Lauterbur, P. C., and McGeer, P. L., 1982. Positron tomography and nuclear magnetic resonance imaging, *Science* 215:619.

Bryant, R. G., 1969. Nuclear magnetic resonance study of calcium-43, *J. Am. Chem. Soc.* 91:1870.

Bubnov, N. N., Bilevich, K. A., Polyakova, L. A., and Okhlobystin, O. Yu., 1972. Electron transfer as the first step in electrophilic aromatic substitution, *J. Chem. Soc., Chem. Commun.* 1972:1058.

Burnett, L. J., and Muller, B. H., 1971. Deuteron quadrupole coupling constants in three solid deuterated paraffin hydrocarbons: C_2D_6; C_4D_{10}; C_6D_{14}, *J. Chem. Phys.* 55:5829.

Bushweller, C. H., Beall, H., Grace, M., Dewkett, W. J., and Bilofsky, H. S., 1971. Temperature dependence of the proton nuclear magnetic resonance spectra of copper(1) borane complexes, $B_3H_8^-$ salts and icosahedral carboranes. Quadrupole-induced spin decoupling. Fluxional behavior, *J. Am. Chem. Soc.* 93:2145.

Bycroft, B. W., 1972. Crystal structure of viomycin, a tuberculostatic antibiotic, *J. Chem. Soc., Chem. Commun.* 1972:660.

Campbell, I. D., Dobson, C. M., Williams, R. J. P., and Xavier, A. V., 1973. Resolution enhancement of protein pmr spectra using the difference between a broadened and a normal spectrum, *J. Magn. Reson.* 11:172.

Campbell, I. D., Dobson, C. M., Moore, G. R., Perkins, S. J., and Williams, R. J. P., 1976. Temperature dependent molecular motion of a tyrosine residue of ferrocytochrome-C, *FEBS Lett.* 70:96.

Campbell, I. D., Dobson, C. M., Ratcliffe, R. G., and Williams, R. J. P., 1978. A method for the accurate measurement of proton spin–spin coupling constants in large molecules, *J. Magn. Reson.* 31:341.

Canet, D., Goulon-Ginet, G., and Marchal, J. P., 1977. Accurate determination of parameters for ^{17}O in natural abundance by Fourier transform nmr, *J. Magn. Reson.* 25:397. (Correction for the original article: *J. Magn. Reson.* 22:537.)

Canet, D., Brondeau, J., Marchal, J. P., and Robin-Lherbier, B., 1982. A convenient method of observing relatively broad nuclear magnetic resonances in the Fourier transform mode, *Org. Magn. Reson.* 20:51.

Carlson, W. D., 1976. X-ray Diffraction Studies of Bovine Pancreatic Ribonuclease-S and *E. coli* Alkaline Phosphatase, Ph.D. dissertation, Yale University.

Carr, H. Y., and Purcell, E. M., 1954. Effects of diffusion on free precession in nuclear magnetic resonance experiments, *Phys. Rev.* 94:630.

Cass, A. E. G., Hill, H. A. O., Smith, B. E., Bannister, J. V., and Bannister, W. H., 1977a. Investigation of the structure of bovine erythrocyte superoxide dismutase by ^{1}H-nuclear magnetic resonance spectroscopy, *Biochemistry* 16:3061.

Cass, A. E. G., Hill, H. A. O., and Smith, B. E., 1977b. The investigation of the proton magnetic resonance spectra of some copper proteins, in *Structure and Function of Hemocyanin (Proceedings of the Fifth European Molecular Biology International Workshop)* (J. V. Bannister, ed.), Springer-Verlag, Berlin, p. 128.

Ceraso, J. M., and Dye, J. L., 1974. Sodium-23 nmr spectrum of the sodium anion, *J. Chem. Phys.* 61:1585.

Chan, T., and Markley, J. L., 1983. Nuclear magnetic resonance studies of two-iron-two-sulfur ferredoxins. 3. Heteronuclear (^{13}C,^{1}H) two-dimensional nmr spectra, ^{13}C peak assignments, and ^{13}C relaxation measurements, *Biochemistry* 22:5996.

Chan, T., Ulrich, E. L., and Markley, J. L., 1983. Nuclear magnetic resonance studies of two-iron-two-sulfur ferredoxins. 4. Interactions with redox partners, *Biochemistry* 22:6002.

Chapman, D., 1967. Liquid crystalline nature of phospholipids, in *Advances in Chemistry Series, No. 63, Ordered Fluids and Liquid Crystals* (R. F. Gould, ed.), American Chemical Society Publications, Washington, D.C., p. 163.

Chapman, G. E., Abercrombie, B. D., Cary, P. D., and Bradbury, E. M., 1978. The measurement of small nuclear Overhauser effects in the proton spectra of proteins, and their application to lysozyme, *J. Magn. Reson.* 31:459.

Chenoweth, M. B., 1949. Monofluoroacetic acid and related compounds, *Pharmacol. Rev.* 1:383.

Chlebowski, J. F., and Coleman, J. E., 1976. Zinc and its role in enzymes, in *Metal Ions in Biological Systems,* Vol. 6 (H. Sigel, ed.), Marcel-Dekker, New York, pp. 1–140.

Civan, M. M., and Shporer, M., 1978. Nmr of sodium-23 and potassium-39 in biological systems, in *Biological Magnetic Resonance,* Vol. 1 (L. J. Berliner and J. Reuben, eds.), Plenum Publishing Corp., New York, pp. 1–32.

Civan, M. M., McDonald, G. G., Pring, M., and Shporer, M., 1976. Pulsed nuclear magnetic resonance study of ^{39}K in frog striated muscle, *Biophys. J.* 16:1385.

Clerc, T., Pretsch, E., and Sternhell, S., 1973. *Methoden der Analyse in der Chemie; Band 16: ^{13}C-Kernresonanzspektroskopie* (F. Hecht, R. Kaiser, E. Pungor, and W. Simon, eds.), Akademische Verlagsgesellschaft, Frankfurt am Main.

Cohen, J. S., 1973. Nuclear magnetic resonance investigations of biomolecules, in *Experimental Methods in Biophysical Chemistry* (C. Nicolau, ed.), John Wiley and Sons, New York.

Cohn, M., and Rao, B. D. N., 1979. Phosphorous nmr studies of enzymic reactions, *Bull. Magn. Reson.* 1:38.

Colman, P. M., Weaver, L. H., and Matthews, B. W., 1972. Rare earths as isomorphous calcium replacements for protein crystallography, *Biochem. Biophys. Res. Commun.* 46:1999.

Cooley, J. W., and Tukey, 1965. An algorithm for the machine calculation of complex Fourier series, *Math. Computation* 19:297.

Cooper, J. W., 1977. *The Minicomputer in the Laboratory*, Wiley-Interscience, New York, p. 211.

Cotton, F. A., and Wilkinson, G., 1972. *Advanced Inorganic Chemistry*, 3rd ed., Wiley-Interscience, New York.

Dadok, J., and Sprecher, R. F., 1974. Correlation nmr spectroscopy, *J. Magn. Reson.* 13:243.

Dalgarno, D. C., Levine, B. A., Williams, R. J. P., Fullmer, C. S., and Wasserman, R. H., 1984. Proton-nmr studies of the solution conformations of vitamin-D-induced bovine intestinal calcium-binding protein, *Eur. J. Biochem.* 137:523.

Dalmark, M., 1976. Effects of halides and bicarbonate on chloride ion transport in human red blood cells, *J. Gen. Physiol.* 67:223.

Davies, D. B., 1978. Discussion Document on Nomenclature. 1. Conformation of Biopolymers. 2. Notations for Descriptions of Conformations. 3. Notations for NMR Parameters Used in Analysis of Nucleotides, *Jerusalem Symp. Quantum. Chem. Biochem.* 11:509.

Dawson, W. H., and Odom, J. D., 1977. Selenium-77 relaxation time studies. Considerations regarding direct observation of selenium resonances in biological systems, *J. Am. Chem. Soc.* 99:8352.

deDuve, C., deBarsy, T., Poole, B., Trouet, A., Tulkens, P., and Van Hoof, F., 1974. Lysosomotropic agents, *Biochem. Pharmacol.* 23:2495.

Derbyshire, W., Gorvin, T. C., and Warner, D., 1969. A deuteron magnetic resonance study of a single crystal of deuterated malonic acid, *Mol. Phys.* 17:401.

Deverell, C., 1969. Nuclear magnetic resonance studies of electrolyte solutions, *Progress in NMR Spectroscopy* 4:235.

Dickerson, R. E., and Timkovich, R., 1975. *The Enzymes*, Vol. 11 (P. D. Boyer, ed.), Academic Press, New York, p. 397.

Donato, H., and Martin, R. B., 1975. Conformations of carp muscle calcium binding parvalbumin, *Biochemistry* 13:4575.

Dunfield, L. G., Burgess, A. W., and Scheraga, H. A., 1978. Energy parameters in polypeptides. 8. Empirical potential energy algorithm for the conformational analysis of large molecules, *J. Phys. Chem.* 82:2609.

Durig, J. R., and Bucy, W. E., 1977. Infrared, Raman, and microwave spectra of ethaneselenol and the determination of the barriers to internal rotation, *J. Mol. Spectrosc.* 64:474.

Dwek, R. A., 1973. *Nuclear Magnetic Resonance in Biochemistry: Applications to Enzyme Systems*, Oxford University Press, England.

Dzidic, I., and Kebarle, P., 1970. Hydration of the alkali ions in the gas phase. Enthalpies and entropies of reactions $M^+(H_2O)_{n-1} + H_2O = M^+(H_2O)_n$, *J. Phys. Chem.* 74:1466.

Earl, W. L., and VanderHart, D. L., 1980. High resolution, magic angle sample spinning ^{13}C nmr of solid cellulose I, *J. Am. Chem. Soc.* 102:3251.

Eggleton, G. L., Jung, G., and Wright, J. R., 1978. The ^1H-nmr spectra of mixed valence complexes of copper with derivatives of 1-amino-2, 2-dimethyl-2-mercaptoethane, *Bioinorg. Chem.* 8:173.

Eichhorn, G. L., and Marzilli, L. G., 1979. *Advances in Inorganic Biochemistry 1*, Elsevier North Holland, New York.

Ellett, J. D., Haeberlen, U., and Waugh, J. S., 1970. High resolution nuclear magnetic resonance of solid perfluorohexane, *J. Am. Chem. Soc.* 92:411.

Ernst, R. R., and Anderson, W. A., 1966. Application of Fourier transform spectroscopy to magnetic resonance, *Rev. Sci. Instrum.* 37:93.

Ernst, C. R., Spialter, L., Buell, G. R., and Wilhite, D. L., 1974. Silicon-29 nuclear magnetic resonance. Chemical shift substituent effects, *J. Am. Chem. Soc.* 96:5375.

Farnell, L. F., Randall, E. W., and White, A. I., 1972. Effect of Paramagnetic Species on the Nuclear Magnetic Resonance Spectra of Nitrogen-15, *J. Chem. Soc., Chem. Commun. (20),* 1159.

Feigon, J., Leupin, W., Denny, W. A., and Kearns, D. R., 1983a. Proton nuclear magnetic resonance investigation of the conformation and dynamics in the synthetic deoxyribonucleic acid decamers d(ATATCGATAT) and d(ATATGCATAT), *Biochemistry* 22:5930.

Feigon, J., Leupin, W., Denny, W. A., and Kearns, D. R., 1983b. Two-dimensional proton nuclear magnetic resonance investigation of the synthetic deoxyribonucleic acid decamer d(ATATCGATAT)$_2$, *Biochemistry* 22:5943.

Fielden, E. M., Roberts, P. B., Bray, R. C., Lowe, D. J., Mautner, G. N., Rotilio, G., and Calabrese, L., 1974. The mechanism of action of superoxide dismutase from pulse radiolysis and electron paramagnetic resonance. Evidence that only half the active sites function in catalysis, *Biochem. J.* 139:49.

Finkel'shtein, A. V., and Ptitsyn, O. B., 1978. Theory of self-organization of protein secondary structure: Dependence of the native globule structure on the secondary structure of the unfolded chain, *Dokl. Akad. Nauk SSSR* 242:1226.

Franklin, K. J., Lock, C. J. L., Sayer, B. G., and Schrobilgen, G. J., 1982. Chemical applications of ^{99}Tc-nmr spectroscopy: Preparation of novel Tc(VII) species and their characterization by multinuclear nmr, *J. Am. Chem. Soc.* 104:5303.

Freeman, R., and Hill, H. D. W., 1971. High-resolution study of nmr spin echoes: J-spectra, *J. Chem. Phys.* 54:301.

Fyfe, C. A., Lyeria, J. R., and Yannoni, C. S., 1978. High resolution ^{13}C nuclear magnetic resonance spectra of frozen liquids using magic angle spinning, *J. Am. Chem. Soc.* 100:5635.

Garson, M. J., and Staunton, J., 1979. Some new nmr methods for tracing the fate of hydrogen in biosynthesis, *Chem. Soc. Reviews* 8:539.

Gerasimova, G. K., and Nakhil'nitskaya, Z. N., 1977. Blood electrolyte levels and erythrocyte potassium ion transport in animals under the effect of a continuous magnetic field, *Kosm. Biol. Aviakosm. Med.* 11:63.

Gerig, J. T., 1981. Fluorine magnetic resonance in biochemistry, in *Biological Magnetic Resonance,* Vol. 1 (L. J. Berliner, and J. Reuben, eds.), Plenum Publishing Corp., New York, Chapter 4.

Glickson, J. D., Mayers, D. F., Settine, J. M., and Urry, D. W., 1972. Spectroscopic studies on the conformation of gramicidin A′. Proton magnetic resonance assignments, coupling constants and H–D exchange, *Biochemistry* 11:477.

Glusker, J. P., 1971. *The Enzymes,* 3rd ed., Vol. 5 (P. D. Boyer, ed.), Academic Press, New York, p. 413.

Gochin, M., James, T. L., and Shafer, R. H., 1984. *In Vivo* ^{19}F-nmr of 5-fluorouracil incorporation into RNA and metabolites in *E. coli* cells, *Biochim. Biophys. Acta.* 804:118.

Goetz, A. M., and Richards, J. H., 1978. Molecular studies of subspecificity differences among phosphorylcholine-binding mouse myeloma antibodies using ^{31}P nuclear magnetic resonance, *Biochemistry* 17:1733.

Goldberg, M., and Gilboa, H., 1978. Sodium exchange between two sites. The binding of sodium to halotolerant bacteria, *Biochim. Biophys. Acta.* 538:268.

Golding, R. M., Pascual, R. O., and Vrbancich, J., 1976. On the theory of nmr screening constants in $t_2{}^1$ and $t_2{}^5$ transition metal ion complexes—the pseudocontact contribution, *Mol. Phys.* 31:731.

Goldman, S., 1948. *Frequency Analysis, Modulation and Noise,* McGraw-Hill, New York.

Gorenstein, D. C. (ed.), 1984. *Phosphorous-31 NMR,* Academic Press, New York.

Gosselin, R. E., Hodge, H. C., Smith, R. P., and Gleason, M. N., 1976. *Clinical Toxicology of Commercial Products,* 4th ed., Williams and Wilkina Co., Baltimore, Maryland, p. 307.

Granger, R. P., 1974. Chemically induced dynamic nuclear and electron polarizations— CIDNP and CIDEP, in *NMR, Basic Principles and Progress,* Vol. 8 (P. Diehl, E. Fluck, and R. Kosfeld, eds.), Springer-Verlag, Berlin.

Griffin, R. G., Powers, L., and Pershan, P. S., 1978. Head-group conformations in phospholipids: A phosphorous-31 nuclear magnetic resonance study of oriented monodomain dipalmitoylphosphatidylcholine bilayers, *Biochemistry* 17:2718.

Gronowitz, S., Johnson, I., and Hornfeldt, A. B., 1975. Selenium-77 nmr studies of organoselenium compounds. I. Selenium-77 nmr parameters of monosubstituted selenophenes, *Chem. Scr.* 8:8.

Gupta, R. K., Ferretti, J. A., and Becker, E. D., 1974. Rapid scan Fourier transform nmr spectroscopy, *J. Magn. Reson.* 13:275.

Gurevich, A. Z., Barsokov, I. L., Arseniev, A. S., and Bystrov, V. F., 1984. Combined COSY–NOESY experiment, *J. Magn. Reson.* 56:471.

Gutmann, V., and Wychera, E., 1966. Coordination reactions in nonaqueous solutions. The role of the donor strength, *Inorg. Nucl. Chem. Lett.* 2:257.

Gutowsky, H. S., and Saika, A., 1953. Dissociation, chemical exchange and the proton magnetic resonance in some aqueous electrolytes, *J. Chem. Phys.* 21:1688.

Harris, R. K., and Kimber, B. J., 1975. ^{29}Si and ^{13}C nuclear magnetic resonance studies of organosilicon chemistry. I. Trimethylsilyl compounds, *J. Magn. Reson.* 17:174.

Harris, R. K., and Mann, B. E., 1979. *NMR and the Periodic Table,* Academic Press, New York.

Hartmann, S. R., and Hahn, E. L., 1962. Nuclear double resonance in the rotating frame, *Phys. Rev.* 128:2042.

Hawkes, G. E., Randall, E. W., and Bradley, C. H., 1975. Theory and practice for studies of peptides by ^{15}N nuclear magnetic resonance at natural abundance: Gramicidin-S, *Nature (London)* 257:767.

Hawkes, G. E., Randall, E. W., Hull, W. E., Gattegno, D., and Conti, F., 1978. Qualitative aspects of hydrogen–deuterium exchange in the ^{1}H, ^{13}C and ^{15}N nuclear magnetic resonance spectra of viomycin in aqueous solution, *Biochemistry* 17:3986.

Herzfeld, J., Griffin, R. G., and Haberkorn, R. A., 1978. Phosphorous-31 chemical-shift tensors in barium diethyl phosphate and urea–phosphoric acid: Model compounds for phospholipid head-group studies, *Biochemistry* 17:2711.

Hinton, J. F., and Briggs, R. W., 1975. Thallium-205 nmr spectroscopy. I.: ^{205}Tl^{1+} chemical shifts. A sensitive probe for preferential solvation and solution structure, *J. Magn. Reson.* 19:393.

Hoekstra, W. G., Suttie, J. W., Ganther, H. E., and Mertz, W. (eds.), 1974. *Trace Element Metabolism in Animals—2, Proceedings of the Second International Symposium,* University Park Press, Baltimore, Maryland.

Hollis, D. P., 1980. Phosphorous nmr of cells, tissues and organelles, in *Biological Magnetic Resonance,* Vol. 2 (L. J. Berliner and J. Reuben, eds.), Plenum Publishing Corp., New York, pp. 1–44.

Horowitz, J., Ofeugand, J., Daniel, W. E., and Cohn, M., 1977. Fluorine-19 nuclear magnetic resonance of 5-fluorouridine-substituted tRNA–Val from *Escherichia coli, J. Biol. Chem.* 252:4418.

Hruska, F. E., Grey, A. A., and Smith, I. C. P., 1970. Nuclear magnetic resonance study of the molecular conformation of β-pseudouridine in aqueous solution, *J. Am. Chem. Soc.* 92:4088.

Hutton, W. C. Yeagle, P. L., and Martin, R. B., 1977. The interaction of lanthanide and calcium salts with phospholipid bilayer vesicles: The validity of the nuclear magnetic resonance method for determination of vesicle bilayer phospholipid surface ratios, *Chem. Phys. Lipids* 19:255.

Inubushi, T., Ikeda-Saito, M., and Yonetani, T., 1983. Isotropically shifted nmr resonances for the proximal histidyl imidazole NH protons in cobalt hemoglobin and iron–cobalt hybrid hemoglobins. Binding of the proximal histidine toward porphyrin metal ion in the intermediate state of cooperative ligand binding, *Biochemistry* 22:2904.

James, T. L., 1975. *Nuclear Magnetic Resonance in Biochemistry: Principles and Applications,* Academic Press, New York.

Jeener, J., Meier, B. H., Bachmann, P., and Ernst, R. R., 1979. Investigation of exchange processes by two-dimensional NMR spectroscopy, *J. Chem. Phys.* 71:1979.

Jefferson, N. A., Beavers, C. R., Beavers, D. A., McElroy, M., and Wright, J. R., 1979. ^{1}H-nmr effects in chloroquine–biopolymer binding interactions, *Physiol. Chem. Phys.* 11:233.

Kanamori, K., and Roberts, J. D., 1983. ^{15}N nmr studies of biological systems, *Acc. Chem. Res.* 16:35.

Kaptein, R., Dijkstra, K., and Nicolay, K., 1978. Laser photo-CIDNP as a surface probe for proteins in solution, *Nature (London)* 274:293.

Karplus, M., 1963. Vicinal proton coupling in nuclear magnetic resonance, *J. Am. Chem. Soc.* 85:2870.

Keepers, J. W., and James, T. L., 1984. A theoretical study of distance determinations from nmr. Two-dimensional nuclear Overhauser effect spectra, *J. Magn. Reson.* 57:404.

Keeton, K. R., and Wright, J. R., 1980. Modifying a spare 60 MHz ^{1}H-nmr detector for ^{23}Na, ^{27}Al and ^{63}Cu, Eighth Annual MBS Biomedical Symposium, Paper P-57 (Spectroscopy), April 10, 1980, Atlanta, Georgia.

Kessler, H., Bermel, W., Friederich, A., Krack, G., and Hull, W. E., 1982. Peptide conformation. 17. *Cyclo*-(L-Pro-L-Pro-D-Pro). Conformational analysis by 270- and 500 MHz one- and two-dimensional ^{1}H-nmr spectroscopy, *J. Am. Chem. Soc.* 104:6297.

Khaled, M. A., Urry, D. W., Sugano, H., Miyoshi, M., and Izymiya, N., 1978. Hydrogen–deuterium substitution and solvent effects on the nitrogen-15 nuclear magnetic resonance of gramicidin-S: Evaluation of secondary structure, *Biochemistry,* 17:2490.

Kidd, G., and Goodfellow, R. J., 1979. The transition metals, in *NMR and the Periodic Table* (R. K. Harris, and B. E. Mann, eds.), Academic Press, London.

Kittel, C., 1968. *Introduction to Solid State Physics,* John Wiley and Sons, Inc., New York, p. 515.

Klemperer, W. G., 1978. ^{17}O-Nmr spectroscopy as a structural probe, *Angew. Chem., Int. Ed. Engl.* 17:246.

Knappenberg, M., Brison, J., Dirks, J., Hallenga, K., Deschrijver, P., and Van Binst, G., 1979. Conformational studies on somatostatin. II. The C-terminal hexapeptide fragment, *Biochim. Biophys. Acta.* 580:266.

Kohler, S. J., and Klein, M. P., 1976. ^{31}P nuclear magnetic resonance chemical shielding tensors of phosphorylethanolamine, lecithin and related compounds: Applications to head-group motion in model membranes, *Biochemistry* 15:967.

Kohler, S. J., and Klein, M. P., 1977. Orientation and dynamics of phospholipid head groups in bilayers and membranes determined from ^{31}P nuclear magnetic resonance chemical shielding tensors, *Biochemistry* 16:519.

Komoroski, R. A., Peat, I. R., and Levy, G. C., 1975. High field carbon-13 nmr spectroscopy. Conformational mobility in gramicidin-S and frequency dependence of ^{13}C spin–lattice relaxation times, *Biochem. Biophys. Res. Commun.* 65:272.

Kumar, A., Ernst, R. R., and Wuthrich, K., 1980. A two-dimensional nuclear Overhauser enhancement (2D NOE) experiment for the elucidation of complete proton–proton cross relaxation networks in biological macromolecules, *Biochem. Biophys. Res. Commun.* 95:1.

Kuntz, I. D., Crippen, G. M., and Kollman, P. A., 1979. Application of distance geometry to protein tertiary structure calculations, *Biopolymers* 18:939.

Kyogoku, Y., and Iitaka, Y., 1966. The crystal structure of barium diethyl phosphate, *Acta Crystallogr.* 21:49.

Lachmann, H., and Schnackerz, C. D., 1984. ^{31}P nuclear magnetic resonance titrations: Simultaneous evaluation of all pH-dependent resonance signals, *Org. Magn. Reson.* 22:101.

LaMar, G., Krishnamoorthi, R., Smith, K. M., Gersonde, K., and Sick, H., 1983. Proton nuclear magnetic resonance investigation of the conformation-dependent spin equilibrium in azide-ligated monomeric insect hemoglobins, *Biochemistry* 22:6239.

Laszlo, P., 1978. Sodium-23 nuclear magnetic resonance spectroscopy, *Angew. Chem., Int. Ed. Engl.* 17:254.

Laszlo, P., 1983. *NMR of Newly Accessible Nuclei: Chemical and Biochemical Applications,* Vol. 1, Academic Press, New York.

Laszlo, P., 1984. *NMR of Newly Accessible Nuclei: Chemically and Biochemically Important Elements,* Vol. 2, Academic Press, New York.

Lauterbur, P., Kramer, D., Hause, W., and Ching-Nien, C., 1975. Zeugmatographic high resolution nuclear magnetic resonance spectroscopy. Images of chemical inhomogeneity within macroscopic objects, *J. Am. Chem. Soc.* 97:6866.

Lee, L., Sykes, B. D., and Birnbaum, E. R., 1979. A determination of the relative compactness of the Ca^{2+}-binding sites of a Ca^{2+}-binding fragment of troponin-C and parvalbumin using lanthanide-induced ^{1}H-nmr shifts, *FEBS Lett.* 98:169.

Lee, M., and Goldberg, W. I., 1965. Nuclear magnetic resonance line narrowing by a rotating radio frequency (RF) field, *Phys. Rev. A* 140:1261.

Lepley, A. R., and Closs, G. L. (eds.), 1973. *Chemically Induced Magnetic Polarization,* John Wiley, New York.

Levi, B. G., 1977. Two magnets set records for field intensity, *Physics Today* 30:20.

Levy, G. C., and Lichter, R. L., 1979. *Nitrogen-15 Nuclear Magnetic Resonance Spectroscopy,* Wiley-Interscience, New York.

Levy, G. C., and Nelson, G. L., 1972. *Carbon-13 NMR for Organic Chemists,* Wiley-Interscience, New York.

Li, S., Johnson, D. L., Gladysz, J. A., and Servis, K. L., 1979. Silicon-29 nmr spectra of metal carbonyl silanes by the selective population transfer method, *J. Organomet. Chem.* 166:317.

Linas, M., and Klein, M. P., 1975. Charge relay at the peptide bond. A proton magnetic resonance study of solvation effects on the amide electron density distribution, *J. Am. Chem. Soc.* 97:4731.

Lindman, B., and Forsen, S., 1976. *NMR Basic Principles and Progress,* Vol. 12, *Chlorine, Bromine and Iodine NMR. Physicochemical and Biological Applications,* (P. Diehl, E. Fluck, and R. Kosfeld, eds.), Springer-Verlag, Berlin.

Lindman, B., and Forsen, S., 1979. The halogens—chlorine, bromine and iodine, in *NMR and the Periodic Table* (R. K. Harris, and B. E. Mann, eds.), Academic Press, London.

Lindstrom, T. R., Ho, C., and Pisciotta, A. V., 1972. Nuclear magnetic resonance studies of hemoglobin M Milwaukee, *Nature (London), New Biol.* 237:263.

Lippmaa, E., Saluvere, T., and Laisarr, S., 1971. Spin–Lattice relaxation of ^{15}N-nuclei in organic compounds, *Chem. Phys. Lett.* 11:120.

Llinas, M., Klein, M. P., and Nielands, J. B., 1970. Solution conformation of ferrichrome, a microbial iron transport cyclohexapeptide, as deduced by high resolution proton magnetic resonance, *J. Mol. Biol.* 52:399.

Llinas, M., Horsley, W. J., and Klein, M. P., 1976. Nitrogen-15 nuclear magnetic resonance spectrum of alumichrome. Detection by a double resonance Fourier transform technique, *J. Am. Chem. Soc.* 98:7554.

London, R. E., Walker, T. E., Kollman, V. H., and Matwiyoff, N. A., 1977. Proton relaxation in ^{13}C-enriched molecules; ^{13}C T_1 and NOE data from proton magnetic resonance measurements, *J. Magn. Reson.* 26:213.

Lowe, I. J., 1959. Free induction decay of rotating solids, *Phys. Rev. Lett.* 2:285.

Lutz, M., Kleo, J., Gilet, R., Henry, M., Plus, R., and Leicknam, J. R., 1976a, Vibrational spectra of chlorophylls *a* and *b* labeled with magnesium-26 and nitrogen-15, in *Proc. 2nd Int. Conf. Stable Isotopes* (1975) (E. R. Klein, and P. D. Klein, eds.), NTIS, Springfield, Virginia.

Lutz, O., Neple, W., and Nolle, A., 1976b. ^{17}O and ^{33}S Fourier transform nmr studies in thiosulfate and thiomolybdate solutions, *Z. Naturforsch.* A31:978.

Lutz, O., Oehler, H., and Kroneck, P., 1978. Copper-63 and -65 Fourier transform nuclear magnetic resonance studies, *Z. Phys. A.* 288:17.

Magnusson, J. A., and Bothner-By, A. A., 1971. ^{25}Mg^{2+} nuclear resonance as a probe of Mg^{2+} complex formation, in *Magnetic Resonances in Biological Research* (C. Franconi, ed.), Gordon and Breach, New York, Paper 35.

Mantsch, H. H., Saito, H., and Smith, I. C. P., 1977. Deuterium magnetic resonance, applications in chemistry, physics and biology, *Progress in Nuclear Magnetic Resonance Spectroscopy* 11:211.

Mariam, Y. H., and Wilson, W. D., 1979. A phosphorous-31 nmr analysis of the helix to coil transition of natural DNA samples: Evidence for the existence of different conformational states, *Biochem. Biophys. Res. Commun.* 88:861.

Marki, W., Opplinger, M., Thanei, P., and Schwyzer, R., 1977. D(−)- and L(+)-γ-Carboxyglutamic acid (Gla): Resolution of synthetic Gla derivatives, *Helv. Chim. Acta.* 60:798.

Mason, J., 1981. Nitrogen nmr spectroscopy in inorganic, organometallic and bioinorganic chemistry, *Chem. Rev.* 81:205.

Mavel, G., 1973. Nmr studies of phosphorous compounds, in *Annual Reports on NMR Spectroscopy,* Vol. 5B (E. F. Mooney, ed.), Academic Press, New York.

Maxfield, F. R., and Scheraga, H. A., 1979. Improvements in the prediction of protein backbone topography by reduction of statistical errors, *Biochemistry* 18:697.

McConnell, H. M., and Robertson, R. E., 1958. Comments on theory of isotopic hyperfine interactions in pi-electron radicals, *J. Chem. Phys.* 28:991.

Mehring, 1976. High resolution nmr spectroscopy in solids, in *NMR, Basic Principles and Progress,* Vol. 11 (P. Diehl, E. Fluck, and R. Kosfeld, eds.), Springer-Verlag, New York.

Meiboom, S., and Gill, D., 1958. Modified spin-echo method for measuring nuclear relaxation times, *Rev. Sci. Instrum.* 29:688.

Metcalfe, J. C., Burgen, A. S. V., and Jardetzky, O., 1968. On the mechanism of binding of choline derivatives to an anticholine antibody, in *Molecular Associations in Biology* (B. Pullman, ed.), Academic Press, New York, p. 487.

Miknis, F. P., Bartuska, V. J., and Maciel, G. E., 1979. Cross-polarization ^{13}C-nmr with magic angle spinning. Some applications to fossil fuels and polymers, *Am. Lab.* 11:19.

Moore, W. J., 1962. *Physical Chemistry,* 3rd ed., Prentice Hall, Englewood Cliffs, New Jersey, p. 231.

Morashima, I., and Hara, M., 1982. High pressure nmr studies of hemoproteins. Pressure induced structural changes in the heme environments of cyanometmyoglobin, *J. Am. Chem. Soc.* 104:6833.

Morishima, I., and Inubushi, T., 1978. ^{15}N nuclear magnetic resonance studies of iron-bound $C^{15}N^-$ in ferric low-spin cyanide complexes of various porphyrin derivatives and various hemoproteins, *J. Am. Chem. Soc.* 100:3568.

Morris, A. T., and Dwek, R. A., 1977. Some recent applications of the use of paramagnetic centers to probe biological systems using nuclear magnetic resonance, *Quart. Rev. Biophys.* 10:421.

Morris, G. A., 1980. Sensitivity enhancement in ^{15}N nmr: Polarization transfer using the INEPT pulse sequence, *J. Am. Chem. Soc.* 102:428.

Moses, V., Holm-Hansen, O., and Calvin, M., 1958. Response of chlorella to a deuterium environment, *Biochim. Biophys. Acta.* 28:62.

Mullen, K., and Pregosin, P. S., 1976. *Fourier Transform NMR Techniques: A Practical Approach,* Academic Press, New York.

Muus, L. T., Atkins, P. W., McLauchlan, K. A., and Pedersen, J. B. (eds.), 1977. *NATO Advanced Study Institute Series,* Vol. C34, *Chemically Induced Magnetic Polarization,* D. Reidel Publishing Co., Dordrecht, The Netherlands.

Muzskat, K. A., and Weinstein, M., 1975. Photochemically induced dynamic nuclear polarization study of reversible hydrogen transfer between dyes and phenolic oxidation inhibitors, *J. Chem. Soc., Chem. Commun.* 1975:143.

Muzskat, K. A., 1977a. Effects of acid–base equilibrium on photo-CIDNP in nitroaromatic compounds, *Chem. Phys. Lett.* 49:538.

Muzskat, K. A., 1977b. Photo-CIDNP in the tyrosyl unit: A new tool in high resolution nuclear magnetic resonance studies of peptides, *J. Chem. Soc., Chem. Commun.* 1977:872.

Muzskat, K. A., Weinstein, M., and Gilon, C., 1978. Biochemical applications of chemically induced nuclear polarization in phenols, peptides, catecholamines and related molecules, *Biochem. J.* 173:993.

Nagayama, K., Wuthrich, K., Bachmann, P., and Ernst, R. R., 1977. Two-dimensional J-resolved ^1H-nmr spectroscopy for studies of biological macromolecules, *Biochem. Biophys. Res. Commun.* 78:99.

Nagayama, K., Kumar, A., Wuthrich, K., and Ernst, R. R., 1980. Experimental techniques of two-dimensional correlated spectroscopy, *J. Magn. Reson.* 40:321.

Nakano, M., Nakano, N. I., and Higuchi, T., 1967. Calculation of stability constants of hydrogen-bonded complexes from proton magnetic resonance data. Interaction of phenol with dimethylacetamide and various ketones. Solvent effects, *J. Phys. Chem.* 71:3954.

Navon, G., and Lanir, A., 1972. Nmr relaxation by intermolecular and intramolecular dipolar interactions in small molecules bound to an enzyme, *J. Magn. Reson.* 8:144.

Navon, G., Ogawa, S., Shulman, R. G., and Yamane, T., 1977. High-resolution ^{31}P nuclear magnetic resonance studies of metabolism in aerobic *Escherichia coli* cells, *Proc. Natl. Acad. Sci. U.S.A.* 74:888.

Njus, D., Sehr, P. A., Radda, G. K., Ritchie, G. A., and Seeley, P. J., 1978. Phosphorous-31 nuclear magnetic resonance studies of active proton translocation in chromaffin granules, *Biochemistry* 17:4337.

Noggle, J. H., and Schirmer, R. E., 1971. *The Nuclear Overhauser Effect, Chemical Applications,* Academic Press, New York.

Noth, H., and Wrackmeyer, B., 1978. *Nuclear Magnetic Resonance Spectroscopy of Boron Compounds,* Springer-Verlag, New York.

Oldfield, E., Gutosky, H. S., Jacobs, R. E., Kang, S. Y., Meadows, M. D., Rice, D. M., and Skarjune, R. P., 1978a. Recent developments in high-field nmr spectroscopy of biological systems, *Am. Lab.* 10:19.

Oldfield, E., Meadows, M., Rice, D., and Jacobs, R., 1978b. Spectroscopic studies of specifically deuterium labeled membrane systems. Nuclear magnetic resonance investigation of the effects of cholesterol in model systems, *Biochemistry* 17:2727.

Orme-Johnson, W. H., 1973. Iron–sulfur proteins: Structure and function, *Annu. Rev. Biochem.* 42:159.

Pardi, A., Wagner, G., and Wuthrich, K., 1983. Protein conformation and proton nuclear-magnetic-resonance chemical shifts, *Eur. J. Biochem.* 137:445.

Passow, H., 1969. *Progress in Biophysics and Molecular Biology,* Vol. 19 (J. A. V. Butler, and D. Noble, eds.), Pergamon Press, New York, p. 425.

Pattnaik, N. M., Kezdy, F. J., and Scanu, A. M., 1976. Kinetic study of the action of snake venom phospholipase A_2 on human serum high density lipoprotein 3, *J. Biol. Chem.* 251:1984.

Pike, M. M., Fossel, E. T., Smith, T. W., and Springer, C. S., 1984. High-resolution ^{23}Na-nmr studies of human erythrocytes: Use of aqueous shift reagents, *Am. J. Physiol.* 246:C528.

Pines, A., Gibby, M. G., and Waugh, J. S., 1973. Proton-enhanced nmr of dilute spins in solids, *J. Chem. Phys.* 59:569.

Pitner, T. P., and Urry, D. W., 1972. Proton magnetic resonance studies in trifluoroethanol. Solvent mixtures as a means of delineating peptide protons, *J. Am. Chem. Soc.* 94:1399.

Pitner, T. P., Glickson, J. D., Rowan, R., Dadok, J., and Bothner-By, A. A., 1975. Delineation of interactions between specific solvent and solute nuclei. Nuclear magnetic resonance solvent saturation study of gramicidin-S in methanol, dimethyl sulfoxide and trifluoroethanol, *J. Am. Chem. Soc.* 97:5917.

Plaush, A. C., and Sharp, R. R., 1976. Ion binding to nucleotides. A chlorine-35 and lithium-7 nmr study, *J. Am. Chem. Soc.* 98:7973.

Poe, M., Phillips, W. D., McDonald, C. C., and Orme-Johnson, W. H., 1971. Pmr and magnetic susceptibility studies on *Clostridium acidiurici* ferredoxin, *Biochem. Biophys. Res. Commun.* 42:705.

Pople, J. A., Beveridge, D. L., and Dobosh, P. A., 1968. Molecular orbital theory of the electronic structure of organic compounds. II. Spin densities in paramagnetic species, *J. Am. Chem. Soc.* 90:4201.

Popov, E. M., and Zheltova, V. N., 1971. Electronic structure and properties of the peptide group, *J. Mol. Struct.* 10:221.

Powers, L., and Clark, N. A., 1975. Preparation of large monodomain phospholipid bilayer smectic liquid crystals, *Proc. Natl. Acad. Sci. U.S.A.* 72:840.

Quin, L. D., Gallager, M. J., Cuncie, G. T., and Chestnut, D. B., 1980. ^{31}P and ^{13}C-nmr spectra of 2-norbornyl phosphorous compounds. Karplus equations for $^3J_{PC}$ in several P(III) and P(IV) derivatives, *J. Am. Chem. Soc.* 102:3136.

Rabiner, L. R., Schafer, R. W., and Rader, C. M., 1969. The chirp Z-transform algorithm, *IEEE Trans. Audio. Electroacoust.* Au-17:86.

Reich, H. J., and Trend, J. E., 1976. Stereochemical dependence of geminal selenium–carbon coupling constants, *J. Chem. Soc., Chem. Commun.* 1976:310.

Reuben, J., Shporer, M., and Gabbay, E. J., 1975. The alkali ion–DNA interaction as reflected in the nuclear relaxation rates of ^{23}Na and ^{87}Rb, *Proc. Natl. Acad. Sci. U.S.A.* 72:245.

Richardson, J. S., Thomas, K. A., and Richardson, D. C., 1975a. Alpha-carbon coordinates for bovine Cu, Zn superoxide dismutase, *Biochem. Biophys. Res. Commun.* 63:986.

Richardson, J. S., Thomas, K. A., Rubin, B. H., and Richardson, D. C., 1975b. Crystal structure of bovine Cu, Zn superoxide dismutase at 3 angstrom resolution: Chain tracing and metal ligands, *Proc. Natl. Acad. Sci. U.S.A.* 72:1349.

Richarz, R., and Wirthlin, T., 1981. CCCP: Carbon–Carbon Connectivity Plots on the XL-200. A New Tool for Structural Analysis of Organic Molecules. Document Z-13, Varian Instruments, Inc.

Righetti, P., and Moeschlin, S., 1971. The therapeutic effect of dithiocarb (DTC) and potassium chloride on experimental thallium poisoning in guinea pigs, *Clin. Toxicol.* 4:165.

Rinaldi, P. L., and Baldwin, N. J., 1982. $^{13}C\{^{2}H\}$ insensitive nuclei enhanced by polarization transfer (INEPT): A new strategy for isotopic labelling studies, *J. Am. Chem. Soc.* 104:5791.

Roberts, G. C. K., Feeney, J., Birdsall, B., Kimber, B., Griffiths, D. V., King, R. W., and Burgen, A. S. V., 1977. Dihydrofolate reductase: The use of fluorine-labeled and selectively deuterated enzyme to study substrate and inhibitor binding, in *NMR in Biology, Proceedings of the British Biophysical Society Spring Meeting* (R. A. Dwek, I. O. Campbell, and R. E. Richards, eds.), Academic Press, London.

Roberts, J. D., 1961. *An Introduction to the Analysis of Spin–Spin Splitting in High Resolution Nuclear Magnetic Resonance Spectra,* Benjamin Press, New York.

Robertson, P., Hiskey, R. G., and Koehler, K. A., 1978. Calcium and magnesium binding to *ν*-carboxylglutamic acid-containing peptides via metal ion nuclear magnetic resonance, *J. Biol. Chem.* 253:5880.

Rodger, C., Sheppard, N., McFarlane, H. C. E., and McFarlane, W., 1979. Group VI-oxygen, sulphur, selenium and tellurium, in *NMR and the Periodic Table* (R. K. Harris, and B. E. Mann, eds.), Academic Press, London.

Rothstein, A., Cabantchik, Z. I., Balshin, M., and Juliano, R., 1975. Enhancement of anion permeability in lecithin vesicles by hydrophobic proteins extracted from red blood cell membranes, *Biochem. Biophys. Res. Commun.* 64:144.

Rybaczewski, E. F., Neff, B. L., Waugh, J. S., and Sherfinski, J. S., 1977. High resolution ^{13}C nmr in solids: ^{13}C local fields of CH, CH_2, and CH_3, *J. Chem. Phys.* 67:1231.

Samoson, A., Kundla, E., and Lippmaa, E., 1982. High resolution MAS-nmr of quadrupolar nuclei in powders, *J. Magn. Reson.* 49:350.

Samuelson, G. L., Obenauf, R. H., and Albright, M. J., 1977. Digital quadrature detection in Fourier transform nmr, *Am. Lab.* 9:85.

Santini, R. E., and Grutzner, J. B., 1976a. A broadband system for the observation of nmr spectra of any resonant nucleus, *J. Magn. Reson.* 22:155.

Santini, R. and Grutzner, J. B., 1976b. Total systematic noise reduction in Fourier transform nuclear magnetic resonance spectrometry, *Anal. Chem.* 48:941.

Schaefer, J., Stejskal, E. O., and McKay, R. A., 1979. Cross-polarization nmr of nitrogen 15-labeled soybeans, *Biochem. Biophys. Res. Commun.* 88:274.

Schwarz, K., 1973. A bound form of silicon in glycosaminoglycans and polyuronides, *Proc. Natl. Acad. Sci. U.S.A.* 70:1608.

Schwarzschild, B. M., 1979. New nmr spectrometer doubles resolving power, *Physics Today* 32:19.

Seelig, J., 1977. Deuterium magnetic resonance: Theory and application to lipid membranes, *Quart. Rev. Biophys.* 10:353.

Shami, Y., Carver, J., Ship, S., and Rothstein, A., 1977. Inhibition of Cl^- binding to anion transport protein of the red blood cell by DIDS (4,4'-diisothiocyano-2,2'-stilbene disulfonic acid) measured by ^{35}Cl-nmr, *Biochem. Biophys. Res. Commun.* 76:429.

Shannon, C. E., and Weaver, W., 1949. *The Mathematical Theory of Communication*, University of Illinois Press, Urbana, Illinois.

Shporer, M., and Civan, M. M., 1974. Effects of temperature and field strength on the nmr relaxation times of ^{23}Na in frog striated muscle, *Biochim. Biophys. Acta* 354:291.

Singleton, R. C., 1969. Algorithm for mixed radix FFT, *IEEE Trans. Audio. Electroacoust.* Au-17:93.

Smith, P. J., and Tupciauskas, A. P., 1978. Chemical shifts of ^{119}Sn nuclei in organotin compounds, in *Annual Reports on NMR Spectroscopy*, Vol. 8, (E. F. Mooney, ed.), Academic Press, New York, pp. 291-370.

Solomon, I., 1955. Relaxation processes in a system of two spins, *Phys. Rev.* 99:559.

Stern, A., Gibbons, W., and Craig, L. C., 1968. A conformational analysis of gramicidin-S-A by nuclear magnetic resonance, *Proc. Natl. Acad. Sci. U.S.A.* 61:734.

Stoesz, J. D., Redfield, A. G., and Malinowski, D., 1978. Cross relaxation and spin diffusion effects on the proton nmr of biopolymers in H_2O, *FEBS Lett.* 91:320.

Strop, P., Cechova, D., and Wuthrich, K., 1983a. Preliminary structural comparison of the proteinase isoinhibitors IIA and IIB from bull seminal plasma based on individual assignments of the ^{1}H nuclear magnetic resonance spectra by two-dimensional nuclear magnetic resonance at 500 MHz, *J. Mol. Biol.* 166:669.

Strop, P., Wider, G., and Wuthrich, K., 1983b. Assignment of the ^{1}H nuclear magnetic resonance spectrum of the proteinase inhibitor IIA from bull seminal plasma by two-dimensional nuclear magnetic resonance at 500 MHz, *J. Mol. Biol.* 166:641.

Stryer, L., 1975. *Biochemistry*, W. H. Freeman and Co., San Francisco, California, Chapter 10, p. 235.

Sudmeier, J. L., and Perkins, T. G., 1977. Studies of single ^{199}Hg(II) ion resonances in the active site of human carbonic anhydrase B by Fourier transform nuclear magnetic resonance, *J. Am. Chem. Soc.* 99:7732.

Suelter, C. H., 1974. Monovalent cations in enzyme-catalyzed reactions, in *Metal Ions in Biological Chemistry*, Vol. 3 (H. Sigel, ed.), Marcel Dekker, New York, p. 201.

Suzuki, M., and Kubo, R., 1964. Theoretical calculation of nmr spectral line shapes, *Mol. Phys.* 7:201.

Swift, T. J., and Connick, R. E., 1962. Nmr-relaxation mechanisms of ^{17}O in aqueous solutions of paramagnetic cations and the lifetime of water molecules in the first co-ordination sphere, *J. Chem. Phys.* 37:307.

Tenforde, T. S. (ed.), 1979. *Magnetic Field Effects on Biological Systems*, Plenum Press, New York.

Terpstra, D., 1979. Two dimensional Fourier transform ^{13}C nmr, in *Topics in Carbon-13 NMR Spectroscopy*, Vol. 3 (G. C. Levy, ed.), Wiley-Interscience, New York, Section VI, p. 62.

Timkovich, R., and Cork, M. S., 1982. Nitrogen-15 nmr investigation of nitrite reductase–substrate interaction, *Biochemistry* 21:3794.

Tomita, S., and Riggs, A., 1970. Effects of partial deuteration on the properties of human hemoglobin, *J. Biol. Chem.* 245:3104.

Urbina, J., and Waugh, J. S., 1974. Proton-enhanced ^{13}C nuclear magnetic resonance of lipids and biomembranes, *Proc. Natl. Acad. Sci. U.S.A.* 71:5062.

Urry, D. W., and Ohnishi, M., 1970. *Spectroscopic Approaches to Biomolecular Conformation* (E. W. Urry, ed.), American Medical Association Press, Chicago, Illinois, p. 263.

Wagner, G., 1984. Two-dimensional relayed coherence transfer-NOE spectroscopy, *J. Magn. Reson.* 57:497.

Wagner, G., and Wuthrich, K., 1982. Sequential resonance assignments in protein ^{1}H nuclear magnetic resonance spectra. Basic pancreatic trypsin inhibitor., *J. Mol. Biol.* 155:347.

Wagner, G., and Zuiderweg, E. R. P., 1983. Two-dimensional double quantum ^{1}H nmr spectroscopy of proteins, *Biochem. Biophys. Res. Commun.* 113:854.

Wasson, J. R., 1984. Nuclear magnetic resonance spectrometry, *Anal. Chem.* 56:212B.

Wasson, J. R., and Corvan, P. J., 1978. Nuclear magnetic resonance spectroscopy, *Anal. Chem.* 50:121R.

Waugh, J. S., Huber, L. M., and Haeberlen, U., 1968. Approach to high resolution nmr in solids, *Phys. Rev. Lett.* 20:180.

Wayland, J. R., and Brannen, J. P., 1977. Absolute reaction rate model for the response of *Bacillus subtillis* to microwave radiation, Sandia Progress Report #Sand-76-9301C.

Weast, R. C. (ed.), 1972. *Handbook of Chemistry and Physics*, 52nd ed., E-57, Chemical Rubber Publishing Co., Cleveland, Ohio.

Williams, R. J. P., 1970. Biochemistry of sodium, potassium, magnesium, and calcium, *Quart. Rev. Chem. Soc.* 1970:331.

Wright, J. R., and Klingen, T. J., 1972. ^{1}H-Nmr spectra of icosahedral carborane polymers in solution: Temperature dependent effects, *J. Inorg. Nucl. Chem.* 34:3284.

Wuthrich, K., 1976. *Nmr in Biological Research: Peptides and Proteins*, Elsevier, New York.

Wuthrich, K., Billeter, M., and Braun, W., 1983. Pseudo-structures for the 20 common amino acids for use in studies of protein conformations by measurements of intramolecular proton–proton distance constraints with nuclear magnetic resonance, *J. Mol. Biol.* 169:949.

Yeagle, P. L., Langdon, R. G., and Martin, R. G., 1977a. Phospholipid–protein interactions in human low density lipoprotein detected by ^{31}P nuclear magnetic resonance, *Biochemistry*, 16:3487.

Yeagle, P. L., Hutton, W. C., Huang, C., and Martin, R. B., 1977b. Phospholipid headgroup conformations; intermolecular interactions and cholesterol effects, *Biochemistry* 16:4344.

Yeagle, P. L., Martin, R. B., Pottenger, L., and Langdon, R. G., 1978. Location and interactions of phospholipid and cholesterol in human low density lipoprotein from ^{31}P nuclear magnetic resonance, *Biochemistry* 17:2707.

Zaner, K. S., and Damadian, R., 1975. Phosphorous-31 as a nuclear probe for malignant tumors, *Science* 189:729.

Zuiderweg, E. R. P., Kaptein, R., and Wuthrich, K., 1983. Sequence-specific resonance assignments in the ^{1}H nuclear-magnetic-resonance spectrum of the *lac* repressor DNA-binding domain 1–51 from *Escherichia coli* by two-dimensional spectroscopy, *Eur. J. Biochem.* 137:279.

Correlation Nmr

The following FORTRAN-80 listing shows the basic features of correlation nmr. Much greater throughput will be realized if the program is rewritten in assembly language in order to take advantage of an inexpensive (~$260) hardware arithmetic unit, e.g., a multiprocessor combination of the Intel 8080 microprocessor with the North Star FBA-A floating point arithmetic processor. This combination will execute a 1024-point FFT in approximately six seconds (32-bit precision). The FFT algorithm also applies to pulsed FT nmr and the crystallographic methods (Chapter 6).

1. The program is easily modified for a 1024-point data array by enlarging the dimension subscripts in lines 150–160 and changing L and N to 10 and 1024, respectively (lines 250–260).

2. Portions of the program are specific for the nmr/computer interface used in our one-of-a-kind design. These are lines 730–1100 (input); lines 2690–3010 and 3080 (graphics); and the ADDA subroutine. Other interfaces may use the program as long as it is realized that the averaged, fast-scan spectrum must be in the X1 array *before* line 1160, and the correlated spectral components are found in arrays AA and BB, respectively, *after* line 2460.

3. For any particular interface, several values of dwell should be entered at line 220 and the sweep times and widths calibrated. *Calibrated* values of sweep width and sweep time are entered at lines 560 and 590. Phase corrections are entered at line 520 only if necessary and positive values at line 620 produce a convolutional improvement in S/N_p with attendant line broadening. The value entered at line 620 is approximately

the linewidth of the narrowest line in the resulting output spectrum. An example dialog is included before the program listing.

Details of the microcomputer/nmr spectrometer interface have been published elsewhere (see Wright, J.R., 1978, Interfacing a microcomputer with the Varian EM-360 nmr spectrometer, *Rev. Sci. Instrum.* 49:1288).

```
EMORY SIZE?
INTERRUPTS? N
HIGHEST DISK NUMBER? 0
HOW MANY DISK FILES? 2
HOW MANY RANDOM FILES? 2

54836E  BYTES AVAILABLE
DOS MONITOR VER 1.0
COPYRIGHT 1977 BY MITS INC
.MNT 0

DIR 0

&DELAY          &BITIN          &ADDA           +DELAY          +BITIN
+ADDA           +CWFT-9         #CWFT-9         &CWFT-9         $CWFT-9

RUN  CWFT-9 0


DDDMMMM    (DWELL/MODE)
   70    1



PHASE ANGLE (DEG.)?000.00


SWEEP WIDTH (HZ)?126.60


SWEEP TIME (SEC)?002.66


CONVOLUTE OR DECONVOLUTE?001.00


SCANS?100


- FORWARD TRANSFORM -

- REORDERING -

- CORRELATING -

- INVERSE TRANSFORM -

- REORDERING -


PLOT REAL(1), IMAGINARY(2) OR REALTRANS(3)?3


NEXT SPECTRUM (1) OR EXIT (2)?2


OS MONITOR VER 1.0
COPYRIGHT 1977 BY MITS INC

+P

00060  CCC - THIS SUBROUTINE CONTROLS SCAN TIME BY INTRODUCING
000070  C        A DWELL BETWEEN SWEEP RAMP INCREMENTS -
000080  C
000100          SUBROUTINE DELAY(J)
000110  .       DO 400 I=1, J
000120  400     CONTINUE
000130          RETURN
000140          END
```

```
*P

000060  CCC
000070  C
000080  C
000100        SUBROUTINE BITIN(I8,N,I5,L)
000110  C
000120  C - FFT BIT INVERSION -
000130  C
000140        I5=0
000150        N1=N
000160        DO 191 I=1,L
000170        N1=N1/2
000180        IF(I8.LT.N1)GOTO191
000190        II=I-1
000200        I5=I5+2**II
000210        I8=I8-N1
000220  191   CONTINUE
000230        RETURN
000240        END


*P

000060  CCC
000070  C
000100        SUBROUTINE ADDA
000120  C - INITIALIZE THE ANALOG/DIGITAL INTERFACE AT OCTAL 070 -
000140        CALL OUT(56,0)
000150        CALL OUT(57,0)
000160        CALL OUT(58,0)
000170        CALL OUT(60,0)
000180        CALL OUT(62,0)
000190        CALL OUT(59,255)
000200        CALL OUT(61,255)
000210        CALL OUT(63,255)
000220        CALL OUT(56,44)
000230        CALL OUT(58,44)
000240        CALL OUT(60,44)
000250        CALL OUT(62,44)
000260        RETURN
000270        END
*

*P

00100   CCC - FORTRAN-80/EM-360 CORRELATION NMR -
000110  C     (C.R. BEAVERS, S.E. GEORGE & J.R. WRIGHT 10/17/79)
000120  C
000130  C
000140        PROGRAM CWFT-9
000150        DIMENSION X1(513),X2(513),AA(513),BB(513)
000160        DIMENSION CO(513),SI(513)
000170        CALL ADDA
000180  C - DEFINE CONSTANTS AND CREATE A TRIG. LOOK-UP TABLE
000190        WRITE(1,1)
000200  C - DWELL DETERMINES SWEEP TIME, MODE TO CORRELATE OR NOT -
000210  1     FORMAT(/,1X,'DDDDMMMM    (DWELL/MODE)',/)
000220        READ(1,661)K17,ISN
000230        WRITE(1,197)
000240  661   FORMAT(2I4)
000250        L=9
000260        N=512
000270        P1=3.14159
000280        V=2.*(P1/N)
000290        RAD=6.283185E0
000300        RADIAN=1.745329E-2
000310        NV=N/2
000320        N5=N-1
000330        N6=NV-1
000340        DO 508 J=0,N5
000350        V1=V*J
000360        CO(J)=COS(V1)
000370        SI(J)=SIN(V1)
000380  508   CONTINUE
000390  C
000400  C - BEGIN DATA INPUT
000410  C
000420  509   CONTINUE
000430        K=K17
000440        DO 703 II=0,N
000450        X1(II)=0.
000460        X2(II)=0.
000470  703   CONTINUE
000480        CALL OUT(61,0)
000490        CALL OUT(63,0)
000500        WRITE(1,702)
000510  702   FORMAT(/,2X,'PHASE ANGLE (DEG. )?')
000520        READ(1,790)PHASE
000530  790   FORMAT(F6.2)
```

119

```
000540        WRITE(1,194)
000550   194  FORMAT(/,2X,'SWEEP WIDTH (HZ)?')
000560        READ(1,790)SW
000570        WRITE(1,196)
000580   196  FORMAT(/,2X,'SWEEP TIME (SEC)?')
000590        READ(1,790)ST
000600        WRITE(1,198)
000610   198  FORMAT(/,2X,'CONVOLUTE OR DECONVOLUTE?')
000620        READ(1,790)W2
000630        NN1=0
000640        WRITE(1,700)
000650   700  FORMAT(/,2X,'SCANS?')
000660        READ(1,701)NN
000670   701  FORMAT(I5)
000680        WRITE(1,197)
000690   197  FORMAT(/)
000700   C
000710   C - UNIT SCAN
000720   C
000730   710  NN1=NN1+1
000740        IX=1
000750        IY=1
000760        I=0
000770   662  CONTINUE
000780        CALL OUT(59,0)
000790        AA(I)=INP(57)
000800        I=I+1
000810        CALL OUT(61,IX)
000820        IY=IY+1
000830        CALL DELAY(K)
000840        IF(IY.LT.0)GOTO720
000850        CALL OUT(59,0)
000860        AA(I)=INP(57)
000870        I=I+1
000880        CALL OUT(63,IX)
000890        IX=IX+1
000900        CALL DELAY(K)
000910        IF(I.LT.N5)GOTO662
000920   720  CONTINUE
000930   C
000940   C - ACCUMULATE SCANS AND LET FLYBACK RINGING STOP -
000950   C - (INCLUDES ABORT FEATURE)
000960   C
000970        CALL OUT(61,0)
000980        CALL OUT(63,0)
000990        DO 785 I=1,20000
001000        III=I
001010   785  CONTINUE
001020        DO 740 II=0,N5
001030        IF(AA(II).LT.0.)AA(II)=AA(II)+256
001040        X1(II)=X1(II)+AA(II)
001050   740  CONTINUE
001060        CALL OUT(59,10)
001070        ITT=INP(57)
001080        IF(ITT.LT.0)ITT=ITT+256
001090        IF(ITT.GT.50)GOTO870
001100        IF(NN1.LT.NN)GOTO710
001110   C
001120   C - AVERAGE NN SCANS -
001130   C
001140        DO 745 II=0,N5
001150        X1(II)=X1(II)/NN
001160   745  CONTINUE
001170   C - DISPLAY CORRELATED (ISN=1) OR UNCORRELATED (ISN=-1)? -
001180        IF(ISN.EQ.-1)GOTO9901
001190   C
001200   C - FFT IN-PLACE ALGORITHM -
001210   C
001220        DO 6 J=0,N5
001230        X1(J)=X1(J)/N
001240   6    CONTINUE
001250        WRITE(1,495)
001260   495  FORMAT(2X,'- FORWARD TRANSFORM -',/)
001270        I1=N/2
001280        I2=1
001290        DO 9 M=1,L
001300        I3=0
001310        I4=I1
001320        DO 8 K=1,I2
001330        I8=I3/I1
001340        CALL BITIN(I8,N,I5,L)
001350        Z1=CO(I5)
001360        Z2=-SI(I5)
001370        I9=I4-1
001380        DO 12 J=I3,I9
001390        A1=X1(J)
001400        A2=X2(J)
```

120

```
001410        J1=J+I1
001420        B1=Z1*X1(J1)-Z2*X2(J1)
001430        B2=Z2*X1(J1)+Z1*X2(J1)
001440        X1(J)=A1+B1
001450        X2(J)=A2+B2
001460        X1(J1)=A1-B1
001470        X2(J1)=A2-B2
001480   12   CONTINUE
001490        I3=I3+2*I1
001500        I4=I4+2*I1
001510    8   CONTINUE
001520        I1=I1/2
001530        I2=2*I2
001540    9   CONTINUE
001550        WRITE(1,300)
001560  300   FORMAT(2X,'- REORDERING -',/)
001570        IU=0
001580  145   IZ=0
001590  146   IS=IU
001600        CALL BITIN(I8,N,I5,L)
001610        AA(IU)=X1(I5)
001620        BB(IU)=X2(I5)
001630        IU=IU+1
001640        IZ=IZ+1
001650        IF(IZ.GT.9)GOTO145
001660        IF(IU.GT.NV)GOTO666
001670        GOTO146
001680  666   CONTINUE
001690   C
001700   C - CORRELATION -
001710   C
001720        WRITE(1,191)
001730  191   FORMAT(2X,'- CORRELATING -')
001740   C
001750   C - THIS IS ONE VERSION OF CORRELATION (THERE ARE VARIANTS) -
001760   C - FOR EXP(X), KEEP X(87; FOR SIN(X), KEEP X<1.5E7
001770   C
001780        PHI=RADIAN*PHASE
001790        CR=-RAD*W2/(2.*SW)
001800        CI=RAD/(2.*SW*ST)
001810        J=0
001820        DO 4 I=1,N6
001830        J=J+1
001840        CJ=CR*J
001850        RE=EXP(CJ)
001860        TH=CI*J*J+PHI
001870        STH=RE*SIN(TH)
001880        CTH=RE*COS(TH)
001890        X=AA(I)
001900        Y=BB(I)
001910   C - COMPLEX MULTIPLICATION -
001920        X1(I)=(X*CTH)-(Y*STH)
001930        X2(I)=(Y*CTH)+(X*STH)
001940    4   CONTINUE
001950   C - CORRELATION DONE, PREPARE FOR FFT -
001960        DO 653 I=NV,N5
001970        X1(I)=0.
001980        X2(I)=0.
001990  653   CONTINUE
002000   C
002010   C - IFFT IN-PLACE ALGORITHM -
002020   C
002030        WRITE(1,505)
002040  505   FORMAT(/,2X,'- INVERSE TRANSFORM -',/)
002050        I1=N/2
002060        I2=1
002070        DO 19 M=1,L
002080        I3=0
002090        I4=I1
002100        DO 18 K=1,I2
002110        I8=I3/I1
002120        CALL BITIN(I8,N,I5,L)
002130        Z1=CO(I5)
002140        Z2=SI(I5)
002150        I9=I4-1
002160        DO 22 J=I3,I9
002170        A1=X1(J)
002180        A2=X2(J)
002190        J1=J+I1
002200        B1=Z1*X1(J1)-Z2*X2(J1)
002210        B2=Z2*X1(J1)+Z1*X2(J1)
002220        X1(J)=A1+B1
002230        X2(J)=A2+B2
002240        X1(J1)=A1-B1
002250        X2(J1)=A2-B2
002260   22   CONTINUE
002270        I3=I3+2*I1
```

121

```
002280          I4=I4+2*I1
002290    18    CONTINUE
002300          I1=I1/2
002310          I2=2*I2
002320    19    CONTINUE
002330          WRITE(1,310)
002340    310   FORMAT(2X,'- REORDERING -',/)
002350          IU=0
002360    155   IZ=0
002370    156   I8=IU
002380          CALL BITIN(I8,N,I5,L)
002390          AA(IU)=X1(I5)
002400          BB(IU)=X2(I5)
002410          IU=IU+1
002420          IZ=IZ+1
002430          IF(IZ.GT.9)GOTO155
002440          IF(IU.GT.N5)GOTO676
002450          GOTO156
002460    676   CONTINUE
002470    C
002480    C - GRAPHIC CHOICE -
002490    C
002500    9901  CONTINUE
002510    689   WRITE(1,350)
002520    350   FORMAT(2X,'PLOT REAL(1), IMAGINARY(2) OR REALTRANS(3)?')
002530          READ(1,1083)IQ
002540    1083  FORMAT(I1)
002550          IF(IQ.EQ.1)GOTO799
002560          IF(IQ.EQ.2)GOTO687
002570          IF(IQ.GT.3)GOTO689
002580          DO 686 I=0,N5
002585          X=AA(I)**2+BB(I)**2
002590          AA(I)=SQRT(X)
002600    686   CONTINUE
002610          GOTO799
002620    687   DO 351 I=0,N5
002630          AA(I)=BB(I)
002640    351   CONTINUE
002650    799   CONTINUE
002660    C
002670    C - GRAPHIC OUTPUT -
002680    C
002690          P=-1.E35
002700          Q=1.E35
002710          IDD=400
002720          IDB=IDD
002730          IDC=8
002740          DO 840 I=0,N5
002750          IF(AA(I).LT.Q)Q=AA(I)
002760    840   CONTINUE
002770          DO 850 I=0,N5
002780          AA(I)=AA(I)-Q
002790    850   CONTINUE
002800          DO 999 I=0,N5
002810          IF(AA(I).GT.P)P=AA(I)
002820    999   CONTINUE
002830          P=255/P
002840          DO 860 I=0,N5
002850          AA(I)=(AA(I)*P)
002860    860   CONTINUE
002870          DO 870 I=0,N5
002880          A=AA(I)
002890          JJ=INT(A)
002900          CALL OUT(63,JJ)
002910          DO 880 J=1,IDC
002920          CALL OUT(61,255)
002930          DO 890 ID=1,IDD
002940    890   CONTINUE
002950          CALL OUT(61,0)
002960          DO 855 ID=1,IDD
002970    855   CONTINUE
002980    880   CONTINUE
002990          DO 900 ID=1,IDB
003000    900   CONTINUE
003010    870   CONTINUE
003020          WRITE(1,910)
003030    910   FORMAT(/,2X,'NEXT SPECTRUM (1) OR EXIT (2)?')
003040          READ(1,915)ILS
003050    915   FORMAT(I1)
003060          WRITE(1,920)
003070    920   FORMAT(/)
003080          CALL OUT(63,0)
003090          IF(ILS.EQ.1)GOTO509
003100          END
```

Mathematical Symbols
Used in Chapter 2

Planck constant	h
$h/2\pi$	$\hbar$
nmr magnetic field at resonance	H_0
nmr radio frequency field (magnetic component)	H_1
nmr radio frequency at resonance	ν_0
Bohr magneton	β
nuclear magneton	β_N
nuclear spin quantum number	I
magnetic dipole moment	μ
nuclear magnetogyric ratio	γ_N
electron magnetogyric ratio	γ_e
capacitive reactance	X_C
inductive reactance	X_L
capacitance (farads); protein/substrate complex	C
inductance (henrys)	L
chemical shift difference; spectral span	$\Delta\nu$
nmr linewidth at half-maximum amplitude	$\Delta\nu_{1/2}$
spin–lattice relaxation time (longitudinal)	T_1
spin–spin relaxation time (transverse)	T_2
correlation time	τ_c
internuclear distance	r
nuclear electric quadrupole moment	Q
proportionality constant between electron spin density and nuclear hyperfine splitting or contact nmr shifts	Q'
electric field gradient	q
asymmetry parameter; viscosity	η

electronic charge	e
substrate or small molecule; electron spin multiplicity	S
protein	P
equilibrium constant	K
Boltzmann constant	k
fraction; probability	f
a pulse-sequence time delay which results in a loss of the nmr resonance	τ_{null}
units of gauss	G
absolute temperature	T
nmr spectrum in the frequency domain, i.e., the spectrum as normally presented	$F(\nu)$
nmr spectrum in the time domain, i.e., the FID signal	$f(t)$
$F(\nu)$ digitized in an array, m	$F(m)$
f(t) digitized in an array, n	$f(n)$
$(-1)^{1/2}$	i
noise power	N_p
nuclear Overhauser factor	ϵ
angle for line narrowing	ϕ
angles measured with respect to molecular coordinates	α, γ', θ'
a component of the chemical shift tensor	σ_{ii}
coupling constant between nucleus x and nucleus y	J_{xy}
components of the electron g-tensor	g_x, g_y, g_z
angles measured from the g_z and g_x axes, respectively	θ, Ω
molecular radius	a

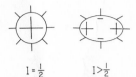

$I = \frac{1}{2}$ $I > \frac{1}{2}$

Figure 3-1. The nuclear quadrupole moment. Nuclei which have spins of zero or one-half may be compared with a charged sphere, as shown in the left-hand example. The electric field strength is equal at all points on the surface of the sphere. Quadrupolar nuclei behave as charged, nonspherical surfaces. Although the net charge is positive, the field intensity is stronger on the more curved regions and weaker on those less curved. These act, respectively, as positive and negative regions; hence, an electric quadrupole. Electric quadrupoles experience an orienting force in an electric field gradient; spherical charges do not. In prolate nuclei, spin is about the long axis; in oblate nuclei, spin is around the short axis.

Table 3-1. Biological Isotopes Which Have Electric Quadrupole Moments

Isotope	% Abundance	Spin	Quadrupole moment (Q) (barns)
^{2}H	1.5×10^{-2}	1	2.73×10^{-3}
^{10}B	19.58	3	7.4×10^{-2}
^{11}B	80.42	3/2	3.55×10^{-2}
^{14}N	99.63	1	1.6×10^{-2}
^{17}O	3.7×10^{-2}	5/2	-2.6×10^{-2}
^{23}Na	100	3/2	0.14–0.15
^{25}Mg	10.13	5/2	0.22
^{33}S	0.76	3/2	-6.4×10^{-2}
^{35}Cl	75.53	3/2	-7.89×10^{-2}
^{37}Cl	24.47	3/2	-6.21×10^{-2}
^{39}K	93.10	3/2	0.07
^{41}K	6.88	3/2	—
^{43}Ca	0.145	7/2	—
^{50}V	0.24	6	—
^{51}V	99.76	7/2	-4×10^{-2}
^{53}Cr	9.55	3/2	—
^{55}Mn	100	5/2	0.55
^{59}Co	100	7/2	0.40
^{61}Ni	1.19	3/2	—
^{63}Cu	69.09	3/2	-0.16
^{65}Cu	30.91	3/2	-0.15
^{67}Zn	4.11	5/2	0.15
^{79}Br	50.54	3/2	0.33
^{81}Br	49.46	3/2	0.28
^{87}Sr	7.02	9/2	0.2
^{95}Mo	15.72	5/2	0.12
^{97}Mo	9.46	5/2	1.1
^{127}I	100	5/2	-0.69
^{135}Ba	6.59	3/2	0.25
^{137}Ba	11.32	3/2	0.2

Nuclear Quadrupole Resonance (Nqr)

This brief chapter has been included in the anticipation that biological applications of nqr spectroscopy will increase as time passes. As in the other spectroscopic methods based on nuclear properties, nqr is able to probe the electronic environment at a specific type of nucleus (Smith, 1978; Semin *et al.*, 1975; Schultz, 1970). It is a solid-state phenomenon and thus has much in common with the Mössbauer effect, while biological nmr is oriented largely toward the liquid and liquid-crystalline states. To date, most nqr investigations have dealt with fairly simple substances. The method is generally applicable to the study of a variety of elements (Biryukov and Vorokonov, 1972), and in spite of certain limitations, biological applications may be expected to increase. For the present, however, small molecules are more appropriate subjects for nqr.

3.1 Properties of Nqr

Nuclei possess a quadrupole moment when the angular distribution of the nuclear electric field is not spherically symmetrical and may be compared with a charged prolate spheroid (positive moment) or oblate spheroid (negative moment), as shown in Figure 3-1. All nuclei with a nuclear spin quantum number $I \geq 1$ meet this condition. Table 3-1 presents the quadrupole properties of biologically interesting isotopes, and it is seen that the list encompasses most of the elements found in biological molecules.

No net force acts on a quadrupolar nucleus if it is in a uniform or spherically symmetrical electric field, as in the case of ^{35}Cl in the simple chloride ion. However, the presence of an electric field gradient along an axis of cylindrical symmetry (e.g., such as that experienced by the ^{35}Cl nucleus in HCl) will create a multiplicity of energy levels, and under these conditions the application of an external force (a radiofrequency field) will cause the nucleus to precess about the field gradient axis. Similar effects occur in electric field gradients of lower symmetry. This calls to mind the resonance frequencies of magnetic dipoles in nmr phenomena (see Chapter 2), but one should note that nqr is an electric effect based on gradients normally present in molecules and does not require the application of a magnetic field, as in the case of nmr.

The absorption frequencies, ν_Q, of a quadrupolar nucleus are proportional to the nuclear quadrupole moment, Q, and the electric field gradient, q, that it experiences:

$$\nu_Q \propto e^2qQ/h \qquad (3\text{-}1)$$

In expression (3-1), e is the electronic charge and h the Planck constant. The right-hand half of expression (3-1) is referred to as the quadrupole coupling constant.

For nuclei in electric gradients with cylindrical symmetry the constant of proportionality for Eq. (3-1) is a function of quantum number I (Das and Hahn, 1958). However, an additional factor η, the asymmetry parameter, is included when the symmetry is lower. Energy levels leading to ν_Q are illustrated in Figure 3-2, which shows the consequences of various nuclear spins and symmetries. This figure also shows that the transitions associated with a high-spin nucleus may be widely spaced in the spectrum. Depending on q and Q, resonances may occur anywhere from zero to 1000 MHz, and typically fall in the range of 200 kHZ to 800 MHz.

An electric gradient at a nucleus reflects the charge distributions of surrounding bonding and nonbonding electrons. The gradient resultant in ^{35}Cl is zero (thus ν_Q is zero), whereas polar bonds of the type $R^{\delta+}$—$Cl^{\delta-}$ would be expected to produce a finite q at the quadrupolar

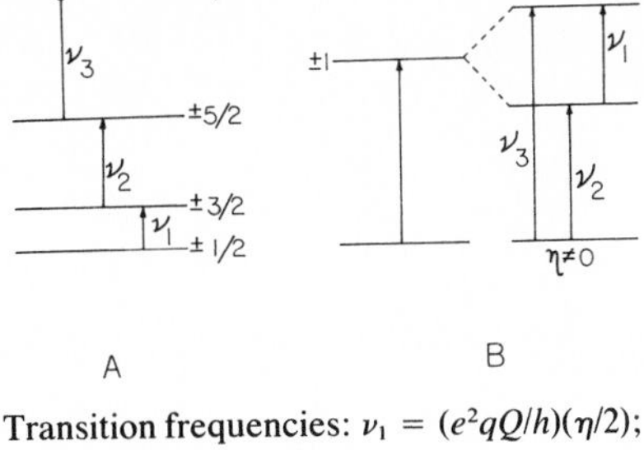

Figure 3-2. Examples of nuclear quadrupole energy levels in inhomogeneous electric fields. The degenerate levels of nuclei with half-integral spins (A) remain degenerate even when the asymmetry factor is non-zero, but the latter factor does alter the spacing between the levels. Transition frequencies for $\eta = 0$: $\nu_1 = (1/14)(e^2qQ/h)$; $\nu_2 = (2/14)(e^2qQ/h)$; $\nu_3 = (3/14)(e^2qQ/h)$. The degeneracy of integral spin nuclei is removed when the asymmetry factor is non-zero, as shown in the right-hand half of B. Transition frequencies: $\nu_1 = (e^2qQ/h)(\eta/2)$; $\nu_2 = (3/4)(e^2qQ/h)[1 - (\eta/3)]$; $\nu_3 = (3/4)(e^2qQ/h)[1 + (\eta/3)]$.

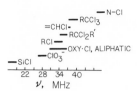

Figure 3-3. Frequencies of analogous ^{35}Cl-nqr transitions in different chemical environments. Spectra were obtained at 77 K in all cases. Correlations with bond type are equivalent to chemical shift correlations in nmr spectrometry. Adapted from Smith (1971b).

nucleus. A decrease in ν_Q thus correlates with increasing bond ionic character as the limit of an isolated, spherically symmetrical ion is approached. Variations in q are analogous to the chemical shift effects observed in nmr. It is possible to correlate ν_Q with bonding environment, as shown in Figure 3-3.

More complex molecules may contain atoms of the same type in different symmetry environments. Barring accidental degeneracy, it is probable that each environment will experience a different q, in which case one would expect to record a count of resonances equal to the number of nonequivalent positions. For example, the molecule in Figure 3-4 has two mirror planes and therefore three unique environments, leading to an expectation of three related ^{35}Cl-nqr absorptions. Measurements on this substance in fact yield a six-line spectrum (Smith, 1971a) since the molecular and crystal lattice symmetries interact to place each chlorine atom in a different electric gradient.

When the molecular rotational diffusion time is short relative to $1/\nu_Q$ and unrestricted, as in the liquid state, electric gradients are averaged to zero. Consequently, nqr in its conventional form necessitates working with solid samples (Borsa and Rigamonti, 1979). Polycrystalline material is often the subject of investigation but single-crystal studies lead to more insight. The example cited in the previous paragraph illustrates the sort of peculiarities one encounters in working with the solid state, and nqr is thus somewhat more difficult to interpret than nmr. In compensation, nqr permits measurements at lower temperatures, where better signal-to-noise ratios are theoretically possible.

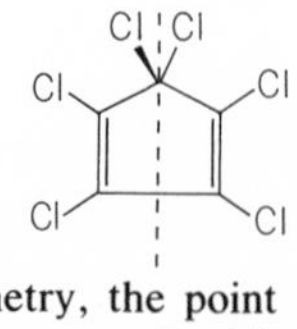

Figure 3-4. Perchlorocyclopentadiene. The point symmetry of the molecule includes a mirror plane (passing perpendicular to the page through the dashed line), and there are three unique chlorine environments (pairs which reflect left to right in the mirror). The ^{35}Cl-nqr spectrum reveals *six* different coupling constants rather than three, indicating a lower symmetry in the crystal, with each atom in a unique environment. In nqr spectrometry, the point symmetry of the molecule cannot be considered separately from crystal symmetry. In this regard the nqr spectrum is more complex than the nmr spectrum.

3.2 Relationship to Mössbauer, Esr, and Nmr Spectroscopy

Nuclear quadrupole coupling with electric gradients leads to hyperfine splitting in Mössbauer spectra (see Chapter 4), from which a quantitative estimate of qQ is possible. The distinction between the two methods lies in the nature of the quadrupolar species, i.e., in the iron Mössbauer effect it is a nuclear excited state while in nqr it is a nuclear ground state. Measurements of coupling constants by nqr are inherently orders of magnitude more accurate than by Mössbauer spectroscopy.

The complete spin Hamiltonian describing the energy levels of the esr spectrum includes a nuclear quadrupole interaction term (Ayscough, 1967) which results in an unequal spacing of hyperfine lines. In this case the nuclear entity is in its ground state. Nuclear quadrupole coupling constants may be determined from esr measurements on single, dilute-spin crystals or sometimes from random polycrystalline samples, especially when qQ is large. As an example, esr spectra of polycrystalline bis(diethyldithiocarbamato)copper(II) yielded e^2qQ values which varied by an order of magnitude, depending on the solvating group present (Liczwek et al., 1983). Esr-derived coupling constants are based on fairly small spectral effects, but the example cited here indicates a probable sensitivity of nqr to biological function, e.g., as in the case of a substrate interacting at a metalloenzyme center.

Values of the nuclear quadrupole coupling constant may also be estimated from nmr measurements in the liquid state. Such measurements depend upon the quadrupolar characteristics of the nmr isotope and the molecular tumbling rate, and while deuterium is suitable in this regard (Schramm and Oldfield, 1983), many other isotopes are not. The presence of a quadrupole moment can drastically increase the rate of nuclear magnetic relaxation (see Section 2.3.2.3), resulting in broadened lines. Most of the heavier biological isotopes suffer from this effect, which further complicates the noise problem associated with their typically low resonant frequencies. By contrast, nqr frequencies generally increase with atomic weight. Nqr thus offers an alternative to nmr for probing the environment of heavier biological elements, though one which has not been much explored.

3.3 Instrumentation

Conventional detection circuits for nqr are relatively less sophisticated than those used in nmr since free running oscillators are applicable. Marginal oscillators are generally used below 5 MHz while superregenerative designs are better at higher frequencies. Polycrystalline materials are enclosed in a thin-walled cuvette (~ 1 cm^3), which forms the core of the tuned circuit inductor. The probe assembly is generally cooled to low temperature, e.g., 77 K. Pulse detection is also possible (Lutz and Oehler, 1978), and coherent pulse instruments may have double resonance capabilities (Ramachandran and Narasimhan, 1983). The latter approaches not only permit access to relaxation phenomena but also lead to better signal-to-noise ratios.

Signal enhancement techniques (see Sections 2.4.4.2, 2.4.4.5, and 7.1.13) are applicable to a variety of spectroscopic methods, including nqr. Computer averaging and Fourier transform (Lutz and Oehler, 1978) modes of data acquisition increase sensitivity limits and can be done with nmr instruments. The major impediment to broader applications of nqr is clearly the difficulty in locating resonances, especially weak ones, within a very wide frequency range. However, nqr has a high potential for automation, and the recent emergence of powerful microcomputers (see Section 1.2.2) has created an economical way to revolutionize the data acquisition process. Fortunately, a frozen sample may be subjected to a lengthy search without risking degradation, something that is not possible in many nmr experiments.

3.4 Applications

Most applications of Mössbauer spectroscopy have involved iron resonances while nmr is still largely oriented to hydrogen and carbon. Similarly, many nqr studies have dealt with chlorine (Smith and Stoessiger, 1978; Smith, 1971b). Obviously, chlorine is not a commonplace covalent entity among biopolymers or intermediary metabolites, and other quadrupolar nuclei such as ^{14}N or ^{17}O are of much greater utility.

3.4.1 ^{17}O-Nqr

Effective physical methods for probing the electronic environment of oxygen in biomolecules are notably few (see ^{17}O-nmr, Section 2.7.1.6), in spite of the importance of this element. Considering the fractional abundance of ^{17}O, 3.7×10^{-4}, and the fact that it is the only isotope of oxygen with a non-zero spin ($I = 5/2$), it is not surprising that much difficulty surrounds ^{17}O-nmr. The conventional nqr superregenerative detection circuits mentioned above in Section 3.3 are also inadequate when the quadrupolar species is so dilute.

Nqr/nmr double resonance techniques developed by several independent investigators (Redfield, 1963; Slusher and Hahn, 1968) circumvent these problems and allow nqr measurements of ^{17}O (or other isotopes) in natural abundance. These interesting methods are applicable to any half-integral-spin nucleus (I = 3/2, 5/2, 7/2, etc.) and are based on the fact that an abundant nuclear spin system, I, present in the molecule (e.g., protons) may be in rapid thermal equilibrium with any dilute nuclear spin systems, S (e.g., ^{17}O). If the S species is irradiated at one of its quadrupole resonant frequencies using a sequence of $\sim 10^4$ 180° pulses, (see Section 2.4.4.4) the I-spin heat reservoir is significantly disturbed from its thermal equilibrium. Typically, the pulses delivered to the S-spin system are spaced at 500-μs intervals, which allows the energy received in each pulse to be equilibrated with the I-spin system (in the specific example cited here, the rate depends on the hydrogen/oxygen internuclear distance). Finally, a 90° pulse is applied to the sample at the I-spin nmr frequency, and the net I nuclear magnetization is quantified on the basis of the intensity of its free induction decay signal.

The actual nqr spectrum is constructed by repeating this process while varying the RF frequency applied to the S-spin system. Obviously, the I-spins must be allowed to equilibrate before each cycle repeats. In this manner, nqr absorptions are detected as reductions in the net nuclear magnetization of the I-spin system, as shown in Figure 3-5 for ^{17}O in $NaHCO_3$ (Cheng and Brown, 1979).

A block diagram of a double resonance nqr spectrometer of this type is presented in Figure 3-6. It will be noted that the S and I irradiations occur in separate compartments, necessitating a mechanical link for moving the sample rapidly back and forth. The sample itself is a powdered solid at 77 K.

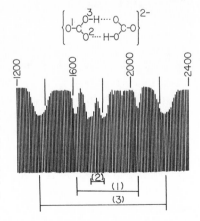

Figure 3-5. ^1H-Nmr/^{17}O-nqr double resonance spectrum of $KHCO_3$. Frequencies are in kilohertz, and the ν_1 and ν_2 resonance positions are indicated with numbers matched to the structure shown at the top. The low-frequency component of each resonance pair is ν_1. Every vertical bar represents one mechanical cycle between the nqr coils (Figure 3-6), and the ^{17}O-nqr absorption spectrum appears in negative contrast along the top of the recording, i.e., the valleys mark resonance positions. Adapted from Cheng and Brown (1979).

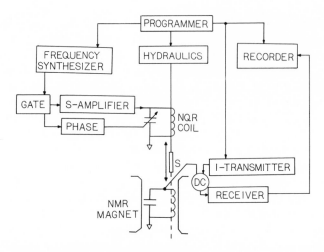

Figure 3-6. Nqr/nmr double resonance spectrometer. The sample (S) is shown midway between the nmr and nqr coils, and the hydraulic unit is able to move it rapidly between those two positions. Each cycle begins with an irradiation (in the nqr frequency range) in the upper position and ends with a measurement of ¹H nuclear magnetization in the lower (nmr) position. "DC" is a directional coupler. Adapted from Cheng and Brown (1979).

It was shown in the cited investigation (Cheng and Brown, 1979) that the ^{17}O quadrupole coupling constant is consistently lower when one of the groups attached to the carbonyl carbon contains lone pair electrons which can enter into π-bond interactions with the carbonyl, as shown in the resonance forms of Figure 3-7. Correlations of this type also have potential for detecting σ-bond inductive effects, hydrogen bonds, and metal coordination which involves the nonbonding electron pairs at oxygen. Dilute spin/abundant spin systems (involving a half-integral I) are common in biological materials; e.g., 1:1 protein complexes of ^{25}Mg, ^{43}Ca, ^{67}Zn, etc.; protein/substrate complexes where the substrate may be a thyronine derivative containing ^{127}I; and ^{33}S-containing structures in proteins, etc. These are the same isotopes which are difficult if not impossible nmr subjects as a result of quadrupolar relaxation effects in covalent binding environments. Thus, nqr/nmr double resonance might prove to be applicable to these elements. One limitation to the method is reduced coupling between the S and I spin systems for ions and atoms of large radii since cross coupling is an inverse cube function of internuclear distance.

$$R\overset{\overset{\displaystyle O}{\|}}{\underset{}{C}}\!\!-\!\!X \longleftrightarrow R\overset{\overset{\displaystyle O^{\ominus}}{\|}}{\underset{}{C}}\!\!=\!\!X^{\oplus}$$

Figure 3-7. Resonance forms of acyl compounds.

The nqr/nmr double resonance method has been used to study nqr correlations in carbonyl compounds (Cheng and Brown, 1979), and the evidence clearly indicates that ^{17}O-nqr spectra are sensitive to both chemical and crystal environment effects. For example, the three oxygens of HCO_3^- each have unique coupling constants and asymmetry factors, the latter being inferred by comparing Figure 3-5 with Figure 3-8, while two of the oxygens should be equivalent in solution. These crystal lattice effects are small in comparison with possible chemical differences; e.g., a peroxide oxygen in benzoyl peroxide experiences a coupling constant of 11.91 MHz, which is much larger than the values of about 7 MHz found in HCO_3^- (see Figure 3-5).

Linewidth also conveys structural information. For example, the broader lines observed for the #3 oxygens in Figure 3-5 are interpreted as saturation effects due to the proximity of their nuclei to the hydrogen-bonded protons, i.e., there is a more efficient coupling of spin energy between the S and I systems.

A major drawback to the use of ^{17}O-nqr is the absence of a good data base from which to draw reliable empirical correlations.

3.4.2 Nqr/Infrared Double Resonance

"Pure" nqr spectra may be obtained from gaseous samples using an nqr/infrared double resonance method (Arimondo *et al.*, 1978) orders of magnitude more sensitive than conventional absorption measurements. The sensitivity of this method is largely due to the narrowness of infrared absorption lines in gaseous phases ($\Delta\nu_{1/2} \sim 25$ MHz). Although gases are of lesser interest in biochemistry it should be noted that optical linewidths may also be quite narrow in the solid state, amounting to as little as 10 kHz at very low temperatures (DeVoe *et al.*, 1979). Analogous solid-state double resonance phenomena have not been explored, but there are potential implications for biochemistry.

The apparatus for nqr/infrared double resonance is shown in Figure 3-9. A sample is placed within the resonant cavity of a tunable infrared

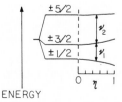

Figure 3-8. A plot showing the effect of the asymmetry factor on the degenerate nqr energy levels of ^{17}O. Frequencies ν_1 and ν_2 are functions of the coupling constant and the asymmetry factor, η: $\nu_1 = (3/20)(e^2qQ/h)f(\eta)$, $\nu_2 = (3/10)(e^2qQ/h)f'(\eta)$.

Figure 3-9. Laser infrared/nqr double resonance spectrometer. The gain tube contains an appropriate infrared laser material and a means for introducing excitation energy. The grating (G) allows tuning of the operating wavelength, and the beam enters a detector (DET.) through an exit mirror (M). The sample (S) acts as a load on the laser and is subjected to a strong, tunable RF field. Partial quenching of the beam occurs when the RF source is tuned to an nqr frequency. "Pure" nqr spectra are obtained when the detector output is recorded as a function of the applied RF frequency. Adapted from Arimondo *et al.* (1978).

laser, which is selected to operate at one of the sample's vibrational absorptions. Surrounding the sample is a coil which delivers tunable RF energy at about five watts. In this configuration the sample acts as a load on the laser and will quench the output if it is not dilute (e.g., a gas or a thin sample). A system of this type is distinctively nonlinear and is able to amplify small effects.

The energy level diagram of Figure 3-10 shows a vibrational transition (Ω) which only affects the populations of levels 2 and 4; in the steady state the load on the laser is determined solely by these populations and their relaxation rate. However, when the sample is irradiated at a quadrupole resonant frequency (ν_{12}) the effective population of the ground state is doubled, i.e., transitions now occur between quadrupole levels 1 and 2. This additional supply of ground state molecules places a heavier load on the laser. Thus, when the RF source is scanned over a range of frequencies, quadrupole resonances appear as sharp attenuations in the laser output, as shown in Figure 3-11. A similar argument applies to the irradiation of ν_{34} (Arimondo *et al.*, 1978).

The last example was included mainly to show that novel approaches to nqr and better sensitivities are possible. It should be noted that nqr/infrared double resonance has a high degree of bond selectivity.

3.4.3 Nqr Studies with ^{14}N and ^{127}I

Applications of nqr in biophysical and biochemical studies involving nitrogen and iodine have been reviewed (Babushkina, 1977). It is thought

Figure 3-10. Energy levels and transitions in infrared/nqr double resonance. *A*: No RF field applied; *B*: RF field applied at the $1 \rightarrow 2$ transition frequency. This randomizes the ground state populations of levels 1 and 2 (nqr) and doubles the load on the laser. *C*: RF field applied at the $3 \rightarrow 4$ transition frequency has a similar effect as in *B*.

Figure 3-11. Laser infrared/nqr double resonance spectrum
of gaseous CH₃I. The time scale of an infrared photon (brief
relative to rotational rate) permits nqr measurements in the
gas phase.

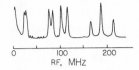
0   100   200
RF, MHz

that nitrogen and iodine nqr could be developed for the study of charge
distributions in biological molecules, provided sufficient empirical correlations are obtained. Tables 3-2 and 3-3 present, respectively, the nqr
coupling constants of ^{14}N in amino acids and ^{127}I in thyroid hormones.
Coupling constants vary among the chemical environments and some
(e.g., imidazole and indole) are distinctively different. In the case of nitrogen the asymmetry factor should be regarded as a source of information
as important as the coupling constant.

3.4.4 Biologically Oriented Studies with ^{35}Cl

Buess *et al.* (1984) found correlations between measured ^{35}Cl-nqr
resonant frequencies and the toxicities, antitumor activities, and Hammett

Table 3-2. ^{14}N-Nqr Coupling Constants and
Asymmetry Factors in Several Amino Acids[a]

Compound	e^2Qq (MHz)	Asymmetry[b]
Glycine	1.25	0.51
L-Alanine	0.97	0.41
L-Alanine	1.24	0.20
Cysteine·HCl	1.22	0.13
Cysteine	1.22	0.47
Tyrosine	1.08	0.41
Phenylalanine	1.35	0.63
Histidine	1.31	0.14
Histidine	1.25	0.11 N—H (imidazole)
Histidine	1.43	0.91 N: (imidazole)
Histidine	3.36	0.13 N: (imidazole)
Tryptophan	1.27	0.55
Tryptophan	3.01	0.18 indole ring
Serine	1.215	0.118
L-Proline	1.623	0.955

[a]Adapted from Babushkina (1977).
[b]α-Amino group unless indicated otherwise.

Table 3-3. Nqr Coupling
Constants of [127]I in Thyroid
Hormones[a,b]

Compound	e^2Qq (MHz)
Thyroxine (T$_4$)	297.85
	295.21
	293.61
	284.65
	283.10
Triiodothyronine (T$_3$)	292.10
	282.45
	290.60

[a]$1/2 \rightarrow 3/2$ transitions.
[b]Adapted from Babushkina (1977).

sigma parameters of several nitrogen mustards. Figure 3-12 shows the relationship between LD$_{10}$ and resonant frequency. While the correlation is excellent for four of the compounds, one differed for reasons which are not clear. These authors attempted to relate their data to C—Cl bond character and an alkylation mechanism involving an aziridinium intermediate (Figure 3-13).

3.4.5 Other Potential Applications of Nqr

Both nmr and nqr are well-suited for studies of phase transitions in condensed phases (Borsa and Rigamonti, 1979). Paramagnetic/antiferromagnetic phase transitions may be detected using nqr (Kawamori, 1978; Guibe and Montabonel, 1978). In materials of this type one is able to observe the nqr resonance of the metal ion in the paramagnetic (higher temperature) phase, but the signal disappears below the Néel temperature (i.e., in the antiferromagnetic phase). For example, the nqr

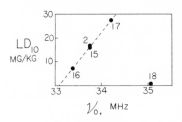

Figure 3-12. The relationship between lethal dose for 10% mortality and the chlorine-35 nqr frequency of nitrogen mustards. NSC compound codes (in parentheses) are keyed to the plotted data as follows: 2 (3088), 15 (16498), 16 (23891), 17 (9698), and 18 (762). The latter compound (R = methyl group in Figure 3-13) does not correlate with the other data, but it also has a considerably lower molecular weight. From Buess *et al.* (1984).

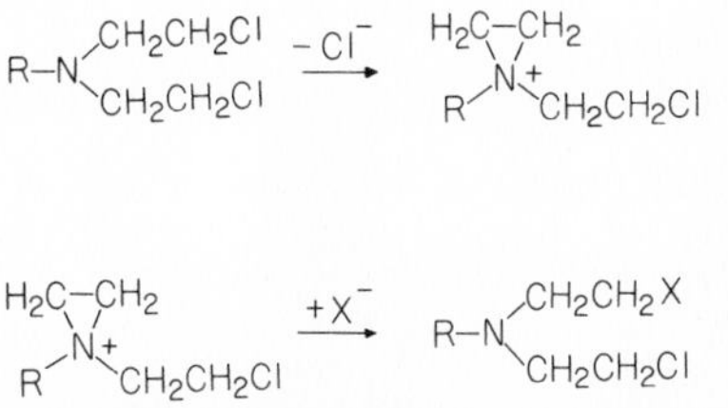

Figure 3-13. A possible mechanism for nitrogen mustard action. One model envisions that the biological activity of nitrogen mustards is related to ability to form an aziridine intermediate, as shown here. Aziridines are efficient alkylating agents, and the "X" shown could be a carboxylate group, a phosphate group, or any other which might donate a pair of electrons in the second ring-opening reaction. From Buess *et al.* (1984).

resonances of ^{63}Cu and ^{65}Cu in $CuBr_2$ are present at room temperature but are absent at 77 K (Guibe and Montabonel, 1978).

Similar magnetic behavior occurs in the polynuclear atom assemblies found in metalloproteins. Nqr thus seems potentially useful in studies of biological forms of the transition metals, in spite of problems associated with the detection of very weak resonance signals over a wide frequency range.

References

Arimondo, E., Glorieux, P., and Takeshi, O., 1978. Radiofrequency spectroscopy inside a laser cavity: "Pure" nuclear quadrupole resonance of gaseous CH_3I, *Phys. Rev. A.* 17:1375.

Ayscough, P. B., 1967. *Electron Spin Resonance in Chemistry*, Methuen & Co. Ltd., London, p. 102.

Babushkina, T. A., 1977. Use of nuclear quadrupole resonance in biophysical studies, *Nucl. Quadrupole Reson.* (Russian) 2:3.

Biryukov, I. P., and Vorokonov, M. G., 1972. *Tables of Nuclear Quadrupole Resonance Frequencies*, Halsted Press, New York.

Borsa, F., and Rigamonti, A., 1979. Nmr and Nqr in fluids, paramagnets and crystals, in *Magnetic Resonance of Phase Transitions* (F. J. Owens, ed.), Academic Press, New York.

Buess, M. L., Bray, P. J., and Sheppard, D. W., 1984. ^{14}N and ^{35}Cl nuclear quadrupole resonance data for nitrogen mustards: Attempted correlations with chemical and biological activities, *Org. Magn. Reson.* 22:67.

Cheng, C. P., and Brown, T. L., 1979. Oxygen-17 nuclear quadrupole double resonance spectroscopy. 1. Introduction. Results for organic carbonyl compounds, *J. Am. Chem. Soc.* 101:2327.

Das, T. P., and Hahn, E. L., 1958. *Nuclear Quadrupole Resonance Spectroscopy*, Academic Press, New York.

DeVoe, R. G., Sazabo, A., Rand, S. C., and Brewer, R. G., 1979. Ultraslow optical dephasing of $LaF_3:Pr^{3+}$, *Phys. Rev. Lett.* 42:1560.

Guibe, L., and Montabonel, M. C., 1978. Temperature dependence of nqr frequencies in copper(II) bromide and magnesium bromide etherate, *J. Magn. Reson.* 31:419.

Kawamori, A., 1978. Exchange interaction in copper acetate monohydrate studied by nuclear quadrupole resonance, *J. Magn. Reson.* 31:423.

Liczwek, D. L., Belford, R. L., Pilbrow, J. R., and Hyde, J. S., 1983. Elevation of copper nuclear quadrupole coupling in thio complexes by complexation of the coordination sphere, *J. Phys. Chem.* 87:2509.

Lutz, O., and Oehler, H., 1978. ^{135}Ba and ^{137}Ba Fourier transform nmr and nqr studies, *Z. Physik. A.* 288:11.

Ramachandran, R., and Narasimhan, P. T., 1983. A coherent nuclear quadrupole pulse and double resonance spectrometer, *J. Phys. E.* 16:643.

Redfield, A. G., 1963. Pure nuclear electric quadrupole resonance in impure copper, *Phys. Rev.* 130:589.

Schramm, S., and Oldfield, E., 1983. Nuclear magnetic resonance studies of amino acids in proteins. Rotational correlation times of proteins by deuterium nuclear magnetic resonance spectroscopy, *Biochemistry* 22:2908.

Schultz, H. D., 1970. Nuclear quadrupole resonance spectroscopy in *Inorganic Chemistry*, Vol. I (C. N. R. Rao and J. R. Ferraro, eds.), Academic Press, New York.

Semin, G. K., Babushkina, T. A., and Yakobson, G. G., 1975. *Nuclear Quadrupole Resonance in Chemistry* (English translation), Halsted Press, New York.

Slusher, E., and Hahn, E. L., 1968. Sensitive detection of nuclear quadrupole interactions in solids, *Phys. Rev.* 166:332.

Smith, J. A. S., 1971a. Nuclear quadrupole resonance spectroscopy, Part Two. Instruments, *J. Chem. Ed.* 48:A77.

Smith, J. A. S., 1971b. Nuclear quadrupole resonance spectroscopy, Part Three. Chemical applications, *J. Chem. Ed.* 48:A243.

Smith, J. A. S., 1978, ed. *Advances in Nuclear Quadrupole Resonance*, Vols. 1–3, Heyden Press, Philadelphia.

Smith, P. W., and Stoessiger, R., 1978. Nuclear quadrupole resonance studies of compounds of the type $A_2(I)M(III)Cl_5(H_2O)$, *J. Magn. Reson.* 31:431.

Mössbauer Spectroscopy

At the time of this writing, the number of published articles involving Mössbauer spectroscopy is approaching 20,000 (Stevens and Bowen, 1984). This figure includes *all* applications, of which biologically relevant contributions represent only a part. Mössbauer spectroscopy is appropriate for this volume as a result of its high degree of element specificity, and Mössbauer spectra have a substantial information content, being comparable to esr spectra. Most Mössbauer studies have focused on compounds containing iron or tin.

Certain modes of nuclear transmutation (e.g., α, β^+, β^-, electron capture, and γ) create daughter nuclei in an excited isomer state. This is illustrated in Figure 4-1, which shows the electron capture decay of ^{57}Co ($t_{1/2} = 270$ days) to the excited isomer of ^{57}Fe. The latter quickly returns to the nuclear ground state by the emission of a 136.46-keV gamma photon or, alternatively (and more significantly), by a two-step process involving an intermediate excited state and a final transition energy of 14.41 keV. Photons emitted by ^{57}Fe isomers are thus approximately resonant with other ground state ^{57}Fe nuclei (about 2% natural abundance in terrestrial iron), and an iron-containing absorber will experience ^{57}Fe excitation and nuclear fluorescence when it interacts with 14.41-keV gamma radiation. By comparison, atomic absorption is the analogous phenomenon involving electronic rather than nuclear energy levels. The Mössbauer effect is enhanced if the nuclear transition is of relatively low energy (which reduces recoil energy) and has a large absorption cross section as well as a long excited state lifetime. From Table 4-1 it will be seen that ^{57}Fe meets these criteria, and this is the reason for iron's preeminence in Mössbauer studies.

Prior to Rudolf Mössbauer's classical experiments (Mössbauer, 1958) it had been assumed that the γ-emitting nucleus would experience a recoil governed by the energy conservation law of Eq. (4-1):

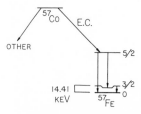

Figure 4-1. Production of ^{57}Fe nuclear isomers. The radio-active isotope ^{57}Co with a half-life of 270 days decays predominantly by electron capture (99.8%), i.e., the nucleus assimilates a K-shell electron, and the atomic number is reduced by one. The resulting iron nucleus is formed in an excited state and may proceed directly to the ground state by emission of a gamma photon. Alternatively, ground may be reached via the intermediate excited state ($I = 3/2$; the indentation indicates the metastable nature of this level, which has a half-life of approximately 98 ns). Further decay (i.e., $3/2 \rightarrow 1/2$) results in the emission of a 14.41-keV gamma photon. The latter resonance is the one commonly used in Mössbauer studies of iron. The same nuclear isomers of ^{57}Fe are produced in the decay of ^{57}Mn.

$$E_T = E_\gamma + E_R \qquad (4\text{-}1)$$

Here the subscripts T, γ, and R denote transition, gamma, and recoiling nuclear energies. The dissipation of E_R in a source means that resonant excitation in an absorber is possible only if the relative thermal motions of the emitting and receiving nuclei can make up the deficit between E_R and E_T. That is to say, nuclear resonant absorption and fluorescence are temperature-dependent events and their occurrence should be expected to increase at higher temperatures. Simple nuclear fluorescence behaves in this manner.

The Mössbauer effect occurs in crystalline or amorphous solids at low temperature (see Section 4.5.3). In the above model it was assumed that γ-photon emission would be accompanied by nuclear recoil. Solids, however, have quantized phonon (lattice vibration) energy levels, which may be related to an oscillator frequency ν and a quantum number n in an Einstein solid, as described in Eq. (4-2):

$$\Delta E_p = h\nu (n + \tfrac{1}{2}); \qquad n = 0, 1, 2 \ldots \qquad (4\text{-}2)$$

Under these conditions a relatively weak nuclear transition, e.g., 150 keV, might not result in a recoil since E_R is comparable to (or even less than) ΔE_p, where $n = \pm 1$. Quantized behavior of this type is dominant below the Einstein temperature, θ_E, defined in Eq. (4-3):

$$\theta_E = h\nu/k \qquad (4\text{-}3)$$

where k is the Boltzmann constant and h the Planck constant.

We thus see that the Mössbauer effect is the result of quantum mechanical restrictions inherent in low-temperature solids, i.e., recoil is possible only when there is enough energy to excite a lattice phonon transition. Barring this mechanism, the emitted γ-photon must contain *all* of the energy of the nuclear transition, i.e., $E_\gamma = E_T$. In practice, both types of events have finite probabilities, and the Mössbauer line may be identified by its shorter wavelength.

Table 4-1. Mössbauer Isotopes Which Have Been or Could Be Used in Biological Investigations

Isotope	% Natural abundance	γ-Energy (keV)	Half-life (ns)	Nuclear spin (h/2π) Excited	Nuclear spin (h/2π) Ground	Cross section (10^{-20} cm^2)	Line width (mm/s)	Recoil energy (10^{-3} eV)
^{40}K	0.0118	29.4	4.26	3	4	28.97	2.184	11.60
^{57}Fe	2.19	14.4125	97.81	3/2	1/2	256.6	0.1940	1.957
^{57}Fe	2.19	136.46	8.7	5/2	1/2	4.30	0.2304	175.4
^{61}Ni	1.25	67.40	5.06	5/2	3/2	72.12	0.8021	39.99
^{67}Zn	4.11	93.31	9150	3/2	5/2	10.12	0.000320	69.78
^{119}Sn	8.58	23.871	17.75	3/2	1/2	140.3	0.6456	2.571
^{127}I	100	57.60	1.9	7/2	5/2	21.37	2.500	14.03

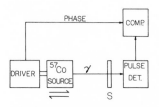

Figure 4-2. Basic features of a Mössbauer spectrometer. The driver is a mechanical or electromechanical device which alternately moves the source (e.g., a quantity of ^{57}Co producing excited ^{57}Fe in the case shown here) toward and away from the absorbing sample (s). The sample container must have provisions for maintaining low temperatures. The ^{57}Fe-resonant gamma photons pass through the sample and are counted in a scintillation detector. A computer or a simple processor sorts and adds the incoming count pulses into a memory array corresponding to Doppler velocities from the most negative to the most positive. Accurate sorting necessitates a rigid phase lock between the driver and the computer. In the type of instrument shown here, Mössbauer lines will appear as *decreases* in the count rate at specific Doppler velocities (see Figure 4-3).

4.1 Instrumentation

Spectral dispersion of Mössbauer lines is achieved by creating a Doppler effect between the source and the sample under investigation. The basic components of a working spectrometer are shown in Figure 4-2. Its source is moved alternately toward and away from the sample detector assembly, causing first a slight blue-shift and then a corresponding red-shift in the gamma source.* There are several ways to create a velocity component which varies systematically, e.g., by mounting the source on a centrifuge wheel (Mössbauer's first experiments made use of a motorized toy!) or by fixing the source to the driven element of a solenoid (see also Section 4.6; piezoelectric drivers have been used in ^{67}Zn spectrometers).

The sample cell in Figure 4-2 lies between the source and a scintillation detector, and spectra recorded in this arrangement will be of the absorptive type. The output of the scintillation photomultiplier is routed first through a pulse height discriminator, which is set to count the energetic Mössbauer events while rejecting others. Pulses from the discriminator are then accumulated in a multichannel scaler which maintains an accurate synchronization between source Doppler velocity and counting channel number. A typical Mössbauer spectrum obtained in this fashion is shown in Figure 4-3, where resonant absorptions appear as downward spikes, i.e., the count rate at the scintillation detector (ordinate) is re-

---

*Velocity modulation may be at constant acceleration or at constant velocity. In the latter, counts are averaged for one velocity value. The spectrum may be constructed sequentially, i.e., velocity is incremented and a new data point is obtained. Constant velocity modulation is useful if one wishes to monitor a particular spectroscopic feature as a function of time. In the constant acceleration mode the whole spectrum array is obtained during each mechanical cycle. Successive spectra are summed into a storage array. Both methods allow for signal averaging.

Figure 4-3. Mössbauer absorption spectra showing polarization effects. Doppler velocities expressed in mm/s on the abscissa are bipolar (the driver moves the source toward and away from the absorbing sample). The ordinate presents scintillation detector count rate as percent transmittance, where 100% T is the count rate away from Mössbauer-resonant absorption lines. The example shown here illustrates another feature of Mössbauer spectrometry based on polarization of the emitted gamma radiation. Polarimetry analogous to that used in optical spectrometry involves applying strong magnetic fields to both the source and the absorbing sample (Faraday effect). The two fields are parallel in the top spectrum and perpendicular in the bottom example. Selective effects are evident. (Adapted from Gonser, 1975.)

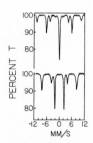

duced. Chemical phenomena lie within a range of only tens of mm/s and are indeed weak modulations of the energetic gamma photon.

Practical Mössbauer spectrometers are more sophisticated than Figure 4-2 indicates. Some instruments rely on backscatter (fluorescent emission) rather than absorption, and most have provisions for applying a magnetic field, measuring polarization, etc. All working spectrometers require cryogenic arrangements for cooling the sample and the source. To be useful the spectrometer should be on-line linked to a minicomputer for the purpose of comparing theoretical spectra with the data. There are even more economical options, e.g., inexpensive microcomputers are entirely adequate for the purpose (Linares and Sundqvist, 1984).

4.2 Conversion Electron Detection

Absorption and backscattering detection require sensors which register γ-ray photons. The latter entities are deeply penetrating and their detection necessitates large scintillator crystals. Also, the direct method may encounter dynamic range problems when the absorption is very small. An alternate process of detection involves counting the conversion electrons produced in a sample itself when the resonant energy is absorbed and then re-radiated (Manuschev *et al.*, 1976).

In many of the resonant absorption events occurring in the specimen the nuclear isomer simply re-radiates a γ-ray photon in a random direction after a brief interval. However, in some nuclear events the excitation energy is imparted to an atomic electron, which departs with the amount of energy given by Eq. (4-4):

$$E_{C.E.} = h\nu_0 - B.E. \tag{4-4}$$

where B.E. is the binding energy. Such electrons may be detected and counted in the manner of β-emitters if the sample is enclosed within a plastic scintillator block. Unlike β-emissions, the conversion electrons originating in a thin sample have better-defined energies. Bonchev *et al.* (1977) were able to detect 10^{-7} g of tin using this approach. Obviously, any method which increases sensitivity is pertinent to biological investigations.

4.3 Sample Considerations

The necessity of low temperatures in Mössbauer spectroscopy precludes any possibility of approximating *in vivo* conditions, and information obtained by this method contains some uncertainties. Nmr and esr are the two methods definitely suited for the biological temperature range. Nqr and the diffraction techniques, in contrast, are restricted to crystalline phases and are more comparable in this respect to Mössbauer spectroscopy. However, the Mössbauer effect is simply a low-temperature phenomenon and does occur in amorphous solids.* Some Mössbauer studies, for example, have been carried out on frozen, whole organisms (Dickson and Rottem, 1979).

Since ^{57}Fe is not abundant (2%) an advantage is gained by replacing the normal iron with this isotope, e.g., by growing the organism in a ^{57}Fe-enriched medium or chemically exchanging it with the active isotope. Alternatively, one may accumulate statistically large counts at the Mössbauer wavelength, from which weak absorptions may be recovered. The latter approach is costly in time, but thermal decomposition of the sample is not a problem at low temperature. Also, the multiscaling counter inherent in the spectrometer's design (Figure 4-2) is a type of signal averager. Gamma radiation damage to the sample is negligible in most measurements, although the possibility should not be overlooked in lengthy measurements.

Sample thickness is an important consideration. The effective absorber thickness is given in Eq. (4-5), where σ is the absorption cross section, f_m is the fraction of Mössbauer events, n' is the number density of Mössbauer active atoms (atoms cm^{-3}), and d is the thickness (cm) presented between the source and the scintillator:

$$t = \sigma f_m n' d \qquad (4\text{-}5)$$

The sample thickness, d, should be adjusted so that t is approximately one. If t is much greater than one, the lines will broaden (due to saturation)

---

*In working with biological specimens it is often advantageous to freeze the material as rapidly as possible since crystal growth can destroy cellular structures. This of course does not apply in simpler systems, e.g., the growth of single crystals of proteins.

and lose their Lorentzian shape. Obviously, absorption problems quickly arise in a dilute sample when the gamma energy is low, as in the case of the 14.41-keV resonance of ^{57}Fe. Under such circumstances it may be preferable to observe *backscattered* photons.

4.4 Mössbauer Effect and the Chemical Environment

The energetic γ-ray photons involved in the Mössbauer effect are capable of breaking hundreds of chemical bonds if photoelectron production occurs. By contrast, nmr, nqr, and esr are all based on transitions with energies far smaller than the weakest chemical bonds, and the effects due to chemical differences are even more diminutive. This is best seen in nmr where chemical modulations are less than a part per thousand of the measuring frequency. Similarly, the variations in isomer shifts seen in the Mössbauer spectra of iron compounds are due to variations in the electron density at the active nucleus and are almost vanishingly small in comparison with the energy of the γ-photon. The ability to observe chemical phenomena by means of the Mössbauer effect is based entirely on the extreme narrowness of Mössbauer lines. Were the lines much broader, the technique would be of little or no interest to chemists.

Mössbauer phenomena which relate to the chemical environment at the resonant nucleus include intensity and linewidth effects, slight differences in the resonant condition between the source and sample (isomer shift), and two conditions which produce a multiplicity of absorptions: one due to electric gradients at resonant nuclei (quadrupole splitting) and a paramagnetic manifestation (magnetic hyperfine splitting) closely related to features observed in esr spectra. In addition, polarization phenomena may be of considerable interest. These sources of chemical information will be discussed briefly in the following section.

4.4.1 Isomer Shifts, δ (Also Known as Chemical Shifts)

Although the inner shell (or core) electrons do not participate directly in chemical bonding, their charge densities are in fact effected to some extent by the atom's environment. In turn, alterations in the electron charge density in the nuclear region may either increase or decrease the energy *difference* between the ground and isomer states, as shown in Figure 4-4, resulting in a shifting of the resonant condition to a slightly higher or lower energy. It is emphasized that the shift involved is *slight*. As a result, the values are reported relative to a reference substance

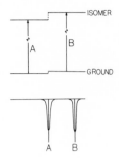

Figure 4-4. Isomer shift (synonymous with chemical shift). Examples A and B show Mössbauer-resonant absorptions involving the same isotope (e.g., ^{57}Fe) in two chemically distinct environments. Changes in the nuclear surroundings produce unequal changes in the ground and isomer state energy levels. As a result, the exact Doppler conditions for resonances A and B will differ slightly. Such changes are extremely small in comparison with the gamma photon energy. Free of other effects, the spectra of A and B will appear as shown in the lower part of the figure. In practice, the shift scale is calibrated with respect to a standard reference substance (e.g., metallic Fe).

(usually stainless steel in the case of iron Mössbauer spectroscopy), as shown in Eq. (4-6):

$$E_a - E_s = \frac{\delta E_\gamma}{c} \qquad (4-6)$$

where c = speed of light, E_a = absorber transition energy, E_s = source transition energy (reference), and E_γ = Mössbauer gamma energy.

A rigorous theory for the isomer shift has not been established; however, the observed values are found to reflect the atom's valence and bonding environment. For example, the spin and oxidation states of iron compounds correlate with isomer shifts (relative to metallic iron) in the decreasing order Fe^{2+} high spin ($\delta \sim 1.3$ mm/s) > Fe^{3+} high spin ($\delta \sim 1$) > Fe^{2+} low spin ($\delta \sim 0.1$) > Fe^{3+} low spin ($\delta \sim 0$). Ligands which are covalent also tend to *decrease* the observed isomer shift. Isomer shifts are relatively insensitive to oxidation state among low-spin forms of iron, whereas in high-spin complexes, the isomer shift is a more reliable indicator of oxidation state.

The isomer shift thus provides a clue about the atomic environment but one that can be ambiguous since much variability between similar environments has been noted (Dwivedi *et al.*, 1979). One should also realize that the information contained in isomer shifts appears to be similar to the chemical shift perturbations found in ESCA and Auger spectra (see Section 7.1.5). The latter phenomena correlate primarily with atomic oxidation state.

4.4.2 Magnetic Effects

4.4.2.1 The Nuclear Zeeman Effects (External Field)

If a magnetic field is applied to a diamagnetic ^{57}Fe atom, the result will be as shown in Figure 4-5. The field causes the nuclear excited state

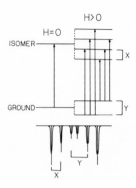

Figure 4-5. The nuclear Zeeman effect in the absence of paramagnetism. Vertical spacing between the ground and isomer
states relative to the nuclear magnetic splitting is enormously
larger than shown. This example is representative of ^{57}Fe,
where the ground state, $I = 1/2$, is split into two levels ($+ 1/2$,
$- 1/2$) and the isomer state, $I = 3/2$, is split into four levels
($+ 3/2$, $+ 1/2$, $- 1/2$, $- 3/2$) when an external magnetic field is
applied. Also, the upper and lower levels are not split to the
same extent (X and Y), leading to the complex spectrum shown
in the lower part of the figure.

spins to assume $2I + 1$ discrete energy levels, and a six-line spectrum
results from the applicable selection rules, i.e., $\Delta m = 0, \pm 1$. This relatively simple pattern is symmetrical, and the differences in amplitude are
due to differences in the transition probabilities.

4.4.2.2 Paramagnetism in the Absence of an Applied Field

If the Mössbauer nucleus is at a center of paramagnetism, as in low
spin Fe^{3+}, the nuclear and electron spins interact to produce a total
angular momentum (coupled) of $F = S + I$, where S and I are, respectively, the electron and nuclear spins. In the case of low-spin Fe^{3+}, which
has a single unpaired electron ($S = 1/2$) and nuclear spin values of $I =
3/2$ (the nuclear excited state) and $I = 1/2$ (the nuclear ground state),
there are four energy levels, as shown in Figure 4-6. The three-line spectrum of this atom will be observed only in the absence of other complicating effects, i.e., those due to an applied field (see Sections 4.4.2.1 and
4.4.2.3) or electric field gradients (see Section 4.4.3).

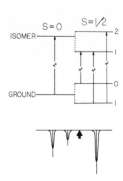

Figure 4-6. Zero-field splitting in paramagnetic environments.
The total angular momentum, F, is the vector sum of S, the
electron spin multiplicity, and I, the nuclear spin quantum number, i.e., $F = S + I$. In the example shown at the right, the
electron spin multiplicity is 1/2 and the value of I is 1/2 and 3/2,
respectively, for the ground and isomer states, as in the case of
^{57}Fe. F values are shown at the right. Selection rules lead to the
three-line spectrum ($\Delta m = 0, \pm 1$ and $\Delta M_e = 0$). The location
of resonance in the absence of electron–nuclear coupling (diamagnetism) is marked with a heavy arrow.

One will note that the spectrum of Figure 4-6 is asymmetric, which is the result of unequal probabilities among the three allowed transitions. The electron's spin–lattice relaxation time must be long if zero-field splitting is to be observed; specifically $\tau_e >> \hbar/A$, where A is the electron–nuclear coupling constant.

4.4.2.3 Paramagnetism in an Applied Field

The application of a magnetic field to a Mössbauer nucleus contained within a paramagnetic atom leads to more complex effects. Figure 4-7 shows the energy level correlation diagram for ^{57}Fe ($S = 1/2$) when an external field is applied. The pattern is now more or less a hybrid of Figures 4-5 and 4-6, but the Mössbauer selection rules include a new condition, $\Delta M_e = 0$. The external field influences both nuclear and electron magnetic dipoles, and since the latter's field may either reinforce or oppose the applied field at the nucleus in question, the result is a set of low and high nuclear spin energy levels separated by the value E, as shown in Figure 4-7. The transition rule simply requires that the electron's spin state is not altered during a Mössbauer absorption.

The energy difference N, Figure 4-7, appears in the esr spectrum as electron–nuclear hyperfine splitting. More correctly, the esr splitting is $2N = 2A/2g\beta$ (see Chapter 5), and for ^{57}Fe (ground state $I = 1/2$) the result will be a doublet separated by about 20 G in the esr spectrum. The corresponding effect of the electron on the nucleus is much larger, typically on the order of 200 kG. Its larger effect is simply a result of the considerably greater strength of the electron magnetic dipole compared with the nuclear magnetic dipole.

In Mössbauer spectra the selection rule $\Delta M_e = 0$ forbids transitions between upper and lower members of the electron splitting sets, and the

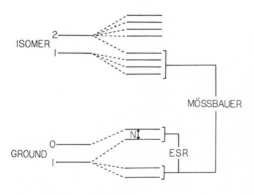

Figure 4-7. Energy levels involving paramagnetism ($S > 0$) and an applied magnetic field. The left part of the figure coincides with the right half of Figure 4-5. By applying a strong field the levels are further split, as shown. Mössbauer transitions between the levels indicated in this figure are closely comparable to those of Figure 4-4. The relationship between esr and the Mössbauer effect is also presented in this figure.

actual spectrum closely resembles the case in Figure 4-5. The energy levels are given by Eq. (4-7), where the first two terms are significant at moderate applied fields and the third becomes significant in an intense field.

$$\mathcal{H} = g\beta \mathbf{H} \cdot S + AS \cdot I - g_n\beta_n\mathbf{H}_{ext} \cdot I \qquad (4\text{-}7)$$

At low fields one may observe a spectral rearrangement from the zero-field condition to the nuclear Zeeman effect. In the latter situation the electron and nuclear spins begin to precess independently about the applied field. It is emphasized that the electron–nuclear interactions appear only when the electron spin state is long-lived, i.e., when $\tau_e \gg \hbar/A$. The probability of observing these effects thus improves at lower temperatures.

The result is somewhat more complex when there is a multiplicity of unpaired electrons ($S > 1/2$). Also, the values of both g and A may be anisotropic, which may lead to severe, uncorrectable line broadening. The cases shown in Figure 4-5, 4-6, and 4-7 are thus idealized and may not apply in many situations.

4.4.2.4 Information Revealed through Mössbauer Magnetic Effects

We have seen that the interaction of an unpaired electron with a magnetic nucleus is observable as hyperfine splitting in the esr spectrum and as zero-field splitting in the Mössbauer spectrum. A closely related effect, the contact shift (see Section 2.7.2.4.3), is observed in the nmr spectra of paramagnetic molecules under certain conditions. Thus, all three methods are somewhat redundant, at least in principle. In practice, Mössbauer studies of iron are possible throughout a wide range of spin and oxidation states, while direct nmr probes of the iron nucleus are limited to the uncommon diamagnetic forms of iron (e.g., Hb·O_2). Similarly, esr is not applicable to all oxidation states of iron (see Section 5.4.3). The Mössbauer method also produces evidence, through coupling constants, of the degree of localization of the unpaired electron at iron, i.e., the coupling is weaker when the electron is more delocalized.

4.4.3 Quadrupole Splitting (ΔE_Q)

If the Mössbauer nucleus is localized in an environment which is diamagnetic and spherically symmetrical (i.e., lacking an electric field gradient), only a single resonance of finite width will be observed. How-

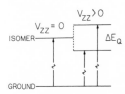

Figure 4-8. Quadrupole splitting. The right-hand example shows the splitting of an isomer state ($I = 3/2$) into two levels when placed in an electric field gradient (V_{zz}), while the ground state ($I = 1/2$) is unaffected. The resulting pair of transitions is a common feature of ^{57}Fe-Mössbauer spectra since bond polarity and asymmetry often lead to an electric field gradient at the iron nucleus.

ever, it will be noted that all of the isotopes collected in Table 4-1 have quadrupole moments in their nuclear excited states (where I is always greater than 1/2). As a result, the presence of charge asymmetry in the atom's bonding environment will cause a splitting of the single resonance into two or more since an electric field gradient will then exist at the Mössbauer nucleus. Figure 4-8 shows the result for ^{57}Fe, where I(ground) = 1/2 and I(excited) = 3/2. Thus, depending on the presence or absence of symmetry at the iron atom, Mössbauer spectra of ^{57}Fe may or may not exhibit the quadrupole doublet structure shown in Figure 4-9. Figure 4-8 is a fairly simple case, and in other instances (e.g., ^{67}Zn) quadrupolar splitting involves both the ground and excited states.

The value of ΔE_Q depends on the intensity of the electric field gradient, $\tilde{V}$, at the nucleus and the nuclear quadrupole moment, i.e., $e^2\tilde{V}Q$. The information provided by Mössbauer quadrupole effects parallels what is observed in nqr spectra and may be used for inferences about the symmetry environment at the Mössbauer atom. However, in ordinary nqr the quadrupole interaction occurs in the nuclear ground state while in Mössbauer resonance the ground and excited states may be involved (e.g., ^{127}I), or in some cases, just the excited state (^{57}Fe). Measurements of quadrupole coupling to electric field gradients are inherently more accurate by the nqr method (see Chapter 3). One should note that the quadrupole splitting will be observed throughout the temperature range of the Mössbauer phenomenon, whereas the magnetic splitting may vanish in the upper temperature range. The two effects may thus be distinguished.

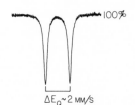

Figure 4-9. Quadrupole doublet with an energy difference of 2 mm/s. The ^{57}Fe = Mössbauer spectrum frequently presents doublet signatures corresponding to transitions of the type shown in Figure 4-8.

4.4.4 Linewidth and Lineshape

The natural linewidth of a Mössbauer line is given by Eq. (4-8), where τ_I is the effective lifetime of the isomer state. Isomers which decay rapidly produce broader lines. The best case in Table 4-1 is ^{67}Zn, which is theoretically able to produce very narrow lines.

$$\Gamma = \frac{\hbar}{\tau_I} \tag{4-8}$$

Other factors which contribute to linewidth include saturation effects, motional disorder near phase transitions (and other types of Brownian motion), and various relaxation processes. These act, when present, to broaden the lines. Delayed coincidence and thermal spikes may produce line-narrowing effects.

Lineshapes are commonly Lorentzian, though occasionally Lorentzian convoluted with Gaussian gives a better fit. Linewidths can be artificially narrowed by Fourier transforming the spectrum to the time domain. After multiplication with an exponentially *increasing* function (deconvolution) the inverse transform yields a spectrum with narrower lines. Deconvolution presents the disadvantage of degrading S/N.

4.5 Biological Applications of ^{57}Fe

The Mössbauer spectral features of isolated iron atoms in porphyrin structures are presented in Table 4-2. A more complex type of behavior has been observed among the four-iron clusters of high-potential iron proteins, where redox changes are accompanied by marked changes in the Mössbauer spectrum. In particular, the cluster iron of these proteins may become diamagnetic due to antiferromagnetic coupling, even in a one-electron redox process, which shows that the electron spins of the cluster act cooperatively. In these clusters, an antiferromagnetic state is indicated by evidence from several sources, including Mössbauer spectra.

Mössbauer measurements of iron in *any* of its compounds yield detailed information about the valence- and spin-states of iron. In biological studies, particularly those involving large aggregate structures or whole organisms, there may be two or more chemical forms of iron present, leading to a spectrum which may be impossible to interpret. The spectral dispersion of the Mössbauer method is relatively low when compared with others such as nmr. Thus, the interpretation of Mössbauer spectra

Table 4-2. The Mössbauer Spectral Features of Isolated Iron Atoms in
Porphyrin Structures

Iron type	Quadrupole splitting	Zero-field splitting	Magnetic hyperfine splitting	Isomer shift
Low-spin Fe(II), $S = 0$	Yes; 2 mm s^{-1}	None; spins paired.	None; spins paired.	~0.2 mm s^{-1}
High-spin Fe(II), $S = 2$	Yes; 2.5 mm s^{-1}	None; electron relaxation rapid.	Hyperfine interaction broadens lines.	~1 mm s^{-1}
Low-spin Fe(II), $S = 1/2$	Yes; 1.5 mm^{-1}		$S > 0$; electron relaxation slows at very low temperature, where hyperfine splitting occurs.	~0.2 mm s^{-1}
High-spin Fe(III), $S = 5/2$	Yes, but small and broadening obscures it at higher temperatures.	No, broadening interferes.	Marked, resolved splitting observed in a field; three doublets.	Broadening is a problem at higher temperature.

depends on the availability of pure protein preparations. Generally speaking, structure-oriented studies must follow the latter route, although the method has been applied to complex entities (see Section 4.5.3).

4.5.1 Iron Porphyrin Structures

Kent *et al.* (1979) conducted studies of Fe(II) forms of deoxymyoglobin, deoxyhemoglobin, and model compounds with a view toward interpreting magnetic hyperfine splitting effects. The substances in question contain high-spin Fe(II) ($S = 2$), which is esr silent. Magnetic susceptibility measurements are not impossible (Nakano *et al.*, 1971; Alpert and Banerjee, 1975) but are difficult and not highly informative. In contrast, Mössbauer spectra of these compounds present magnetic hyperfine interaction effects, which should yield a much more detailed picture of the electronic structure of the iron atom. In practice the interpretation of hyperfine splitting may not be straightforward in many cases. Mössbauer spectra of oxyhemoglobin are complex due to non-Lorentzian lineshapes, and various interpretations are possible, including the effect of a low-lying excited state (Cianchi *et al.*, 1984). Nevertheless, Mössbauer may

Figure 4-10. High-field ^{57}Fe-Mössbauer spectra of 1,2-dimethylimidazole-*meso*-tetraphenylporphyrin Fe(II). Measured spectra appear as dots (discrete counting channels) while Hamiltonian-derived (theoretical) spectra are shown as solid lines. Legend: *A*, 77 K with no field applied; *B*, 16 K and a 6-T field; *C*, 48 K and a 6-T field; *D*, 140 K and a 6-T field. The agreement is generally good over a broad range of conditions. Also, closely similar results were obtained for other porphyrins and heme proteins not shown here. Adapted from Kent *et al.* (1979).

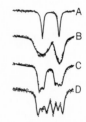

be the best approach, as in the cited example of high-spin Fe(II), where esr is useless and magnetic susceptibility measurements have limitations. Similarly, the iron nucleus is a poor nmr subject due to its long nuclear spin relaxation time (see Section 2.7.2.4).

In the studies of Kent *et al.* (1979) the iron was enriched to 80% ^{57}Fe. By using a constant acceleration velocity modulator and a 30-mCi ^{57}Co source, usable protein spectra could be built up in approximately 24 h. A 6-T superconducting magnet was employed to obtain high-field spectra.

In the zero-field condition the cited compounds produced Mössbauer spectra which were dominated by a quadrupole doublet splitting of ΔE_Q = 2.2 mm/s at a shift near 0.9 mm/s. The hemoglobin spectrum presented an additional quadrupole doublet at 0.46 mm/s with a splitting of ΔE_Q = 1.07 mm/s. The latter was attributed to a hemochrome impurity, i.e., hemoglobin in which the distal histidine is bound to iron. Linewidths did not vary significantly over the temperature range 4.2–195 K. The isomer shifts were weakly dependent upon temperature, thus indicative of 3*d*-electron delocalization. These shifts are probably due to a second-order Doppler effect.

In high applied fields the spectrum becomes more complex due to hyperfine splitting, shown over a temperature range for the 1,2-dimethylimidazole-*meso*-tetraphenylporphyrin Fe(II) complex in Figure 4-10. Field-dependent spectra may be calculated using the Hamiltonian* of Eq. (4-9) (Kent *et al.*, 1977).

$$\mathscr{H} = \left\{ \frac{eQV_{\hat{z}\hat{z}}}{12} \left[3I_{\hat{z}}^2 - \frac{15}{4} + \eta(I_{\hat{x}}^2 - I_{\hat{y}}^2) \right] \right\}$$
$$- [g_N^* a \beta_N \mathbf{H}^{\text{app}} \cdot (1 + \bar{\omega}) \cdot \mathbf{I}] \tag{4-9}$$

*In classical mechanics the Hamiltonian is the sum of kinetic and potential energies ($H = T + V$). Quantized systems restrict the Hamiltonian to a set of discrete solutions (eigenvalues). Given a set of calculated line positions and amplitudes, a reasonable simulation of a spectrum is obtained if an appropriate lineshape (e.g., Lorentzian) is imposed on each eigenvalue. In the Hamiltonian presented here, $V_{\hat{z}\hat{z}}$, the electric field gradient tensor, replaces q in $\Delta E = eqQ$.

The left-hand term accounts for the nuclear quadrupolar/electric field gradient interaction while the right-hand term describes the interaction of the nucleus with the applied field H^{app}, augmented by $\bar{\omega}\cdot H^{app}$, the effect due to an induced electronic moment. $\bar{\omega}$ is intimately related to the symmetry of the orbitals containing unpaired spins, and should vary as $1/T$ in the upper temperature (>20 K) range. Equation (4-9) contains several parameters, which should be self-consistent over wide temperature variations. Best fit was found when the Euler angles were set at zero and $\eta = 0.7 \pm 0.1$ (temperature independent) with the sign of the electric field gradient, $V_{\hat{z}\hat{z}}$, negative. The linewidth, Γ, was ~0.34 mm/s. Axial symmetry occurs with $\omega_{\hat{x}\hat{x}} = \omega_{\hat{y}\hat{y}} = \omega_{\perp}$ and $\omega_{\hat{z}\hat{z}} \neq \omega_{\perp}$, where $\bar{\omega}$ varies as $1/T$ above 20 K. The dependence of the tensor components of $\bar{\omega}$ upon temperature is shown in Figure 4-11. This choice of parameters leads to Hamiltonian-derived spectra (solid lines in Figure 4-10) which agree closely with the experimentally observed spectra.

Spectra below 20 K are much more difficult to interpret. Hemoglobin spectra resembled those of myoglobin but were complicated by the previously noted hemochrome impurity. Otherwise, the spectra, including those of model compounds, yielded similar parameters.

If it may be assumed that the selected parameters are valid, then the Mössbauer method is particularly informative. The negative sign of $V_{\hat{z}\hat{z}}$, the large values for η, and the axial symmetry of $\bar{\omega}$ are indicative of a prolate ground orbital for the sixth $3d$-electron, assuming that the valence electron dominates the electric field gradient. Also, the negative values of $\omega_{\perp}$ and $\omega_{\hat{z}\hat{z}}$ are consistent with hyperfine interaction dominated by a contact term. Computer simulation and parameterization, as applied here, are potentially powerful methods, particularly if the derived parameters may be compared with theory or measured by other methods.

Studies on Fe(III) forms of myoglobin have been conducted by Harami (1979). These also made use of Hamiltonian-derived spectra, and involved oriented single-crystal measurements of metmyoglobin and its fluoride and azide derivatives. The azide form of myoglobin is low spin ($S = 1/2$) while the other two are of the high spin ($S = 5/2$) type. The

Figure 4-11. Internal field coefficients (ω) of myoglobin as a function of temperature. The upper curve is ω_z while the lower curve is $\omega_{\perp}$. Quantitatively similar results were obtained for other heme-containing compounds. Ordinate and abscissa scales are logarithmic and equivalent. The plots tilt at a 45° angle (arrow) in the high-temperature limit, indicating a slope of -1, hence a $1/T$ dependence. Behavior becomes complex in the low-temperature range. Coefficients of this type were used to construct the hypothetical spectra of Figure 4-10. Adapted from Kent *et al.* (1979).

Hamiltonian describing high-spin forms is shown in Eq. (4-10), which presents more terms than Eq. (4-9).

$$\mathcal{H} = D[S_z^2 - S(S + 1)/3] - \frac{eQV_{\hat{z}\hat{z}}}{4I(2I - 1)} [3I_z^2 - I(I + 1)]$$
$$- g_N\beta_N \mathbf{H}^{\text{app}}\cdot I + 2\beta_e \mathbf{H}^{\text{app}}\cdot S - A I \cdot S \qquad (4\text{-}10)$$

The first term of Eq. (4-10) describes zero-field splitting. The second gives the nuclear quadrupole/electric field gradient interaction. Terms three and four are, respectively, the nuclear and electron Zeeman interactions, and the last term expresses the electron–nuclear hyperfine coupling interaction, i.e., an effect also measurable in esr spectra.

For low-spin Fe(III) the Hamiltonian for hyperfine interaction is:

$$\mathcal{H} = \frac{eQV_{\hat{z}\hat{z}}}{4I(2I - 1)} [3I_z^2 - I(I + 1) + \eta(I_x^2 - I_y^2)]$$
$$- g_N\beta_N I \cdot \mathbf{H}^{\text{eff}}(\theta_H, \phi_H) \qquad (4\text{-}11)$$

Single-crystal measurements produce more detailed geometric information, in this case for a compound with a known crystallographic structure. This compensates in part for the relatively large number of parameters in Eq. (4-10). Line positions are functions of the angle between the applied field and a symmetry axis of the molecule in question. Figure 4-12 shows the Hamiltonian-derived line positions of metmyoglobin as a function of the angle between the applied field and an axis perpendicular to the heme plane. The actual spectrum from a single crystal will consist of a superpositioning of line groups from the small number of crystallographically nonequivalent heme sites. In contrast, a polycrystalline powder is randomly oriented, causing the discrete lines to be smeared out into a continuum. The single-crystal method thus yields more information.

Comparing Hamiltonian-derived spectra with those measured for the cited compounds (Harami, 1979), the best-fit values for the electric field

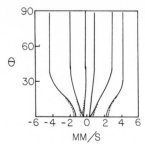

Figure 4-12. ^{57}Fe-Mössbauer line positions as a function of the angle (θ) between the normal to the heme plane and applied field vector. Experimental data shown as dots while calculated spectra are solid lines. The experimental data were obtained from single-crystal measurements. Adapted from Harami (1979).

gradient were found consistent with a symmetry less than orthorhombic about the iron atom, e.g., one nearer monoclinic.

Standard computer programs may be used to calculate Mössbauer spectra when the tensors for g, A, and electric field gradient are known (Münck *et al.*, 1973). The tensors may be systematically varied, using the computer method, until the calculated and measured spectra agree closely. This type of approach was used in the interpretation of the Mössbauer spectra of both oxidized and reduced forms of *Pseudomonas aeruginosa* cytochrome C-551 (Dwivedi *et al.*, 1979). Excess line broadening was encountered in the case of cytochrome C-551 and was attributed to an intrinsic protein heterogeneity, i.e., the binding site for iron may vary from molecule to molecule as a result of conformational irregularities, leading to a distribution of g tensors and hyperfine interactions. Good agreement between measured and simulated spectra are obtained if the theoretical values are based on crystal field calculations in which Δ and R conform to Gaussian distributions.* Better resolution was observed in protoporphyrin IX Fe(II) solutions containing sulfur ligands, and these model compounds could be compared with cytochromes P-450 (Silver and Lukas, 1984).

Kinetic studies with Mössbauer are also possible. Marcolin *et al.* (1979) used a constant velocity detection system to study the recombination of photodissociated myoglobin–CO. Time-dependent Mössbauer spectra were observed over a 4.2–46 K temperature range. Recombination was noted well below 46 K, possibly due to quantum mechanical tunneling of the relatively large CO molecule.

4.5.2 Nonheme Iron

Nonheme iron clusters present a more complex situation. This is particularly true of the metal-rich proteins involved in nitrogen fixation, e.g., the molybdenum/iron nitrogenases. Mössbauer and esr spectroscopic methods have proved valuable in the characterization of these enzymes, and the reader is referred to a series of elegant studies by Orme-Johnson and associates (Huynh *et al.*, 1979; Zimmerman *et al.*, 1978).

---

*The symbol Δ stands for the crystal field splitting energy. Distortions in the geometry and radius of the first coordination sphere surrounding the metal ion directly affect the value of Δ. Random distortions of the crystal field may thus produce a Gaussian distribution of Δ-values, which is the basis of the cited calculations. Changes in Δ may be accompanied by changes in d-electron paramagnetism and are thus observable in Mössbauer spectra. The relative values of Δ and the electron pairing energy determine whether the complex is of the high or low spin type, i.e., the spin multiplicity, S, depends on Δ.

The MoFe holoprotein of *Azotobacter vinelandii* appeared to consist of four protomeric subunits, containing two molybdenum atoms and 30 ± 2 of iron. This rather large assembly of iron appears to be further subdivided into two distinct spin-coupled iron centers, each containing about six iron atoms and (probably) one molybdenum. These are referred to as the cofactor or M-type clusters. The remaining iron appears to be organized in four distinct 4Fe–4S clusters similar to those of the ferredoxins. The latter comprise the P-type clusters. Both classes of substructure exist in a multiplicity of redox states, and some (e.g., $S = 0$) are esr silent. In contrast, Mössbauer should permit the observation and characterization of *all* redox states. The inherent difficulty of the Mössbauer method lies in analyzing the relative contributions of each substructure toward the overall spectrum. A summary of MoFe protein characteristics is given in Table 4-3.

The cited investigators used a computerized method for merging and comparing experimental and theoretically derived spectra. Simulated spectra for specific cluster types could be subtracted from the measured spectra, allowing an assessment of contributions from the different kinds of iron present. Also, redox states could be changed readily, e.g., thionine and molecular oxygen convert both cluster types to the most oxidized state (Table 4-3).

Figure 4-13 shows Mössbauer spectra of thionine-oxidized MoFe protein from *Azotobacter vinelandii* at different field strengths. The central features of the spectrum are assigned to the M^{ox} cofactor component, and the simulated spectrum for $S = 0$ agrees well (solid line). Outer signatures in this spectrum show a magnetic pattern even in the absence of an applied field. This component has been assigned as originating in iron clusters, particularly since it appears during a 4-electron oxidation in titration experiments. Its high-field signature is consistent with antiferromagnetic coupling and half-integral spin. An attempt at simulation is shown in the upper solid curve.

The Mössbauer spectrum of protocatechuate 4,5-dioxygenase is so uniquely different from the cluster and heme iron forms described in the preceding sections that it has been presumed to represent a hitherto unrecognized kind of metalloenzyme iron (Arciero *et al.*, 1983).

4.5.3 More Complex Entities

Mössbauer spectroscopy may be used with more complex preparations. For example, Figure 4-14 shows a marked similarity between a crude membrane preparation (A) isolated from *Chlorogloea fritschii* and

Table 4-3. Iron Components Observed in the MoFe Protein[a]

Cofactor cluster M

			Proposed character
	$M^{ox} \overset{1e^-}{\rightleftharpoons}$	$M^N \overset{1e^-}{\rightleftharpoons}$	M^R	
Oxidation state			
Electronic spin	$S = 0$	$S = 3/2$	$S = $ integer	Two spin-coupled iron clusters, each containing most likely six iron atoms (and possibly molybdenum). State M^N is observed in the native enzyme and yields the prominent epr signals around $g = 4.3$, 3.6, and 2. M^R is observed in the steady state under N_2-fixing conditions. M^{ox} is attained after oxidation with thionine or oxygen. Redox reactions are completely reversible at low thionine concentrations.
Mössbauer component	Doublet M^{ox}	Magnetic Mössbauer spectrum at 4.2 K	Doublet M^R	

P-Cluster

			Proposed character
	$P^{ox} \overset{1e^-}{\rightleftharpoons}$	P^N	
Oxidation state			
Electronic spin	$S \geq 3/2$, half-integer	$S = 0$	Most likely four 4Fe–4S clusters. The state P^N is observed in the native enzyme and in the steady state under N_2-fixing conditions. Components D and Fe^{2+} are diamagnetic and in the ratio of 3:1. State P^{ox} is attained by thionine and air oxidation. High-field Mössbauer studies prove spin coupling of iron atoms. Oxidation with low concentrations of thionine is completely reversible.
Mössbauer component	M1 at 4.2 K; doublets I–IV at 170 K	D and Fe^{2+}	

[a]Adapted from Zimmerman et al. (1978).

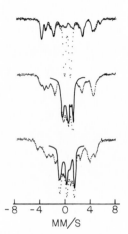

Figure 4-13. ^{57}Fe-Mössbauer spectra of thionine-oxidized MoFe protein from *Azotobacter vinelandii*. More than one type of iron is present; the spectral simulations (solid lines) account for these separately. The top curve was obtained at 0.9 kG, and shows features (solid line) not associated with the M^{ox} component. Assuming diamagnetism ($S = 0$), on the part of the M^{ox} component, the lower two simulations show how the central features of the spectrum are accounted for over a range of conditions (and are thus presumed due to M^{ox}). The middle and bottom spectra were obtained at 30 kG and 54 kG, respectively, and the solid lines mark the calculated absorption spectra of diamagnetic M^{ox}. All spectra were measured at 4 K. Adapted from Zimmerman *et al.* (1978).

its purified photosystem I fraction (B) (Evans *et al.*, 1979). This is indicative of the same form of iron predominating in each type of preparation, although one should not draw the conclusion that the spectrum represents only one iron environment.

Whole, frozen bacterial cells have been the objects of scrutiny (Dickson and Rottem, 1979). Large quantities of iron were observed to be incorporated and stored in cultures of *Proteus mirabilis*, particularly during the stationary growth phase. Also, the iron was distributed throughout both soluble and insoluble fractions. It might be argued reasonably that essentially the same information is available through atomic absorption measurements or simple chemical analyses. However, Mössbauer spectra offer a unique signature which may be compared with known compounds, leading to some insight if a storage form predominates (Danson *et al.*, 1977). More information is produced if spectra are recorded as functions of temperature and applied magnetic field.

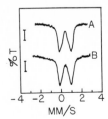

Figure 4-14. ^{57}Fe-Mössbauer spectra of fractions from *Chlorogloae fritschii*. A: Crude membrane fraction (vertical bar marks 1.5% absorption). B: Photosystem I fraction (vertical bar marks 0.25% absorption). From Evans *et al.* (1979).

4.6 Mössbauer Isotopes Other than ^{57}Fe

In analogy with multinuclear nmr, Mössbauer spectroscopy would have broader applicability to biological problems if active elements other than iron were available. Table 4-1 shows that a limited multinuclear approach is feasible. In practice, however, studies with ^{57}Fe have completely dominated biological applications of Mössbauer spectroscopy.

In part this is a problem of economics and logistics since isotopes which decay to appropriate excited isomers may be expensive to produce, and half-lives may be prohibitively short. In addition, the daughter nucleus may not be abundant in nature. Other factors which mitigate against practical Mössbauer spectroscopy include high photon energies, which decrease the probability of recoil-free transitions and low nuclear cross sections, which reduce the sensitivity of the method [Eq. (4-5)]. Isomer transition lifetimes may be so short that Heisenberg uncertainty broadening of the lines degrades spectral resolution. This may be seen by comparing transition lifetimes and linewidths of the two excited isomers of ^{57}Fe (Table 4-1).

The ideal location of a multinuclear Mössbauer spectrometer is near an isotope-producing accelerator or reactor. Studies with a variety of isotopes will have a difficult beginning, as in the case of nmr; however, the Mössbauer effect offers a method for probing specific atoms and often yields evidence unavailable by other spectroscopic methods.

About 60% of the total Mössbauer literature involves ^{57}Fe (Stevens and Bowen, 1978) and ^{119}Sn follows in second place at about 15%. The only other elements for which some experience exists and which are potentially applicable to biological problems are ^{129}I, ^{61}Ni, and ^{67}Zn.

Forster et al. (1977) have described a ^{67}Zn-Mössbauer spectrometer which uses a piezoelectric crystal to generate sample Doppler velocities. In using the piezoelectric effect, an accurately modulated electric field is applied to a quartz crystal, producing proportional modulations in the crystal's dimensions and, thus, fore and aft Doppler modulations in a sample attached to the crystal. Piezoelectric drivers are completely electronic, and it is relatively easy to maintain a high degree of coherence between the driving signal and the detector's counting channels.

The major advantage associated with ^{67}Zn is its long-lived isomer state, which leads to potentially narrow linewidths. As shown in Table 4-1, these can be as narrow as 3.2×10^{-4} mm/s. Disadvantages include a smaller cross section than ^{57}Fe and a much higher recoil energy, which decreases the rate of recoil-free events. The ^{67}Zn isomer may be produced using a ^{67}Ga source (Vetterling and Pound, 1977).

In view of the number of enzymes and proteins which are known to contain Zn, the isotope ^{67}Zn would appear to be particularly applicable and one which should attract attention.

The isotope ^{137}La may be of some interest to the biological sciences in view of the ability of ions of the type Ln^{3+} to replace Ca^{2+}. Preliminary nonbiological studies with ^{137}La have been reported by Gerdau et al., (1978). It is interesting to note that Mössbauer studies have been carried out on antileukemic diorganotin(IV) compounds interacting with 2-thiouracil (Stocco et al., 1984), in which the spectrum of ^{119}Sn was recorded. In this example, the Mössbauer-active compound was being used as an extrinsic probe substance.

Stevens and Bowen (1978, 1984) have reviewed the applications of several Mössbauer isotopes. The reader is referred to some important references of a general nature (Gonser, 1975; Bancroft, 1974; Stevens and Stevens, 1969 to present; Gruverman and Siedel, 1976).

4.7 Mössbauer Emission Spectra

Mössbauer isotopes may also be examined in the emission mode. To accomplish this, one uses a single line absorber, e.g., stainless steel in the case of ^{57}Fe, for which one would use a compound of ^{57}Co as the source. This method has been applied in studies of vitamin B$_{12}$ (Nath et al., 1970), although it must be realized that one is observing the K-capture-produced nuclear isomer of ^{57}Fe, not ^{57}Co, at the site of interest.

4.8 Bragg Scattering Effects

One potentially useful though undeveloped application of the Mössbauer phenomenon involves Bragg scattering from single crystals. The 14.4-keV photon from ^{57}Fe has a wavelength of 0.86 Å, which is comparable to the atomic spacings in crystals, and such radiation is indeed Bragg scattered on interacting with atomic electrons. The unique feature of the Mössbauer method lies in its ability to produce phase information using velocity modulation. The offsetting problem with this method is the extreme weakness of the scattered radiation, which necessitates inordinately long measurement intervals. Mössbauer's group (Parak et al., 1970, 1976) developed the diffraction method (a more general account of X-ray diffraction phase methods is presented in Section 6.4.1).

References

Alpert, A., and Banerjee, R., 1975. Magnetic susceptibility measurements of de-oxygenated hemoglobins and isolated chains, *Biochim. Biophys. Acta.* 405:144.

Arciero, D. M., Lipscomb, J. D., Huynh, B. H., Kent, T. A., and Munck, E., 1983. Epr and Mössbauer studies of protocatechuate 4,5-dioxygenase. Characterization of a new Fe(II) environment, *J. Biol. Chem.* 258:14981.

Bancroft, G. M., 1974. *Mössbauer Spectroscopy: An Introduction for Inorganic Chemists and Geochemists,* Halsted Press, New York.

Bonchev, Ts., Manuschev, B., and Vuchkov, M., 1977. Application of the Mössbauer effect to microgram samples with resonance sample scintillators, *Anal. Chim. Acta.* 92:23.

Cianchi, L., Pieralli, F., Del Giallo, F., Mancini, M., Spina, G., and Fiesolo, L., 1984. Interpretation of the Mössbauer spectra of oxyhemoglobin in terms of a low lying excited state, *Phys. Lett. A.* 100:57.

Danson, D. P., Williams, J. M., and Janot, C., 1977. Particle size distribution in ferritin, *Proc. Int. Conf. Mössbauer Spectrosc.* 1:307.

Dickson, D. P. E., and Rottem, S., 1979. Mössbauer spectroscopic studies of iron in *Proteus mirabilis, Eur. J. Biochem.* 101:291.

Dwivedi, A., Toscano, W. A., and Debrunner, P. G., 1979. Mössbauer studies of cytochrome c-551. Intrinsic heterogeneity related to g-strain, *Biochim. Biophys. Acta.* 576:502.

Evans, E. H., Rush, J. D., Johnson, C. E., and Evans, M. C. W., 1979. Mössbauer spectra of photosystem-I reaction centres from the blue–green alga *Chlorogloae fritschii, Biochim. J.* 182:861.

Forster, A. A., Potzel, W., and Kalvius, G. M., 1977. Piezoelectric Mössbauer spectrometer with fast channel advance rates, *AIP Conf. Proc.*, Vol. 38, *Workshop New Directions Mössbauer Spectroscopy,* pp. 29–34.

Gerdau, E., Winkler, H., and Sabathil, F., 1978. Mössbauer effect with lanthanum 137, *Hyperfine Interactions* 4:630.

Gonser, U. (ed.), 1975. *Mössbauer Spectroscopy. Topics in Applied Physics Series:* Vol. 5, Springer-Verlag, Berlin.

Gruverman, I. J., and Siedel (eds.), 1976. *Mössbauer Effect Methodology,* Vol. 10, Plenum Press, New York.

Harami, T., 1979. Mössbauer spectroscopic studies of ferric myoglobin single crystals in a small applied magnetic field, *J. Chem. Phys.* 71:1309.

Huynh, B. H., Munck, E., and Orme-Johnson, W. H., 1979. Nitrogenase XI: Mössbauer studies on the cofactor centers of the MoFe protein from *Azotobacter vinelandii* OP, *Biochim. Biophys. Acta.* 576:192.

Kent, T., Spartalian, K., Lang, G., and Yonetani, T., 1977. Mössbauer investigation of deoxymyoglobin in a high gradient and magnetic tensors, *Biochim. Biophys. Acta.* 490:331.

Kent, T. A., Spartalian, K., Lang, G., Yonetani, T., Reed, C. A., and Collman, J. P., 1979. High magnetic field Mössbauer studies of deoxymyoglobin, deoxyhemoglobin, and synthetic analogues, *Biochim. Biophys. Acta.* 580:245.

Linares, J., and Sundqvist, T., 1984. Mössbauer data collection using an Apple II micro-computer, *J. Phys. E* 17:350.

Maeda, Y., 1979. Mössbauer studies on the iron–ligand binding in hemoproteins and their related compounds, *J. Phys. (Orsay), Colloq.* 1979:514.

Manuschev, B., Bonchev, Ts., Ivanov, C., and Condeva, N., 1976. Methods of determining the recoilless absorption probability of Mössbauer radiation, *Nucl. Instrum. Methods* 136:267.

Marcolin, H., Reschke, R., and Trautwein, A., 1979. Mössbauer spectroscopic investigations of photodissociated myoglobin–CO at low temperatures, *Eur. J. Biochem.* 96:119.

Mössbauer, R. L., 1958. Nuclear resonance absorption of gamma rays in [191]Ir, *Naturwissenschaften* 45:538.

Münck, E., Groves, J. L., Tumolillo, T. A., and Debrunner, P. G., 1973. Computer simulations of Mössbauer spectra for an effective spin $S = 1/2$ Hamiltonian, *Comput. Phys. Commun.* 5:225.

Nakano, N., Otsuka, J., and Tasaki, A., 1971. Fine structure of iron ion in deoxymyoglobin and deoxyhemoglobin, *Biochim. Biophys. Acta.* 236:22.

Nath, A., Klein, M. P., Kuendig, W., and Lichtenstein, D., 1970. Mössbauer studies of after-effects of Auger ionization following electron capture in cobalt complexes, *Radiat. Eff.* 2:211.

Parak, F., Mössbauer, R. L., and Hoppe, W., 1970. Experimental solution of the phase problem for protein structure determination by interference between electron and nuclear resonance scattering, *Ber. Bunsenges. Phys. Chem.* 74:1207.

Parak, F., Mössbauer, R. L., Hoppe, W., Thomanek, U. F., and Bade, D., 1976. Phase-determination of a diffraction peak of a myoglobin single crystal by nuclear γ-resonance scattering, *J. Phys. (Paris), Colloq.* 1976:703.

Silver, J., and Lukas, B., 1984. Mössbauer studies on protoporphyrin IX iron(II) solutions containing sulphur ligands and their carbonyl adducts. Models for the active site of cytochromes P-450, *Inorg. Chim. Acta* 91:279.

Stevens, J. G., and Stevens, V. E., 1969 to present (a continuing series). *Mössbauer Effect Data Index*, Plenum Press, New York.

Stevens, J. G., and Bowen, L. H., 1978. Mössbauer spectroscopy, *Anal. Chem.* 50:176R.

Stevens, J. G., and Bowen, L. H., 1984. Mössbauer spectroscopy, *Anal. Chem.* 56:199R.

Stocco, G. C., Pellerito, L., Girasolo, M. A., and Osborne, A. G., 1984. Organotin(IV) derivatives of the ambidentate ligand 2-thiouracil. Infrared, Mössbauer, ^{1}H and ^{13}C nmr studies, *Inorg. Chim. Acta* 83:79.

Vetterling, W. T., and Pound, R. V., 1977. The quadrupole interaction in zinc metal, *AIP Conf. Proc.*, Vol. 38, *Workshop New Directions Mössbauer Spectroscopy*, pp. 27–28.

Zimmerman, R., Münck, E., Brill, W. J., Shah, V. K., Heuzl, M. T., Rawlings, J., and Orme-Johnson, W. H., 1978. Nitrogenase X: Mössbauer and epr studies of reversibility oxidized MoFe protein from *Azotobacter vinelandii* OP, *Biochim. Biophys. Acta.* 537:185.

Electron Spin Resonance Spectroscopy (Esr)

5

5.1 Introduction

Electron spin resonance (esr) (also known as electron paramagnetic resonance, epr) has been successfully applied to biological problems since the late 1950s (Commoner *et al.*, 1957). The method compares closely with nmr in being essentially nondestructive since the measuring wavelengths fall within the microwave portion of the spectrum, corresponding to photon energies, $h\nu$, much too weak to break chemical bonds. A magnetic field must be applied in order to detect esr phenomena, but in general, the field is less intense than for nmr (see Chapter 2). The distinctive advantage of esr (compared with nmr) is its sensitivity, permitting measurements at very low concentrations. In some cases this can be a disadvantage, i.e., interfering impurities may be confused with the entity one is attempting to observe (see an example in the last paragraph of Section 5.2.5.4).

Esr spectra are measurable only when molecules or ions in the specimen contain unpaired electrons and are thus paramagnetic. In most stable substances the electrons occur pairwise in bonding (sometimes also antibonding) molecular orbitals or paired in nonbonding orbitals localized on individual atoms. Substances of this type are diamagnetic and cannot exhibit esr spectra. In biological systems there are two major categories of molecular (or atomic) entities which have unpaired electrons. The first category includes all organic free radicals produced by homolytic bond cleavage or other processes, e.g., one-electron reductions or oxidations. These are usually highly reactive intermediates intrinsic to normal metabolic processes but may also originate in deleterious reactions such as those produced by ultraviolet light or ionizing radiation. In some cases an organic free radical may be long-lived, as in the example of the oc, oc-diphenyl-β-picrylhydrazyl free radical, which is stable over a period of years. The second category is made up of atoms or ions which have partially filled inner shells, notably those of the transition elements, which contain unpaired d-orbital electrons. Under conditions which preclude redox changes, the paramagnetism associated with unpaired inner shell electrons is a persistent feature of the substance in which it occurs.

Esr is a method which generates moderate amounts of information. The detection of an esr spectrum indicates the presence of unpaired electrons, while complexities within the spectrum (e.g., hyperfine splitting, resonance position, linewidth) reflect the environment of the orbital, be it localized or delocalized, which contains the unpaired electron. For example, the esr spectrum of the free radical species shown in Figure 1-2 will be strongly dependent on the aromatic ring and its immediate environment (e.g., attached methyl and methylene groups) since the unpaired spin is delocalized over that region, but it will be fairly indifferent to the more remote parts of the molecule (e.g., the isoprenoid side chain). With this limitation in mind, the esr spectrum of a transition metal ion located within a large protein molecule would seem to contain a rather limited amount of information about the protein, i.e., it would be influenced mostly by the metal ion's immediate coordinating ligands, a region measuring no more than a few angstroms across. However, the importance of the esr spectrum is better appreciated when one considers that the metal ion is likely to be intimately involved in the protein's function, e.g., at the catalytic (active) site of an enzyme or as the entity mediating in a specific binding function, such as hemoglobin's Fe(II) in oxygen transport. The esr spectrum is then found to be sensitive to redox events, substrate binding, and other features associated with function.

Readers seeking more detailed information on the chemical and biological applications of esr spectroscopy are referred to several excellent treatises on the subject (Pake, 1962; Ingram, 1969; Poole, 1982). The reader should also be aware of computer programs which simulate esr spectra (Bartl and Sommer, 1984). Methods for measuring static paramagnetism have also been reviewed (Haberditzl, 1973). The latter falls into the low-information category of techniques and will not be considered in this volume.

5.2 Theory

According to electromagnetic theory, a spinning, charged particle is analogous to a current flowing in a loop of wire, as shown in Figure 5-1. In both cases a magnetic dipole will result, oriented along the spin axis of the charged particle [Figure 5-1(B)], or in the example of an electric

Figure 5-1. Spinning charges and magnetism.
(A) Loop of wire viewed from above. A clockwise flow of negatively charged electrons
produces a magnetic field with the north pole
directed upwards. (B) In a similar manner, a
negatively charged sphere rotating clockwise
generates a magnetic field with north at the top.
(C) Shown here is a more massive, positively
charged sphere rotating counterclockwise with
the same angular momentum as the smaller
sphere. In this case the magnetic field is *weaker*

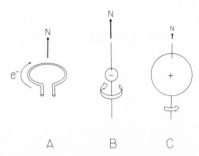

due to a slower rotation rate. The middle and right-hand examples are analogous to electrons
and protons, respectively, but these examples ignore the wave-mechanical properties of
fundamental particles.

current, perpendicular to the loop of wire [Figure 5-1(A)]. This rule may
be applied to charged fundamental particles such as electrons and protons.
The latter two entities appear to "spin" with angular momenta constrained
by quantum mechanical rules to a value of $h/2\pi$ (h = the Planck constant,
6.625×10^{-27} erg s); thus both must possess magnetic dipoles. Because
these particles have the same angular momentum but differing masses
they will have unequal magnetic moments, i.e., the proton is more massive
than the electron by a factor of 1,836 and thus possesses a correspondingly
weaker magnetic moment [Figure 5-1(B) and (C)].

The behavior of a spinning, charged particle in an external field is
illustrated in Figure 5-2, which applies equally to electrons or protons.
Quantum rules require that the particle's spin be either parallel or antiparallel to the applied field vector, representing, respectively, the lower
and upper energy levels in Figure 5-2. The energy gap between these
quantized states is field dependent, and for the electron it is determined
by $g\beta H$, where H is the field strength, β the Bohr magneton for electrons,
and g the Landé splitting factor, which has a value of 2.0023 for an electron
unassociated with an atom. As a result of the electron's larger magnetic
moment, the energy gap between parallel and antiparallel states increases

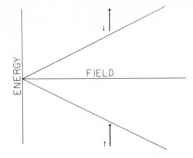

Figure 5-2. Magnetic field-induced energy levels
(Zeeman effect). The large arrows represent the
applied or instrument field, and the small arrows
denote the electron magnetic dipole. In the high-
energy state the electron dipole opposes the field
(top) while it augments the field in the low-energy
state (bottom).

much more markedly with applied field than in the case of the proton. Consequently, the resonance condition for electrons, $h\nu = g\beta H$, occurs in the gigahertz (microwave) frequency range when the applied field is on the order of 1–10 kilogauss, while for protons the corresponding resonance falls in the megahertz (radio) part of the spectrum. The proportionality between field and resonance frequency is such that at any applied field H_0 a unique resonance frequency ν_0 must exist. Under these conditions the sample will *absorb* microwave energy.

5.2.1 Esr and Molecular Structure

A featureless esr resonance of the type described above would be completely uninteresting to chemists, but in most cases the esr spectra of paramagnetic molecules include one or more of several modulating effects due to the structure of the molecule. These include (1) hyperfine splitting of lines caused by interactions between the electron and nuclear dipoles; (2) linewidth effects; and (3) modulations of the g-factor based on the nature of the atom or atoms carrying the unpaired spin. These features should now be treated in some detail.

5.2.2 Hyperfine Splitting Effects

5.2.2.1 Interaction of the Unpaired Electron with Magnetic Nuclei

The Hamiltonian expressing quantized energy states in a system containing an unpaired electron interacting with a nuclear spin and an applied external field may be written as follows:

$$\mathcal{H} = g\beta \mathbf{S} \cdot \mathbf{H}_0 + \hbar A \mathbf{S} \cdot \mathbf{I} - g_I \beta_I \mathbf{I} \cdot \mathbf{H}_0 \tag{5-1}$$

This equation could describe a neutral hydrogen atom with its lone electron in the $1s$ atomic orbital. It should be noted that Eq. (5-1) is a simplified expression based on the assumption that the factors g and A are scalar quantities (in practice these are usually tensors and the Hamiltonian is more complex). The reader will recognize the first term, which defines the two energy levels resulting from the interaction of electron magnetic dipoles with the external field. Here $g\beta \mathbf{S}$ is the electron magnetic moment, and $\mathbf{S}$ is the electron spin ($+\frac{1}{2}$ and $-\frac{1}{2}$ for a single, unpaired electron). The first term describes the largest energy difference. The third term is the analogous interaction between the nuclear dipole and the applied field, where I is the nuclear spin ($\pm 1/2$ for a proton) and $g_I \beta_I \mathbf{I}$ the nuclear magnetic moment. As noted earlier, the latter quantity is small; thus the third term defines weak interactions. The second term is also weaker than the first, and it describes the interaction between electron and nuclear spins, where A is the hyperfine coupling constant.

In the example given, A is assumed to be scalar. Scalar coupling occurs when the orbital (atomic or molecular) has a finite amplitude, ψ, at the nucleus in question. The relationship between A and ψ is as follows:

$$A = -8\pi g g_I \beta \beta_I / 3 \; |\psi^2(0)| \tag{5-2}$$

in analogy to charge density, $\psi^2(0)$ is the "spin density" at the nucleus. This type of coupling is also referred to as contact interaction (see Section 2.7.2.4.3). It follows from the relationship between A and ψ that a molecule containing a multiplicity of nuclei may, depending on symmetry, have more than one A-value, i.e., different nuclei may experience different spin densities. Clearly, the observed hyperfine coupling is structure dependent, and the esr spectrum can be used to resolve questions of free radical or metal complex structures.

Figure 5-3 illustrates the solutions to the Hamiltonian given in Eq. (5-1). It is apparent that a unique resonance condition does not exist. Rather, if the microwave frequency, ν_0, is kept constant and the magnetic field H is swept through the region where H_0 would normally occur (for the electron unperturbed by nuclei), two resonances will now be observed.

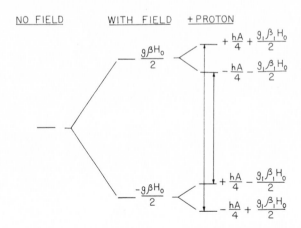

Figure 5-3. Hyperfine splitting effects. In the no-field condition (left) isolated, unpaired electrons are at the same energy level. Applications of a magnetic field (middle) creates two levels (see Figure 5-2). Finally, the introduction of a nucleus of spin 1/2 results in four different energy levels (right). Quantum selection rules permit only two transitions between these.

These correspond to electron spin-state transitions which involve no change in the orientation of the nuclear dipole, i.e., $\Delta m_I = 0$; $\Delta m_s = \pm 1$ (the two vertical arrows in Figure 5-3). These absorptions will appear to be equal in intensity since the field-parallel and -antiparallel nuclear spin states are almost identically populated.

The hyperfine splitting pattern given in this simple example is a doublet, and the general expression for hyperfine multiplicity is $2I + 1$. Spectra are more complex when the nuclear spin is not $1/2$. For example, $I = 1$ for the ^{14}N nucleus, and an unpaired electron in contact with a single ^{14}N nucleus produces an esr spectrum consisting of three absorption lines of equal intensity. Similarly, the multiplicity of the Mn(II) ion is six ($I = 5/2$), and aqueous solutions of the Mn(II) ion show a characteristic six-line esr spectrum.

The hyperfine splitting pattern becomes still more complex when the unpaired electron interacts with two or more nuclei. Figure 5-4(A) depicts the spectrum resulting from contact interaction involving one proton and a ^{14}N nucleus, where $A_H > A_N$. Contact interaction with the proton produces a doublet and each of these lines is further split into triplets by the ^{14}N nucleus. Another hypothetical case is shown in Figure 5-4(B), where three protons (e.g., a methyl group) interact equally with the unpaired electron. The successive addition of each proton to the structure results in a splitting of all preexisting lines into doublets, and the net effect is a $1:3:3:1$ quartet.

These structural concepts may be applied to the simple free radical shown in Figure 1-2. Homolytic cleavage of the phenolic O—H results in a phenoxy free radical carrying a delocalized unpaired electron. Molecular orbital calculations predict that the spin density is greater at the

Figure 5-4. Hyperfine splitting effects with two or more nuclei. (A) Hyperfine splitting due to the electron interacting with a single proton, and to a lesser extent with a single ^{14}N nucleus. The resulting hyperfine pattern has six lines of equal intensity. (B) Successive splitting with one, two, and three protons, all equally coupled to the unpaired electron. One proton produces a doublet with each component half the intensity of the original line. Coincident lines occur with two or more protons, and the result with all three protons is a $1:3:3:1$ quartet.

oxygen atom and also (to a lesser extent) at the *ortho* and *para* carbons, while minimum densities occur at the *meta* carbons. The esr spectrum (7 major lines, each split 6-fold) is consistent with these expectations (Kohl *et al.*, 1969).

It should be noted that the delocalized spin of the phenoxy free radical occupies a π molecular orbital. The symmetry of π orbitals requires that spin amplitude be zero in the ring plane, i.e., at the node surface. Non-zero contact interaction with *ortho* and *para* protons (e.g., a hydrogen atom substituting for one of the methyl groups) thus cannot be explained by direct interaction with the electron spin; rather it is envisioned that spin polarization exists between the π-system and the σ-bonding frame-work. In practice it is found that unpaired electrons in *p*-, *d*-, and *f*-type orbitals (and their molecular orbital equivalents) experience non-zero *A*-values even though the nucleus in question is in a node plane, as shown in Figure 5-5.

Nuclear hyperfine interactions may be observed in covalent structures when linewidths (usually expressed in gauss or milligauss units) are narrower than the isotropic *A*-values. Hyperfine splitting may be obscured if line broadening is excessive, or if significant anisotropies of *g* and *A* exist. Anistropy of the latter two variables can produce marked broadening if the paramagnetic entity is frozen in a random solid. However, these may average to a constant value if the paramagnetic molecule tumbles rapidly, as in a liquid solution. Thus, a complete resolution of hyperfine lines present is more likely in liquid specimens.

Any structure written for the paramagnetic entity must be consistent with the observed hyperfine splitting pattern. Molecular orbital calculations can be used to estimate *A*-factors at the various nuclei, from which a theoretical spectrum can be constructed. Computer simulations may also be based on empirically derived coupling constants, e.g., from collections of reference spectra (Poole, 1982). These methods resemble those used in nmr.

If the precursor of the free radical species is in question, then isotopic

Figure 5-5. Spin density and orbital symmetry. A nucleus at position 1 is in the node plane of the *p*-atomic orbital, and with no other effects taken into consideration it should experience zero spin density from the unpaired electron. Thus it should not cause hyperfine splitting. A nucleus at position 2 is in a region of finite spin density.

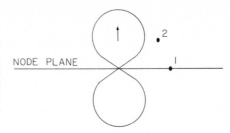

labeling of a proposed precursor can confirm or disprove its identity. For example, the phenoxy radical of Figure 1-2 prepared from ordinary carbon can have no ^{12}C hyperfine splitting since this isotope has zero nuclear spin. However, if ^{13}C is introduced specifically at a region of high spin density, e.g., an *ortho* substituent, then the spectrum should reveal twice as many lines (presuming they are resolved) due to doublet splitting with the ^{13}C nucleus ($I = 1/2$). As an example of a biological application of this principle, the hyperfine splitting of a chloroplast esr signal described as "Signal II" was assigned to protons since the spectrum of organisms grown in heavy water lacked the structure, i.e., the spectrum was narrower (Commoner, 1961). This was the expected result since A-values for deuterons are less than those for protons at the same spin density. Similar manipulations of the coordinating ligands of paramagnetic complexes, including those of transition metal ions in proteins, are possible. The presence of hyperfine splitting due to ligand nuclei is direct evidence of covalent bonding between the paramagnetic metal ion and its ligands. Isotope substitutions are often difficult, requiring one or more steps of synthesis and purification.

5.2.2.2 Anisotropic Hyperfine Splitting Effects

The type of hyperfine splitting described above was strictly isotropic, i.e., it does not depend on the orientation of the molecule in the applied field. There exists another type of hyperfine coupling which originates in the through-space interaction of electron and nuclear magnetic dipoles. This coupling depends on the angle, ϕ, between the line joining the dipoles and the dipolar axis, and on the distance between dipoles, r:

$$A' = gg_I\beta\beta_I\left\langle\frac{1 - 3\cos^2\phi}{r^3}\right\rangle_{average} \qquad (5-3)$$

The expression containing r and ϕ is an averaged value since the electron is not in a fixed position but delocalized over the orbital that it occupies. For systems with spherical symmetry, e.g., s-orbitals, the averaged value is zero, but in p- and d-orbitals, this cannot be so, as illustrated in Figure 5-6.

The anisotropic contribution, A, can be expressed as a tensor and the net hyperfine coupling, A'', is a summation of both isotropic and anisotropic terms.

$$\begin{vmatrix} A_x'' & & \\ & A_y'' & \\ & & A_z'' \end{vmatrix} = \begin{vmatrix} A & & \\ & A & \\ & & A \end{vmatrix} + \begin{vmatrix} A_x' & & \\ & A_y' & \\ & & A_z' \end{vmatrix} \qquad (5-4)$$

Figure 5-6. Coupling constants and
molecular orientation. The largest
arrow is the instrument field, and
the smaller two are the electron
and nuclear dipoles. Owing to the
symmetry of the p-orbital and Eq.
(5-3), the two orientations shown

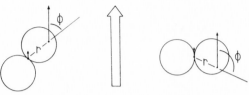

here will not have identical values of A', e.g., the averaged value of ϕ will be larger in the
right-hand configuration (the value decreases as the p-orbital axis approaches a vertical
configuration).

The ability to detect anisotropic hyperfine effects of the type described
here depends on the relative absence of other complicating effects
(linewidth, anisotropic g). This is not always the case, but when the
situation is favorable A' gives an indication of orbital symmetry.

Consistent with the emphasis of this volume, hyperfine coupling constants can provide a picture of atom connectivities around the paramagnetic center, particularly when isotope substitution experiments are carried out. The other features of the esr spectrum (i.e., g-value and linewidth
effects) relate more to the oxidation state of the paramagnetic entity and
the symmetry of its surrounding ligands.

5.2.3 The g-Tensor

It was noted earlier that the value of g (in $h\nu_0 = g\beta H_0$) is 2.00232
for the free, relativistic electron spin. Departures from this value occur
when the electron occupies an atomic orbital, where spin and orbital
angular momenta may couple and zero-field splitting exists. The latter
effects are minimal for organic free radicals, for which g-values close to
2 are generally observed. It should be noted that divergent g-values *can*
be observed in organic species, notably in sulfur-containing radicals, where
unoccupied sulfur $3d$-orbitals may be involved in the bonding. Coupling
is especially important for the unpaired d-electrons of transition metal
ions. The g-value is sensitive to the spin state and the surrounding coordination field, and the observed g (i.e., resonance position) is often
dependent on the angle between the applied field and the molecular coordinate system. Thus, g can be (and often is) anisotropic and is treated
as a tensor.

Complexities observed in the esr spectra of transition metal ions often
result from the ion having more than one unpaired electron. To gain insight
into this problem it is helpful to write the spectroscopic term for the ion
in question using the vector model of the atom. First, the resultant spin,

S, of the entity is the sum of the individual electron spin moments, S_i, where each electron has a value of 1/2. Four electrons could combine to produce $S = 2$, 1, or 0. Three spins would yield $S = 3/2$ or 1/2. It is seen that S must have integral or half-integral values. The spin multiplicity is $2S + 1$. Similarly, L is the vector sum of the orbital angular momenta, and since individual moments have only integral values, L must also be a whole number. The term symbol is of the form:

$$^{2S+1}L_J$$

where J (referred to as the inner quantum number) is the vector sum of L and S, and L-values of 0, 1, 2, 3, etc. are represented, respectively, by symbols S, P, D, F, G, etc.

To illustrate a simple case, one can apply these rules to V(IV) which has just one $3d$-electron (i.e., $3d^1$). Thus S must be 1/2 and $2S + 1$ is 2. This electron is in a d-orbital ($l = 2$) and L must also be 2, represented by symbol D. For the ground state, $J = L - S = 3/2$; thus the ground state term symbol is $^2D_{3/2}$.

A closely related situation is found in the $3d^9$ configuration [e.g., Cu(II)], for which the term symbol is $^2D_{5/2}$. Since this ion requires one electron to attain a filled d-subshell, it can be treated as having an electron vacancy; thus the net angular momentum is that of a half-filled d-orbital.

The remaining special case is found in ions which have five d-electrons. In the free ion these will spread out so that each electron occupies a different orbital (Hund's rule). The result is spherical symmetry, i.e., the L_i-vectors cancel so that $L = 0$. The term for this special case is thus $^6S_{5/2}$, and the ions Mn(II) and Fe(III) are biologically important examples of S-states. Ions of this type present a special problem in that further splitting due to the presence of surrounding ligand groups cannot be accounted for on the basis of crystal field theory (Pake, 1962), i.e., there can be no first-order interaction between ligands and the spherically symmetrical $3d$-subshell. Mathematical treatment of the *observed* spectral effects is complex. The cited $3d^1$ and $3d^9$ configurations are the relatively simple cases among transition metal ions.

Table 5-1 lists the configurations and ground state terms of transition metal ions which either are or could be encountered in biological entities. It should be noted that, neglecting J, the terms have a symmetry about the half-filled condition (d^5).

On introducing a metal ion into a field of coordinating ligands it is found that the d-orbitals behave differently depending on the symmetry of the ligands around the ion, i.e., a splitting of d-orbital energies occurs. This is illustrated in Figure 5-7. Electrostatic repulsion between the incoming ligand electron pairs and the d-electrons is greatest (for octahedral

Table 5-1. Ions and Term Symbols

Ions	d-Configuration	Term (ground state)
V^{4+}, Mo^{5+}	$3d^1$, $4d^1$	$^2D_{3/2}$
V^{3+}	$3d^2$	$^3F_2$
Cr^{3+}, V^{2+}, Mo^{3+}	$3d^3$, $4d^3$	$^4F_{3/2}$
Cr^{2+}, Mn^{3+}	$3d^4$	$^5D_0$
Mn^{2+}, Fe^{3+}	$3d^5$	$^6S_{5/2}$
Fe^{2+}	$3d^6$	$^5D_4$
Co^{2+}	$3d^7$	$^4F_{9/2}$
Ni^{2+}	$3d^8$	$^3F_4$
Cu^{2+}	$3d^9$	$^2D_{5/2}$

fields) along the x, y, and z coordinate axes. As a result, the $d_{x^2-y^2}$ and d_{z^2} orbitals, which are directed along these approaches, become the higher energy levels. In the octahedral environment the $d_{x^2-y^2}$ and d_{z^2} levels are equal in value, i.e., degenerate, while the lower energy level is the three-fold degenerate set of d_{xy}, d_{xz}, and d_{yz} orbitals. The D-state ions are split by the field into a doublet and a triplet, as described above. Configurations d^1 and d^6 are similar in having one d-electron beyond spherically sym-metric configurations (d^0 and d^5), and for these the higher energy level is the doubly degenerate one. Configurations d^4 and d^9 are also comparable since they are one electron short of spherically symmetric cases (d^5 and d^{10}). Note that the energy levels are inverted, with the doubly degenerate set ($d_{x^2-y^2}$ and d_{z^2}) now the lower level. These four cases differ in spin multiplicity and electron populations.

F-state ions have an L-multiplicity of seven. Thus, the incoming octahedral ligands act upon sevenfold degeneracy, producing the triplet, triplet, singlet of Figure 5-8. The configurations d^2 and d^7 have two elec-trons beyond a spherically symmetric state while d^3 and d^8 lack two.

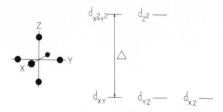

Figure 5-7. Octahedral symmetry and d-orbital energy levels. Six ligand groups are arranged octahedrally on the x, y and z coordinate axes. Electrostatic repulsion between the ligands and two d-orbitals ori-ented along these axes is greatest, resulting in those two orbitals being degenerate high-energy states. The other three orbitals in-terleaf with the ligands and are low-energy states. Δ is the octahedral splitting energy. It is assumed here that the configuration is $3d^1$ or $3d^6$. Ions which are nearly half-filled ($3d^4$) or nearly filled ($3d^9$) may be treated as containing a ''hole'' (i.e., a positive charge which attracts the ligands) and the order of the energy levels is inverted. See Figure 5-9.

F ——

Figure 5-8. Octahedral symmetry and *F*-state energy levels. The effect of octahedral symmetry on the energy levels (orbital and spin) of an *F*-state. The order shown here is for $3d^3$ and $3d^8$ ions. For $3d^2$ and $3d^7$ the order is inverted.

Again the order of orbitals is inverted between d^2, d^7 and d^3, d^8. Table 5-1 shows the symmetry about the *S*-state. Further effects upon splitting and the positions of energy levels arise from reduced symmetry, spin–orbit coupling, and an applied magnetic field. These will now be described.

Environments of less symmetry remove degeneracy, as can be seen in the tetragonal example for Cu(II) in Figure 5-9 (compare with Figure 5-7). In this case the $d_{x^2-y^2}$ orbital experiences less repulsion and has the lower energy, while d_{z^2} is higher. Further removal of degeneracy is possible as a result of the Jahn–Teller effect. The latter allows ligands to shift from positions of perfect symmetry if the system's energy is thereby minimized. Kramer's rule should be considered in conjunction with this effect. The rule distinguishes between ions with odd and even numbers of unpaired electrons, requiring that degeneracy cannot be reduced below twofold by the crystal field if the ion has an odd number of unpaired electrons, while those with even numbers can become fully nondegenerate.

Asymmetry is the general rule in biomolecules. The free ion can be treated theoretically, but the introduction of a symmetric coordinating field complicates the problem. Further difficulties occur when all symmetry is removed.

5.2.3.1 Zero-Field Splitting and Fine Structure

Returning to the configurations for ground state ions (Table 5-1), the simple cases ($3d^1$, $3d^9$) have only one unpaired spin, and while the spin-

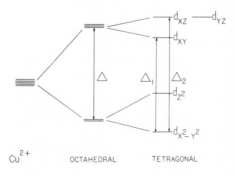

Figure 5-9. Energy levels of copper(II). The fivefold degenerate *d*-orbitals of a free copper(II) ion are subjected first to an octahedral environment, then to tetragonal distortion along the *z*-axis, as in a square planar complex. The lowest orbital contains the electron "hole," but some quantum mechanical mixing with the degenerate upper level (d_{xz}, d_{yz}) and the next highest (d_{xy}) occurs.

containing orbital is perturbed by the coordinating field, the spin degeneracy is 2 (with $m_s = \pm 1/2$). On applying a magnetic field the parallel $(-1/2)$ and antiparallel $(+1/2)$ spin states are resolved and a resonance condition exists (as in the case of an organic free radical). In contrast, the ions with $S > 1/2$ (^3F, ^4F, ^5D) generally experience splitting by the ligands (the crystal field) in the absence of a magnetic field.

The progression from high to low symmetry is: spherical (free ion) > octahedral or tetrahedral > tetragonal > rhombic. Further distortions occur in protein-bound metal ions, e.g., the ligands can, in some cases, be distorted away from the x, y, and z coordinate axes. Orbital and spin degeneracy may be completely lifted by spin–orbit coupling and a crystal field of low symmetry (the ions which have half-integral values of S are exceptions due to Kramer's Theorem, to be discussed shortly). Since crystal field energy differences often correspond to transitions in the infrared and visible portions of the spectrum, while conventional esr instruments operate in the less energetic microwave region, it is often impossible to observe the esr spectrum. This effect is illustrated in Figure 5-10.

To interpret these examples bear in mind the rule for esr transitions, i.e., $\Delta m_s = \pm 1$. In Figure 5-10(A), which could be a V(III) ion, the effects of cubic symmetry, lower symmetry, and finally spin–orbit coupling are shown as successive splittings of the energy levels. Numbers written above the levels denote spin degeneracy. In this example it is seen that the second lowest level is doubly degenerate in spin. It is referred to as a Kramer's doublet and it consists of the spin states $m_s = +1$ and -1. The lowest level is a spin singlet, $m_s = 0$, and the energy difference between this singlet and the Kramer's doublet is the zero-field splitting, D. Zero-field splitting has completely lifted spin degeneracy in the three levels immediately above the doublet.

The application of a magnetic field removes all remaining spin degeneracy. A vertical dashed line on the left marks the zero-field condition while the dashed line on the right is the highest field that the spectrometer can produce. Being more specific, in a 9000-MHz spectrometer the "reach" between levels (the spectrometer's photon energy) is approximately 0.8 cm^{-1}, but in the case of a V(III) ion the zero-field splitting is on the order of 5 cm^{-1}. Applying the selection rule, transitions between the $+1$ and -1 levels are forbidden. Also, the allowed transitions between the 0 and $+1$ or -1 and 0 levels cannot occur because the zero-field splitting exceeds the reach of this particular spectrometer. The reader will note that a transition between the -1 and 0 levels would occur if the magnetic field could be further increased. Alternatively, a higher-frequency instrument would permit measurements on this system.

Although ions with integral S-values are difficult, there are cases for

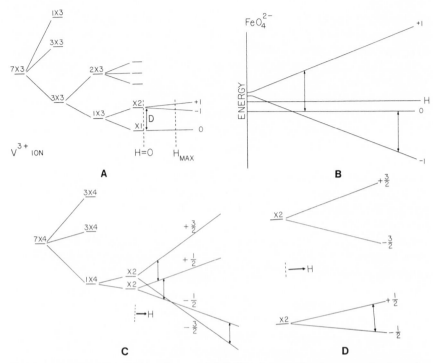

Figure 5-10. Zero-field splitting effects. (*A*) Splitting of energy levels progressing from the free ion through octahedral symmetry, lower symmetry, zero-field splitting (D), and finally the effect of an applied magnetic field. The example given here is V(III), an *F*-state ion, for which $L = 3$. Thus the *L*-multiplicity is $2L + 1$ or 7. For this ion, $S = 1$, and the number of spin levels, $2S + 1$, is 3. The combined *L* and *S* multiplicity is indicated above the free ion as 7×3. (The diagram shows how these are affected as degeneracy is removed.) (*B*) In FeO_4^{2-}, *D* is small relative to magnetic splitting. The two vertical arrows mark values of field strength (*H*) which satisfy the resonance condition. Thus, the spectrometer will record two absorptions. (*C*) Splitting of energy levels beginning with Cr(III) ($S = 3/2$) free ion and progressing through octahedral symmetry, lower symmetry, and finally the effect of an applied field. In this case the magnetic field effect is larger than the zero-field splitting and selection rules allow resonances to occur at three different field strengths (vertical arrows). (*D*) Same as (*C*), except the zero-field splitting is larger than the field effect. Only the lowest Kramer's doublet has a permissible esr transition.

which the zero-field splitting is small enough to permit measurements. Figure 5-10(B) shows the levels of FeO_4^{2-} ($S = 1$), which are nondegenerate in the zero-field condition. If the magnetic field is swept through its full range, two resonances will be observed. This type of spectral multiplicity is called fine structure. From the observed resonance positions one may obtain effective *g*-values (g_{eff}), but these are not easily converted to actual *g*-values.

Figure 5-10(C) illustrates the effect of a weak field on a system with half-integral S, in this case 3/2 [e.g., Cr(III)]. The low-symmetry crystal field leaves two Kramer's doublets, one consisting of the degenerate pair $+1/2$, $-1/2$ (the lowest) and another of the pair $-3/2$, $+3/2$ (the next higher). In this case the zero-field splitting is small relative to the reach of the esr spectrometer, and the selection rule allows three separate transitions at markedly differing field strengths.

In a stronger crystal field the two Kramer's doublets become more widely separated [Figure 5-10(D)]. As a result, two of the transitions are no longer possible, leaving only the transition between the field-separated levels of the lower doublet. This example illustrates what has been said about the relative tractability of half-integral S systems. One or more Kramer's doublets must occur if S is 1/2, 3/2, or 5/2, and of these, one must consist of the pair $+1/2$, $-1/2$. Even if zero-field splitting produces widely separated levels, the selection rule $\Delta m_s = \pm 1$ permits a transition between the latter two levels, and the spectrum can be observed (granting other factors will allow it, e.g., narrow linewidths). In such cases the effective S is 1/2. On the other hand, systems with $S = 1$ or 2 can have doublets $+1$, -1 and $+2$, -2 but not $+1/2$, $-1/2$, and the typically large crystal field splittings render these cases intractable (with a few exceptions, such as the FeO_4^{2-} example given here). It should be noted that esr absorptions deriving from fine structure are frequently of unequal intensity.

Applications of electron spin resonance spectroscopy at zero field have been reviewed, including biologically relevant aspects (Bramley and Strach, 1983). Extension of the technique into the far-infrared region of the spectrum opens the possibility of determining zero-field parameters for most transition metal complexes and paramagnetic metalloproteins.

5.2.3.2 Variations in g and the Relationship to Structure

The Landé g applies only to free ions, where LS coupling dominates:

$$g = 1 + \left[\frac{J(J + 1) + S(S + 1) - L(L + 1)}{2J(J + 1)} \right] \qquad (5-5)$$

Actual g-values obtained from esr spectra are measured quantities which may or may not agree with this equation. For example, the d-electrons of the transition elements are near the surface of the atom and participate in bonding; hence, in the case of transition metal ions with strong ligands attached, d-orbital angular momentum may be quenched. The effective Bohr magneton number (μ_{eff}) for these ions in fact agrees with $\mu_{eff} \cong g[S(S + 1)]^{1/2}$, where $g = 2$. This is the "spin only" value expected if LS coupling is disrupted or "quenched."

In contrast, the lanthanide ions are paramagnetic due to unpaired electrons in f-orbitals. Since these are located deep within the atom and their angular momenta cannot be quenched by a bonding mechanism, LS coupling would be expected. With some exceptions, the measured μ_{eff} agrees with $\mu_{\text{eff}} = g[J(J + 1)]^{1/2}$, where g is the value calculated by the Landé equation (Pake, 1962), suggesting that LS coupling prevails in these ions.

The expressions "free ion" and "spin only" describe the extremes: complete LS coupling or none at all (respectively). Transition elements of the iron group have g-factors which are not exactly equal to the free-spin value. These differences are explained as *a reintroduction of orbital angular momentum when spin–orbit coupling causes the ground state to undergo quantum mechanical mixing with excited states.* The Hamiltonian for spin–orbit coupling is:

$$\mathcal{H} = \lambda \mathbf{L} \cdot \mathbf{S} \tag{5-6}$$

where λ is the spin–orbit coupling constant. If λ is less than Δ_1 or Δ_2 (see below), a first-order perturbation may be applied to the ground state wave function:

$$\psi_0' = \psi - \sum_n \psi_n \frac{\langle n | \lambda \mathbf{L} \cdot \mathbf{S} | 0 \rangle}{E_n - E_0} \tag{5-7}$$

where n are the excited states with L the same as the ground state.

A set of d-orbital states is illustrated in Figure 5-9, which shows first the effect of an octahedral field on a $3d^9$ ion [e.g., Cu(II)] with splitting Δ, and then the further effect of tetragonal distortions along the z-axis. The upper level is now an orbital doublet (d_{xz}, d_{yz}) and the next highest is a singlet (d_{xy}). New energy separations Δ_1 and Δ_2 are defined in Figure 5-9. This arrangement has axial symmetry, and it can be shown that the components of the g-tensor are:

$$g_z \cong g_{\text{free}} - \left(\frac{4\lambda}{\Delta_1} \right) \tag{5-8}$$

and:

$$g_x = g_y \cong g_{\text{free}} - \left(\frac{\lambda}{\Delta_2} \right) \tag{5-9}$$

where g_{free} is the free electron value, 2.0023. Since g_z is parallel to the symmetry axis it is usually represented by the symbol $g_{\parallel}$, while $g_{\perp}$ is the transverse component in the xy-plane. Thus it is seen that the anisotropy of g is a function of λ and the crystal field splitting energies. The anisotropy of g is more pronounced when ligands are weaker because quantum mechanical mixing is greater when energy levels occur close to each other. In the case of Cu(II) the ground state is the upper orbital doublet (Figure 5-9), and the excited states are lower levels, i.e., it takes energy to move the unpaired spin "hole" downwards since an electron must move up to fill its "vacancy." The spin–orbit coupling constant λ is a quantity associated with the free ion. It is positive when the d-subshell is less than half-filled, negative when more than half-filled.

On inspecting Figure 5-9 for octahedral complexes, $\Delta = \Delta_1 = \Delta_2$. Also, the constant 4 in Eq. (5-8) is specific for tetragonal symmetry. The g-tensor is isotropic for strict octahedral geometry. On the other hand it is not uncommon to find complexes which lack axial symmetry, and all three g-tensor components are found to be unequal. The latter situation is described as rhombic.

5.2.3.3 g-Values and Polycrystalline (Randomly Oriented) Samples

The esr spectrum recorded for an amorphous or polycrystalline specimen depends on the degree of anisotropy of g. If the system is isotropic, then the resulting spectrum will consist of a single line or possibly a single group of hyperfine lines. Lower symmetry will result in up to three distinct g-values (provided the substance has only one kind of paramagnetic center).

Figure 5-11 shows absorption spectra and their first derivatives (as they are usually recorded) for cases of axial and rhombic symmetry. Axial symmetry [Figure 5-11(A)] may occur in square planar complexes, for example, with the result that two of the g-tensor components (in the ring plane, $g_\perp$) are identical but one is different (along the axis, $g_\parallel$). The result is a ramp-like signal spread over a region of the spectrum. The spreading is due to randomly oriented paramagnetic centers. It should be remarked that the spectrum of a single crystal would show orientation-dependent resonances. In the rhombic case [Figure 5-11(B)] all three g-tensor com-

AXIAL   RHOMBIC

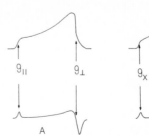 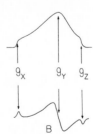

Figure 5-11. g-Anisotropy. Effects of anisotropic g on the esr signal from a randomly oriented specimen. The upper two curves are absorption signals while the lower two are derivatives. These spectra are not complicated by hyperfine splitting effects.

ponents are different, and a tent-like spectrum results. The outer lines typically show only positive or negative slopes, as required by the form of the absorption signal.

An actual spectrum may be complex if the substance contains more than one kind of paramagnetic center. Even so, an inspection of the number of lines present (and their slopes) may suggest a count of the different kinds of centers present in the sample along with their symmetry environments.

The spacing between g-components is increased in high-field instruments, whereas hyperfine coupling remains constant. Thus, esr measurements at different frequencies allow one to distinguish between the two effects.

5.2.4 High-Spin and Low-Spin Complexes

The strength of the ligands surrounding the metal ion can, in certain cases, determine the value of S. This is visualized by imagining an octahedral field surrounding Fe(III) (which contains five d-electrons, i.e., $S = 5/2$). In Figure 5-7, orbitals $d_{x^2-y^2}$ and d_{z^2}, directed toward the coordinating ligands, form the doubly degenerate, high energy level, while d_{xy}, d_{xz}, and d_{yz} make up the triply degenerate lower level. The value of Δ will be small or large depending on the strength (or affinity) of the coordinating ligands. If Δ is smaller than the energy required to pair electrons (this is an electrostatic effect; electron repulsion must be overcome if two electrons are to occupy the same orbital), no pairing will occur, and each of the five d-orbitals will be occupied by a single, unpaired electron. The value of S will be 5/2, as in the free ion. However, if Δ is large relative to the pairing energy, the two higher-energy electrons will be forced into the lower levels. The ion will now have only one unpaired electron, with $S = 1/2$. The former case is referred to as a high-spin ion while the latter is low spin.

It can be seen that S is constant for some ions, e.g., octahedral Cr(III) with three d-electrons. Also, intermediate cases are possible, and the effective S becomes temperature dependent. The latter systems are generally difficult from the perspective of esr.

5.2.5 Linewidth Effects

Several factors work either to narrow or broaden esr spectra. Linewidth effects are often encountered as problems, e.g., as excessive broadening which obscures hyperfine structure. In other situations, linewidth measurements may become a source of information. These features of esr spectra will now be examined.

5.2.5.1 Spin State Lifetime and the Uncertainty Principle

According to Heisenberg's Uncertainty Principle, the uncertainty in energy of a quantum state is a function of the time a particle exists in that state. Given that Δt is the lifetime of the quantum state and ΔE the energy uncertainty, a fundamental relationship exists:

$$\Delta E \cdot \Delta t \geq h/(4\pi) \qquad (5\text{-}10)$$

This effect is illustrated in Figure 5-12. A transition takes place between upper and lower energy levels (both long-lived) in Figure 5-12(A). The difference between these two states has a definite value, resulting in the sharp spectral line shown to the right. However, in Figure 5-12(B) the energy levels are short-lived with an accompanying uncertainty in their energy states. The original quantum ($h\nu_0$) is now surrounded by a distribution of possible transitions ($h\nu_0 \pm \Delta h\nu_0$), and the resulting spectral line, shown to the right, appears broadened.

Any factor which acts to decrease the electron spin state lifetime will broaden esr spectra. These factors can be classified into the following: relaxation mechanisms which determine the spin–lattice relaxation time (T_1), a mechanism which affects the spin system internally by dipole–dipole interaction, leading to a characteristic transverse relaxation time (T_2), and chemical exchange effects in which the lifetime is made brief by rapid chemical reactions. These are the same factors which cause line broadening in nmr spectra (see Chapter 2).

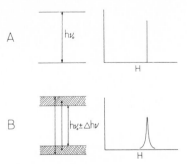

Figure 5-12. The Uncertainty Principle and linewidth. In (A) the quantum states are long-lived and their energy uncertainty is small. Thus the resonant absorption, shown to the right, is a narrow line. In (B) the quantum states are short-lived and their energy is uncertain (shaded area). Resonant absorption has a distribution of possible values, resulting in the broadened line shown to the right.

5.2.5.2 Spin–Lattice Relaxation Time

Transitions between magnetically separated energy levels, such as those upon which esr is based, involve energy differences which are low relative to the available thermal energy. As a consequence the low- and high-energy states (respectively, field augmenting and field opposed) are almost equally populated, i.e., the system behaves as if it were approaching an infinite temperature. It is only because the lower state has a slightly greater population that the esr spectrum can be observed at all.

If a paramagnetic specimen is irradiated at its resonant frequency using a strong microwave source, the two levels may become equally populated. The system is then said to be saturated. Saturation effects are seen by conducting a series of scans through an esr signal and increasing the microwave power between each successive scan. With sufficient power the signal broadens and diminishes in amplitude. Further applied power may cause the signal to disappear entirely. Any critical esr study should include an evaluation of saturation characteristics because the amount of power required to cause saturation varies from one chemical entity to another.

If the irradiation is stopped, the spin system begins to relax toward its equilibrium condition with an excess population in the lower energy state. Thus, energy must be dissipated to the surroundings (or lattice). The time constant of this decay is known as the spin–lattice relaxation time or T_1. Since this process involves changes in the relative populations of field-opposed and field-augmenting spins, T_1 is related to the rate of change of magnetization along the axis of the applied field:

$$\frac{dM_z}{dt} = \frac{M_z - M_z'}{T_1} \tag{5-11}$$

Here M_z' is any instantaneous magnetization along the field axis and M_z the corresponding equilibrium value. If T_1 is long, the dissipation of excess spin energy proceeds inefficiently and the esr signal is easily saturated. When T_1 is not masked by other effects and the relaxation rate can be described by a simple exponential law, lineshapes resemble the Lorentzian distribution shown in Figure 5-13. However, in real paramagnetic substances one may expect to encounter a multiplicity of relaxation mechanisms. Equation (5-11) is not a strictly correct representation of T_1 in that case.

It is customary to relate the linewidth at half-maximum amplitude ($\Delta\nu_{1/2}$) to a generalized relaxation time, T_2. If the lineshape is Lorentzian, the equation $\Delta\nu_{1/2} = 1/(2\pi T_2)$ applies, whereas for Gaussian lines the relationship $\Delta\nu_{1/2} = 1.476/(2\pi T_2)$ is used. The observed T_2 may be deter-

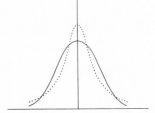

Figure 5-13. A comparison of Gaussian (solid) and Lorentzian (dotted) distributions.

mined by spin–lattice relaxation, and this is often the case among transition metal ions. However, other modes may dominate, e.g., a spin–spin relaxation mechanism (T_2') to be discussed in Section 5.2.5.4.

5.2.5.3 Mechanisms of Spin–Lattice Relaxation

Vibrations of the surrounding lattice can induce spin energy dissipation when the characteristic frequency matches the resonance frequency. However, at temperatures above 4 K the predominating lattice frequencies are much higher than ν_0. The relatively less populated modes corresponding to ν_0 induce spin flips by motional modulations of neighboring dipoles (i.e., other electron spins). This is known as the direct mechanism, and it is field dependent, with linewidth varying in proportion to H_0^4. Another mechanism known as the Raman process involves inelastic interactions between lattice phonons (vibrational quanta) and the unpaired electrons. In the latter case the scattered phonon has a higher frequency than the incident phonon when the encounter flips an electron dipole from its high-energy state to the lower one. The latter process is *not* field dependent. Both mechanisms should correlate with temperature, with linewidth varying as a power of the absolute temperature, and this is in agreement with the observed effects. However, theoretical treatments which include only these modes predict linewidths which are orders of magnitude in error (specifically, the predicted values are too small).

Experimentally, linewidth is found to be correlated closely with spin–orbit coupling where broadening is more pronounced with increasing λ. A theoretical treatment by van Vleck (1939), which includes spin–orbit coupling, is much more consistent with the experimental facts. This model notes the effects of lattice vibrations on crystal field energy levels. It will be recalled from Section 5.2.3.2 that spin–orbit coupling is quenched when the metal ion is bound to the incoming ligand groups, but that quantum mechanical mixing with higher states reintroduces some coupling [Eq. (5-7)]. Thus, vibrational modulation of the upper levels (i.e., by the lattice) leads to an interaction between the lattice and the spin system, resulting in a mechanism for spin energy dissipation. One should think of spin–orbit coupling as an augmenting effect since the direct and Raman mechanisms are still applicable.

The presence of low-lying energy levels thus increases both spin–orbit coupling and the T_1-determined linewidth. For some species, e.g., Fe(II) with an orbital only ~ 100 cm^{-1} away from the spin-containing level, broadening can be extreme. In such cases it may be impossible to record esr spectra except at very low temperatures. T_1 is best determined by pulsed methods similar to those used in nmr spectroscopy (see Section 2.4.4.4 and Poole, 1982).

5.2.5.4 Spin–Spin Relaxation and Dipolar Broadening

The relaxation mechanisms described in the preceding sections act even if the paramagnetic species is dilute, i.e., when neighboring dipoles are far apart. A biological specimen is dilute, for example, if a single paramagnetic metal ion is imbedded within a large protein molecule. On the other hand, a protein (or any other structure) might contain several paramagnetic centers relatively close to one another. In that case the sample would not be dilute.

An additional relaxation mechanism may exist if two or more paramagnetic centers are separated by a few angstroms, especially if these entities have approximately the same resonance frequency. The field experienced by any one dipole will now be the sum of the external (instrument) field plus the local field due to neighboring dipoles. The local field is given by:

$$H_{\text{local}} = \Sigma \mu_i (1 - 3 \cos^2\theta_i)/r_i^3 \qquad (5\text{-}12)$$

where μ_i is the z-component (i.e., parallel to the applied field) of the ith neighboring dipole, r_i the interdipolar distance, and θ_i the angle between the applied field and the line connecting the two dipoles. If the local field is sufficiently strong, the dipoles may interact such that when one flips to the low-energy state, the other is flipped to the high-energy state. The reader will note that this mechanism does not alter spin populations, but it does decrease spin lifetime and thus increases linewidth according to the Uncertainty Principle. This mechanism is known as spin–spin (or transverse) relaxation and the symbol T_2' is used to represent the time constant associated with the mode.

Dipolar broadening will also occur under these conditions, and it does not require that the neighboring dipoles have similar resonance frequencies. As illustrated in Figure 5-14, a resonance which is inherently narrow (as determined by spin lifetime) may occur over a range of con-

Figure 5-14. Dipolar broadening. In the absence of nearby electron dipoles the narrow Lorentzian absorption signal would occur at H_0. This figure shows one population of unpaired spins with the absorption signal (solid line) displaced upfield due to dipoles opposing the applied field. Other spin populations exist in different environments so that the net absorption is the broad, Gaussian envelope (dashed line).

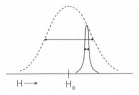

ditions when random local fields augment or oppose the applied field. Dipolar broadening thus results from fluctuations in the apparent resonance condition rather than a characteristic lifetime.

Spin–spin and dipolar broadening are comparable effects, and the associated lineshapes are Gaussian. It should be noted that these properties are detected by observing the effect of concentration on the esr spectrum. This may be difficult in biological specimens if the broadening is due to several paramagnetic centers within, for example, a protein molecule.

Dipolar and spin–spin relaxation effects can become dominant in complexes which contain a multiplicity of spins. A mechanism may then exist for rapid electron spin–spin (transverse) relaxation. For example, red-violet clusters of the type $Cu(I)_8Cu(II)_6L_{12}Cl^{5-}$ (L = penicillamine) present extremely broad esr spectra due to spin–spin and dipolar interactions between six Cu(II) ions. These paramagnetic centers are known to interact ferromagnetically (van Kempen *et al.*, 1981). This example differs from most simple copper(II) complexes since the latter typically produce well-resolved spectra, even at ambient temperatures, reflecting an isolated, unpaired electron with a relatively long spin state lifetime. Nmr spectra are usually broadened beyond detection in simple copper(II) complexes. In contrast, the red-violet clusters present broadened, paramagnetically shifted ^1H-nmr spectra (Eggleton *et al.*, 1978; see also Section 2.3.1 and Figure 2-4). This possibility should be considered when approaching completely novel metalloproteins, and an effect of this type has been observed in a two-iron–two-sulfur ferredoxin (see Section 2.7.2.4.3).

Early work with these clusters illustrates the need for care with respect to possible interfering paramagnetic species. The first attempts at structural characterization were biased toward electroneutral species of the type $Cu(II)_1Cu(I)_2L_2$ (Sugiura and Tanaka, 1970), but the structure wãs later shown to be the much larger anionic species described above, based on X-ray crystallography (Birker and Freeman, 1977). Very broad esr absorptions are associated with these cluster ions (Kroneck *et al.*, 1971; Birker and Freeman, 1977) but the counterion can be $Cu(II)_{aq}$! In this case the interfering impurity is easily removed by passing an aqueous solution of the cluster ion through a Chelex-100 (Biorad) column. It is easy to be misled if a composite spectrum of two completely different entities is presumed to derive from only one (similarly, the isolation of a compound containing counterion copper led to erroneous conclusions about the structural formula in the early work on these clusters). Caution about the possible presence of paramagnetic impurities or multiple species cannot be overstressed.

5.2.5.5 Exchange Narrowing

At very high concentrations, for example, in the pure crystalline forms of low molecular weight paramagnetic substances, one may observe line narrowing due to direct overlap of the spin-containing orbitals. Linewidths at half-maximum are given by:

$$\Delta \nu_{1/2} = \left(\frac{\pi}{2}\right)\left(\frac{\nu_d^2}{\nu_e}\right) \tag{5-13}$$

where ν_d is the dipolar frequency, equal to $g^4\beta^4/r^6$, and ν_e is an exchange frequency derived from the quantum mechanical exchange integral, J:

$$\nu_e = J/h \tag{5-14}$$

The exchange integral is a function of ψ_i and ψ_j, the spin amplitudes of overlapping orbitals, and the extent of line narrowing thus depends on how well the two orbitals overlap.

In simple systems the presence of exchange interaction may be revealed by dilution. Initially, the line broadens as dipolar effects begin to dominate. Further dilution leads to line narrowing, where the dominating mechanism is T_1. It may be quite difficult to detect exchange narrowing in the heterogeneous environment of biological systems. Exchange-narrowed lines have Lorentzian shapes.

5.2.5.6 Chemical Exchange Broadening

The lifetime of a paramagnetic species may be markedly reduced if it participates in a rapid chemical reaction. Enzyme-controlled reactions (at least some of them) belong in this category. There are numerous cases of chemical exchange broadening in simple systems. For example, the napthalene anion radical reacts with napthalene in a "no-reaction reaction" (Atherton and Weissman, 1961) i.e., the unpaired electron jumps from molecule to molecule, reducing the mean lifetime of its molecular existence. Similar one-electron oxidations and reductions occur among the metalloenzymes.

5.2.5.7 *Motional Narrowing Effects*

If the spin state lifetime is long at biologically meaningful temperatures and if the dipolar and exchange contributions are negligible, motional effects may be observed. Most metal ion spectra are badly broadened at higher temperatures, although there are exceptions [e.g., Cu(II)]. Dilute solutions of organic free radicals are more likely to meet these criteria.

Small free radicals in liquid solution reorient so rapidly (i.e., their correlation time, τ_c, is sufficiently short) that the anisotropies of g and A become averaged. The observed g-value is the mean of its tensor components, and hyperfine splitting patterns appear symmetric. Alcohol solutions containing the vitamin E phenoxy free radical behave in this manner (to observe organic radicals in solution it is usually necessary to create them in a flowing reactor, i.e., the necessary reagents meet in a confluence immediately before passing through the esr sample cell). However, if the vitamin E radical is frozen in a solid matrix, the reorienting framework is now that of the solid, which has a much longer τ_c. The anisotropic effects are no longer averaged, and a distorted spectrum is the result, as shown in Figure 5-15(B) (Kohl *et al.*, 1969).

Maximum anisotropic effects are assessed by freezing the radical in a solvent glass, usually by plunging the sample cuvette into liquid nitrogen,

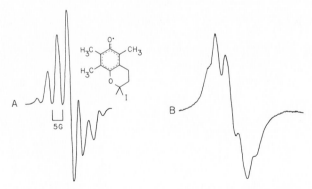

Figure 5-15. Motional averaging effects. (*A*) The signal from the vitamin E free radical shows seven major hyperfine lines originating in the six *ortho* methyl protons. The modulation amplitude (3G) prohibits detection of *meta* substituents (i.e., each major line should resolve into a sextet). (*B*) The same radical frozen at 77 K. Anisotropic *g*- and *A*-effects produce a distorted spectrum.

or by irradiating a solid matrix containing the free radical precursor, i.e., the radical is generated *in situ*. Knowledge of the full range of possible effects between free rotation and complete restriction can be applied to actual spectral effects for an estimate of the kind of motion a radical species is experiencing. This is the basis of spin labeling.

5.2.5.8 Spin Labels

Spin labels are usually esr-active nitroxide free radicals (stable) which may be covalently attached to proteins, lipids, and other biochemical structures using cross-linking agents and as such are used as probes of molecular motion (or lack of it). Spin-labeled molecules show a narrow-line nitroxide esr spectrum if the latter group is in rapid motion, but the spectrum will be broadened and distorted if the nitroxide group experiences restricted rotational motion. The extent of broadening can be correlated with the extent of restriction. Nmr spectra of spin-labeled molecules also show modification, specifically in the form of paramagnetic line broadening when the resonance source is located near the spin label (see Section 2.7.2.4.3). This type of effect can be used to determine the kind of groups located near the labeled site. The structural and dynamic aspects of spin-labeled biomolecules have been reviewed (Buchachenko and Wasserman, 1982; Poole, 1982). Nitroxyl groups must be covalently cross-linked to the biological molecule in order to function as a label, and methods for synthesizing labeled molecules are diverse and well-developed (Keana, 1978).

Spin label methods are among the most frequently used biological applications of esr, and some examples will be cited in Section 5.4.9. The reader should be aware of possible unique applications. For example, changes in spin label line broadening due to dissolved O_2 have been used to obtain quantitative measurements of oxygen uptake by cell cultures (Lai *et al.*, 1982). The sensitivity of esr qualifies this as a true micro-method.

5.3 Esr Instrumentation

The development of esr followed World War II and was stimulated by a surplus of microwave gear originating in that conflict. Most esr spectrometers resemble the example shown in Figure 5-16. The heart of the detection system is the magic tee (an assembly of waveguides and tuned cavities) shown in Figure 5-16, or a variant which behaves similarly. Microwave power at frequency ν_0 originates in the klystron oscillator, and this frequency must be stable. An attenuator (not shown) controls the amount of microwave power reaching the sample and is equivalent to lamp intensity in an optical spectrometer. A second arm of the tee terminates in the sample cavity located between the magnet pole faces. This is also sandwiched between a pair of field modulation coils. For biological work the sample cavity is designed so that the electric field component of ν_0 is minimal (i.e., nodal) at the sample while the magnetic field component of ν_0 is maximal. This avoids undue dissipation of microwave energy in water or other polar solvents.

The detector crystal and tuning cavities on the other tee terminations complete the detector. Tuning is accomplished by varying the cavity geometry, e.g., by moving a slug or disc in or out. The crystal is current biased and is in the input of a radio frequency amplifier tuned to the field modulator frequency. Often, the latter is 100 kHz but other values may be chosen. To understand the operation of this detector, the essential feature to remember is that a change in microwave absorption in the sample arm modifies the tuning of the tee. This is reflected to the crystal, which experiences a corresponding change in current (the arrangement is analogous to a simple Wheatstone bridge).

The operation of this detector is illustrated in Figure 5-17. With the ν_0 fixed, the applied field is swept through the resonance condition. At the left position there is little absorption of microwave energy and the crystal current averages the noise value. However, at the right position the sample absorption varies considerably between the low- and high-field swings of the modulator. This results in an oscillating current in the detector crystal. The latter occurs exactly at the modulation frequency, and as a result, this signal can be amplified and autocorrelated with the modulating signal. Autocorrelation takes place in the phase detector shown

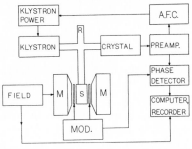

Figure 5-16. Basic features of an esr spectrometer. The waveguide "tee" terminates in the klystron (microwave source), a resistive load (R), a detector crystal, and the tunable sample cavity (S). An automatic frequency control (A.F.C.) maintains a stable klystron frequency. The sample cavity is nested within a pair of Helmholtz coils driven by the modulator (MOD.) and is between the pole faces of an electromagnet (M). Crystal output and a reference from the modulator are autocorrelated in the phase detector. Signals are detected when the sample absorbs energy, which unbalances the tee and changes the crystal current at a frequency determined by the modulator. Information arriving at a recorder or computer consists of the phase detector signal voltage and field information in the form of a voltage from the field sweep generator.

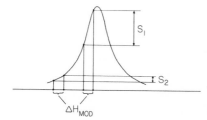

Figure 5-17. Slope detection in an esr spectrometer. Field modulation (ΔH_{mod}) at constant amplitude is shown at two positions relative to an esr absorption line. Since crystal current depends on the degree of unbalance of the tee, the signal produced in the crystal (at the modulator frequency) is greatest on the steeply sloping portion of the absorption line (S_1).

in Figure 5-16. The phase detector output is clearly dependent on the slope of the esr absorption. Consequently, the spectrum will be recorded as a derivative.

Inspection of Figure 5-17 will reveal that the instrument's sensitivity is improved by increasing the modulation amplitude. For optimum results, field modulation amplitude is kept substantially smaller than the natural linewidth. Other methods may be used to enhance the signal-to-noise ratio, e.g., signal averaging (see Section 2.4.4.3). Usually, this involves sweeping field rather than frequency.

Practical esr spectrometers may differ from the example of Figure 5-16. Some studies involve measurements at widely different field and microwave frequencies, necessitating at least one magnet system with two different microwave detector assemblies (waveguide dimensions depend on the operating frequency). Other methods require sophisticated sample cavities, e.g., for particle irradiation, flash photolysis, double resonance, and others. Double resonance will be considered in the section immediately following.

5.3.1 Double Resonance Methods: ENDOR

The method of ENDOR (electron nuclear double resonance) is used to obtain accurate measurements of hyperfine splitting, and it is applicable even when such structure is not clearly resolved in the esr spectrum. ENDOR measurements involve strongly saturating the esr signal (at fixed field and frequency) while scanning an irradiating signal across the nmr frequency range of the isotope causing the hyperfine effects. The esr signal becomes unsaturated at specific nmr frequencies, and the ENDOR spectrum is simply a plot of nmr frequency versus esr signal output. Readers should note that the unique conditions of field and frequency associated with nmr-active isotopes impart a high degree of element specificity to this method (Dorio and Freed, 1979; see Chapter 2).

To illustrate the characteristics of ENDOR, one may consider the

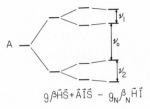

$$g\beta\vec{H}\vec{S} + \vec{A}\vec{I}\vec{S} - g_N\beta_N\vec{H}\vec{I}$$

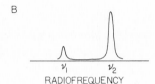

Figure 5-18. ENDOR. (*A*) Energy levels responsible for ENDOR signals. (*B*) An ENDOR doublet. The components are usually of unequal amplitude.

simple case of a single unpaired electron interacting with one proton ($S = 1/2$, $I = 1/2$). The applicable selection rules are: $\Delta m_s = \pm 1$; $\Delta m_I = 0$. This system is described by Eq. (5-1) (see Section 5.2.2.1). These effects are shown graphically in Figure 5-18(A), where the major splitting (caused by electron spin–field interaction) is described by the first term of the Hamiltonian. The second splitting is due to electron–nuclear hyperfine coupling (second term). Finally, interaction between the applied field and the nuclear spin (third term) decreases the separation between the upper two states while increasing the spacing of the lower two. One of the two permissible esr transitions is indicated as ν_0. In ENDOR measurements, the field is held constant with the klystron frequency matching ν_0. Also, the klystron power must be sufficient to saturate this resonance, resulting in the two levels being equally populated.

To obtain the ENDOR spectrum the esr output is recorded as a function of time while a strong radio frequency source is swept progressively through the appropriate nmr frequency range, in this case for the isotope ^{1}H. At ν_1 the esr saturation condition will be upset because part of the population of the second highest level ($m_s = +1/2$; $m_I = -1/2$) will now be transferred to the highest level ($m_s = +1/2$; $m_I = +1/2$). The esr detector will thus register a signal deflection. A similar deflection will occur at ν_2. It should be noted that these two deflections may have different amplitudes, as in Figure 5-18(B). Thus, it is seen that the ENDOR spectrum is a doublet. The spacing between these two lines is the hyperfine coupling constant, A (in Hz or gauss, depending on how the scan is calibrated). Also, the midpoint of the doublet is the nmr frequency of the nucleus in question, i.e., as measured by an nmr spectrometer. This interpretation is usually applicable to hydrogen hyperfine coupling, where the nuclear spin/field interaction exceeds A. When the converse is true, the doublet spacing is twice the nuclear resonance frequency, and the position of the doublet midpoint is $A/2$.

If the unpaired electron interacts with two or more identical nuclei in the same chemical environment, the result will still be a simple doublet. In the special case of $I = 1/2$, the number of ENDOR doublets should match the number of unique chemical environments in the molecule. However, these may or may not be fully resolved in the ENDOR spectrum. A more complex situation exists with $I > 1/2$. For example, if I is 1, the Hamiltonian must include a quadrupolar term, and all of the levels (six) will be unequally spaced. The simplest case will exhibit *two* ENDOR doublets.

5.3.2 ELDOR

The method known as ELDOR (electron–electron double resonance) involves a spectrometer with two klystrons and a bimodel sample cavity. One of the klystrons operates at constant frequency, and the field is set so that a specific absorption line is being monitored, i.e., at ν_0. The second klystron provides a high power input and is frequency-swept over the regions away from ν_0. When the latter source irradiates a line belonging to the set being observed (e.g., lines originating from a unique chemical environment and hyperfine coupling constant), the esr detector records a signal deflection. One may thus unscramble a complex, overlapping esr spectrum. This method resembles nmr decoupling (Chapter 2).

ELDOR may be used to determine *large* hyperfine coupling constants, which can exceed the scan capabilities of an ENDOR spectrometer. The method is also used to detect motional effects.

5.4 Applications

Biological studies using esr reveal naturally occurring organic radicals in a variety of subcellular components. Also, many of the metalloproteins contain paramagnetic metal ions, and these may produce esr signals. Indigenous radicals are usually difficult to characterize since the source of the observed esr signal must be isolated. One must then determine the structural component responsible for its paramagnetism (e.g., metal ion, coenzyme, bound substrate, etc.). One may also add known paramagnetic entities to biological materials. Examples include spin labels (see Sections

5.2.5.8 and 5.4.9) which assess molecular motion and metal ion replacements in which the biological function is retained.

5.4.1 Naturally Occurring Ions and Radicals

Many organic free radicals have only one unpaired electron ($S = 1/2$), and thus are amenable to study by esr. The selection among transition metals is not as broad since S may have integral values. In principle, it is possible to detect esr spectra for $S = 1, 2, 3$, etc. provided the ion is in a symmetric coordinating field (see Section 5.2.3.1) but the latter condition is improbable if the ion is bound to an inherently chiral protein molecule. Among the difficult cases are Ni(II), for which $S = 1$, and Fe(II), with $S = 2$ (high spin). Low-spin forms of ferrous hemoglobin should have $S = 0$, although the NO-ligated derivative has $S = 1/2$ and is therefore useful. Ions which have half-integral values of S are more tractable (if linewidth effects are not a problem). Biological metal ions best suited for esr are therefore Cu(II) and Mo(V) ($S = 1/2$), Co(II) ($S = 3/2$ for the high-spin configuration, and $S = 1/2$ for the low-spin case), and Fe(III) and Mn(II) ($S = 5/2$ for high spin and $S = 1/2$ for low spin), assuming that the coordinating field is octahedral. The lower oxidation states of vanadium are paramagnetic, while Cr(III) ($S = 3/2$) may be useful, since this element appears to have a biological role (Schwarz and Merz, 1961). Other oxidation states of these elements may exist in biological materials.

5.4.2 Copper Proteins

There are at least three types of copper found in biological molecules. The first type is Cu(I), which is diamagnetic and cannot be detected by esr. Copper present as the esr-detectable Cu(II) is further classified as type 1 or 2. The major hyperfine splitting seen in Cu(II) ions is a four-line pattern due to the copper nucleus ($I = 3/2$). If the hyperfine coupling constant for this interaction is less than 100×10^{-4} cm^{-1}, the Cu(II) ion is classed as type 1, whereas a value greater than 140×10^{-4} cm^{-1} is associated with type 2. The relatively smaller coupling constant of type 1 Cu(II) reflects a lower spin density at the copper nucleus (see Section 5.2.2.1) and suggests that the unpaired spin is delocalized onto the ligands. Also, type 1 Cu(II) is associated with large extinction coefficients in the visible spectrum; these are the so-called blue copper proteins. Representative examples are the oxidative enzymes ascorbate oxidase (squash) and ceruloplasmin (human serum), which remove electrons from their substrates and reduce O_2 to water (it should be noted that the latter two examples also contain type 2 copper).

The blue copper proteins are thought to contain Cu(II) in a distorted tetrahedral coordination field. An early quantitative model relating symmetry, optical absorption, esr g-factors, and catalytic activity was proposed by Blumberg (1966). X-ray crystallographic investigations of plastocyanin confirm the presence of a distorted tetrahedral field at Cu(II) (Coleman *et al.*, 1978) with coordinating groups of two histidines and two links to sulfur, one sulfur being of cysteine and the other of methionine. Simple complexes with a close relationship to the type 1 proteins include bis(imidotetraphenyldithiophosphino-S,S')copper(II), which has an intense blue color and an esr spectrum resembling that of a type 1 copper protein (Bereman *et al.*, 1976). The strong absorption at visible wavelengths has been assigned as a copper–sulfur charge transfer interaction. Similarly, intense visible absorptions (probably originating in charge transfer(are noted in mixed valence clusters of copper and penicillamine (and related ligands); however, these have a square planar N, N′, S, S′ geometry around Cu(II), are of structural type $Cu(I)_8Cu(II)_6L_{12}Cl^{5-}$ (L = penicillamine), and are red-violet. Also, their nmr and esr characteristics differ markedly from the blue proteins as a result of magnetic interaction between the cluster copper ions (Birker and Freeman, 1977; Eggleton *et al.*, 1978).

N,N'-Ethylenebis(trifluoroacetylacetoniminato)copper(II), a planar complex with rhombic symmetry around copper, more closely resembles type 2 copper (Giordano and Bereman, 1974). Copper ions of types 1 and 2 are often found together in the same metalloprotein. However, the enzyme galactose oxidase contains only a single type 2 copper ion and is uniquely suited for studies of this type of copper.

An esr spectrum of galactose oxidase (Bereman and Kosman, 1977) obtained with a high-field (34.758 GHz or Q-band) spectrometer is shown in Figure 5-19(A) and is of the kind expected for an environment with axial symmetry (see also Figure 5-9). An arrow marks $g_\parallel$, the component of the g-tensor parallel to the symmetry axis. This absorption is split into four hyperfine lines due to the copper nucleus ($I = 3/2$). A second arrow marks $g_\perp$ (g-tensor components directed equatorially), and there is no evident hyperfine splitting of this feature. This example shows the extent to which hyperfine coupling may be anisotropic. Clearly, $A_\parallel(Cu)$ is much larger then $A_\perp(Cu)$.

The low-field (9.09 GHz or X-band) esr spectrum of galactose oxidase is presented in Figure 5-19(B), and it is seen that the $g_\perp$ feature overlaps with the right-most hyperfine component of $g_\parallel$. Superhyperfine splitting

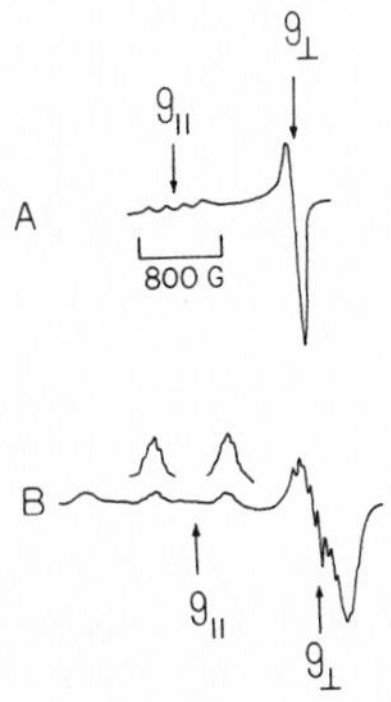

Figure 5-19. Esr spectra of galactose oxidase. (*A*) High field. (*B*) Low field.

originating in ligand atoms is evident in this spectrum (inset and arrows), both in the perpendicular and parallel components, and it appears to consist of a five-line pattern. Nitrogen atoms are potential nearest neighbors, and since ^{14}N has a nuclear spin of 1, the five-line pattern could result from interaction with two nitrogen nuclei. The reader will note that ligand superhyperfine splitting constants appear to be considerably more isotropic than those of copper.

Bereman and Kosman (1977) attempted to clarify ligand identity by observing changes in the spectrum caused by the introduction of exogenous ligands. It was found that various strong ligands could produce marked spectral changes up to a ligand:protein molar ratio of 1:1 but no further changes could be detected beyond this ratio. The evidence is thus indicative of a single, strong coordination site available to exogenous ligands. Results for the ligands ^{19}F (anion) and imidazole (containing ordinary ^{14}N) are collected in Table 5-2 and features of the $g_\perp$ region are shown for the ^{19}F anion interacting in Figure 5-20. The "overshoot" feature is due to unique orientations of the g- and A-tensors intermediate between the parallel and perpendicular conditions (specifically, $0 < \theta < \pi/2$). These perpendicular regions of X- and Q-band spectra are virtually superimposable, ruling out the possibility that the fluoride ligand has introduced rhombic symmetry; thus the overshoot feature is correctly assigned and does not represent a third g-tensor component. (This question would not be resolved with a fixed-frequency instrument; see Section 5.2.3.3.)

Galactose oxidase coordinated to fluoride (^{19}F, $I = 1/2$) presents an esr doublet which is further split into five lines similar to those of the native enzyme. These are characteristics of both $g_\perp$ and the overshoot feature, as shown in Figure 5-20. The galactose oxidase/imidazole complex

Table 5-2. Spin Hamiltonian Parameters
for Cu(II)[a]

Galactose oxidase

$g_\parallel$ = 2.277	$A_\parallel$ = 175.0	$A_N\parallel$ = 14.5
$g_\perp$ = 2.055	$A_\perp$ = small	$A_N\perp$ = 15.1
		A_NO = 14.8

Galactose oxidase–$^{19}F^-$ complex

$g_\parallel$ = 2.305	$A_\parallel$ = 159.7	$A_N\parallel$ = 11.2
$g_\perp$ = 2.050	$A_\perp$ = small	$A_N\perp$ = 14.3
		A_NO = 13.3
		$A_F\parallel$ = 41.0
		$A_F\perp$ = 175.4
		A_FO = 128.1

Galactose oxidase–imidazole complex

$g_\parallel$ = 2.254	$A_\parallel$ = 167.5	$A_N\parallel$ = 12.1
$g_\perp$ = 2.041	$A_\perp$ = small	$A_N\perp$ = 15.7
		A_NO = 13.4

[a]A values are given in gauss; no second-order correction terms are included in the g values. A_NO and A_FO represent ligand hyperfine splittings on the "overshoot" line. From Bereman and Kosman, 1977.

also exhibits an increase in the number of superhyperfine lines (not shown), and the pattern is consistent with three nitrogens interacting with the unpaired electron. The authors (Bereman and Kosman, 1977) reason that the copper(II) site in galactose oxidase could consist of a pseudo square planar arrangement of two histidines, one exchangeable ligand (fluoride, imidazole, H_2O, etc.), and a fourth ligand which could be a carboxylate group.

Later studies, which included chemical modifications of tryptophan and histidine residues (Winkler and Bereman, 1980), suggest that an organic radical rather than Cu(III) is formed when the enzyme is reversibly oxidized. Loss of the esr signal on oxidation could be due to coupling between the radical and copper(II) spins rather than the intractable $S = 1$ state of copper(III). Some chemically modified forms of the enzyme

Figure 5-20. Modified hyperfine splitting pattern of galactose oxidase interacting with ^{19}F [compare with Figure 5-19(B)]. Assignments: (1) perpendicular feature: pentets originating in nitrogen, $A_N\perp$ = 14.3 G; (2) perpendicular feature: doublet due to fluorine, $A_\perp(F)$ = 175.4 G; (3) overshoot feature: pentets originating in nitrogen, $A_O(N)$ = 13.3 G; (4) overshoot feature: doublet due to fluorine, $A_O(F)$ = 128.1 G.

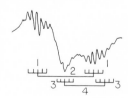

exhibited the normal Cu(II) signal and a radical signal at $g = 2.00$, which might be the result of structural changes that increased the distance between an organic radical, e.g., a tyrosine phenoxy group, and the Cu(II) center.

This example was chosen because it illustrates some commonly encountered limitations of the method. An unambiguous assignment of the five-line superhyperfine pattern as originating in two equivalent nitrogen atoms could be obtained from an enzyme biosynthesized in a medium enriched in ^{15}N ($I = 1/2$). This should decrease the superhyperfine lines to three. Also, an enzyme derived from ^{15}N-labeled histidine would be even more informative. One could argue that the assignments in Figure 5-20 are arbitrary, and close inspection will reveal unassigned hyperfine lines which may or may not originate in the high-field line group of $g_{\parallel}$. Similarly, studies using ^{17}O-enrichment could decide the nature of the remaining ligands. These methods are expensive but necessary in many cases. A good example of isotopic enrichment will be presented in Section 5.4.6. The reader should also note that esr is not directly applicable to some oxidation states, e.g., Cu(I) and Cu(III). Finally, ENDOR may be used to identify the ligands, although the necessary instrumentation is not widely accessible (see Section 5.4.8).

5.4.3 Esr Studies of Iron: Heme Iron

Some (but not all) oxidation states of iron are paramagnetic and esr active, and esr is well-suited for studies of this important biological element. Iron-containing proteins divide roughly into heme and non-heme categories. A detailed review of esr and one class of heme proteins, the hemoglobins (Blumberg, 1981), identifies the following oxidation states and their potential for producing esr spectra: The first is Hb(I), which would be a $S = 1/2$ system, but apparently cannot be prepared due to the destructive nature of the required strong reductant. Hb(II) is the physiological ferrous oxidation state. The deoxy form is an intractable $S = 2$ system, but axial coordination with ligands like CO and OCN$^-$ induces an esr silent $S = 0$ state. The NO-ligated form has $S = 1/2$ and can be used to obtain structural information about the protein, e.g., allosteric effects and other protein interactions. Hb(III) is the nonphysiological ferric oxidation state found in methemoglobin. Strong axial ligands give rise to the low-spin state ($S = 1/2$) while weaker interactions result in a high-spin ($S = 5/2$) state. These are relatively straightforward subjects for esr. Hb(IV) is the ferryl state and has a single oxygen atom at an axial position, for which $S = 1$. As is expected of integral spin cases, this form is intractable by esr (Kramer's rule). The Fe(V) oxidation state appears to occur in peroxidases rather than hemoglobins, although it could apply by analogy. Since $S = 1/2$, this type of heme is appropriate for esr. The Hb(VI) form occurs as oxyhemoglobin and is esr silent, i.e., $S = 0$.

Ferredoxins and related redox proteins contain the Fe_4S_4 cluster prosthetic group. These undergo reversible redox changes which do not appear to alter cluster geometry, and geometry is of the cubane type for Fe_4S_4 structural units:

Other iron–sulfur proteins may have different structures, e.g., Fe_2S_2. The oxidized four-iron cluster is esr silent while the reduced form behaves as a "simple" spin system with $S = 1/2$, even though the unit consists of four interacting spin centers. Reduced Fe_4S_4 usually produces an esr signal at $g \cong 1.94$.

5.4.4 Esr Properties of an Oxidoreductase Containing Both Heme and Nonheme Iron

Complex spectra are observed when the enzyme contains both Fe_4S_4 clusters and heme iron interacting. The hemoprotein subunit of *E. coli* NADPH-sulfite reductase (SiR-HP) consists of one Fe_4S_4 cluster (oxidized) and one siroheme [high-spin Fe(III)] per polypeptide. Janick and Siegel (1982) report several changes in the esr spectrum when the enzyme is subjected to progressive photoreduction using visible light and 5'-deazaflavin with EDTA as the electron donor. These investigators used absorbances in the optical spectrum for quantitative measurements of the extent of reduction of this system.

Figure 5-21(A), (B), and (C) show, respectively, the esr spectra of fully oxidized SiR-HP and subsequent 0.85 and 2.00 electron reductions. The signals at $g = 1.98, 5.24$, and 6.63 originate in the oxidized siroheme group ($S = 5/2$), and at 7 K the double integral corresponds to 0.92 electron spins per heme. From Figure 5-21(A), (B), and (C) it is clear that the first electron reduces the siroheme moiety [Fe(III) to Fe(II)], resulting in a ferroheme product which is probably paramagnetic ($S = 1$ or 2; see Christner *et al.*, 1981). Introduction of the second electron reduces the Fe_4S_4 cluster; however, from Figure 5-21(C) it is seen that the signals

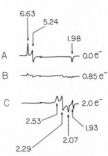

Figure 5-21. NADPH-sulfite reductase at different degrees of reduction (moles electrons per mole protein). The *g*-values are marked.

usually associated with cluster iron (g = 1.91, 1.93, and 2.04) are too
weak to account for the amount present while new resonances at g =
2.07, 2.29, and 2.53 add a combined integrated area similar to that lost
by the siroheme structure. Additional weak resonances were found at
g = 2.80, 3.39, 4.82, and 5.23. Similar signals were found in spinach
assimilatory ferredoxin-SiR (Krueger and Siegel, 1982), which has the
same kinds of prosthetic groups.

These results were interpreted as evidence of exchange coupling between the two reduced prosthetic groups (Janick and Siegel, 1982), and
Christner *et al.* (1981) demonstrated this type of interaction in the fully
oxidized enzyme by means of Mössbauer spectroscopy. The g = 2.29
and 1.93 signal groups show closely comparable microwave power saturation and temperature dependence characteristics. Also, the ferroheme
optical spectrum changes during the second reductive step, i.e., after the
heme Fe(III) has already been reduced to Fe(II). The strong ligands CN^-
and CO are specific for ferroheme and are expected to induce a low-spin
state (S = 0). These agents induce dramatic effects in the esr spectrum,
i.e., the g = 2.29 signal group vanishes with a concomitant increase in
the classical g = 1.93 signal group associated with Fe_4S_4.

Figure 5-22 shows the observed fractional signal changes as a function
of the degree of reduction of the enzyme. If it is assumed that the g =
2.29 and 1.93 signal groups are due to the S = 1/2 (Fe_4S_4) spin interacting
with the ferroheme structure (signal at g = 6.63), e.g., via a bridging

Figure 5-22. Distribution of electrons between the two kinds
of prosthetic groups in NADPH-sulfite reductase (SIR-HP).
The ordinate (f) is the fraction of the maximum signal change
between oxidized and reduced forms. The abscissa gives the
extent of reduction of the enzyme. Data for specific signal
features are labeled as follows: □, signal at g = 6.63; ●, signal
at g = 2.29; ○, signal at g = 1.93.

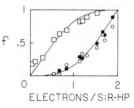

group, then the observed signal intensities are consistent with thermal distributions between the two prosthetic groups, if the reduction potential of the Fe$_4$S$_4$ cluster is 65 mV more negative than that of the heme center. Theoretical values are shown as solid lines in Figure 5-22, obviously in good agreement with the data.

Esr (and Mössbauer) evidence thus provides strong support for interacting groups in this system, but certain features lack a thorough explanation. For example the g = 2.80, 3.39, 4.82, and 5.23 group is enhanced at the expense of the g = 2.29 group when guanidinium sulfate is present in the medium, and the nature of these signals is not clear.

5.4.5 Manganese

This element appears to be intimately involved in the oxygen-producing reactions of photosynthesis (System II), and a characteristic six-line spectrum of Mn(II) is frequently observed in chloroplast preparations. The six lines result from hyperfine splitting with the Mn nucleus (I = 5/2), and this esr-active form, which may be observed at room temperature, is due to water-associated Mn(II). Under these conditions, protein-bound Mn(II) produces no detectable esr signal.

Roughly 2/3 of the chloroplast-associated manganese is loosely bound, and the remaining 1/3 of the manganese is tightly bound. Khanna *et al.* (1981) found that part of the loosely held pool of Mn(II) could be released by increasing Mg(II) ion concentration up to 20 mM. This correlated with an increase in the water-associated Mn(II) signal and a decrease in water-phase proton relaxation rate while Hill activity remained constant, as shown in Figures 5-23 and 5-24. By further increasing Mg(II) concentration to >100 mM, more Mn(II) was lost, but in this case with concomitant loss of Hill activity. This sort of evidence merely establishes correlations, and changes in the Mn(II) signal due to added Mg(II) may be incidental

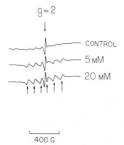

Figure 5-23. Mn(II) signals from Mg(II)-treated thylakoid membranes. The top recording is the untreated control while the lower two were obtained with the added magnesium ion concentration at 5 and 20 mM. The six-line hyperfine splitting from the manganese nucleus is noticeably asymmetric.

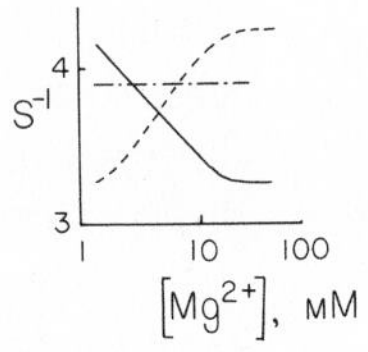

Figure 5-24. Effects of added Mg(II) on thylakoids. Oxygen evolution (horizontal line) is constant, indicating no loss of function below 100 mM added Mg(II). The solid line is the water-phase proton relaxation rate, while the dashed line is the epr detectable signal intensity of Mn(II). Units on the ordinate axis are in reciprocal seconds (proton relaxation rate) or arbitrary esr signal amplitude units.

to the mechanism whereby Mg(II) disrupts System II, e.g., relatively large changes in ionic strength may alter the conformations of a variety of proteins and thus modify function.

Other signals attributed to manganese have been observed in chloroplast preparations. Hansson and Andreasson (1982) report a System II-associated signal with characteristics similar to those of a model compound containing antiferromagnetically coupled Mn(III)/Mn(IV) pairs. This multiline signal was enhanced when chloroplast preparations were treated with System II electron acceptors and chilled to 10 K for observation. Conversely, the signal could not be detected in manganese-extracted chloroplasts.

5.4.6 Isotope Substitutions Which Identify Elements Associated with Paramagnetism

The biosynthesis of hydrogenase in *Methanobacterium thermoautotrophicum* requires nickel, and partially purified preparations of this enzyme are found to contain approximately one mole of nickel per mole of protein (Graf and Thauer, 1981). Esr measurements below 80 K reveal three lines at $g = 2.0$, 2.2, and 2.3 (Albracht *et al.*, 1982), as shown in Figure 5-25(A), and all three signals disappear when the preparation is reduced with molecular hydrogen [Figure 5-25(B)]. The signals in Figure 5-25(A) do not show hyperfine splitting, but with the exception of [61]Ni

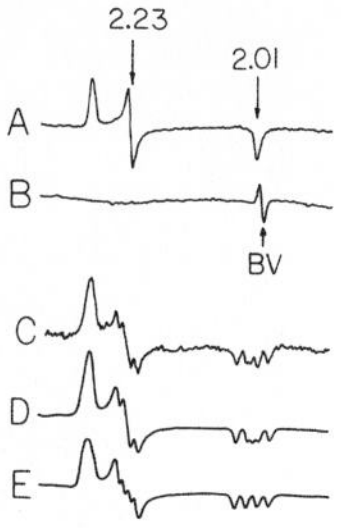

Figure 5-25. Esr spectra of the Ni-containing hydrogenase of *Methanobacterium thermoautotrophicum*. (A) purified natural enzyme; (B) same as (A) but reduced with molecular hydrogen. BV marks the benzyl viologen reference signal; (C) hydrogenase isolated from medium enriched in [61]Ni ($I = 3/2$); (D) theoretical spectrum on the basis of 80% [61]Ni enrichment; (E) theoretical spectrum with 100% [61]Ni enrichment. The spectrum in (D) is a composite of (A) and (E).

(1.19% natural abundance), the normally occurring isotopes of nickel have no nuclear spin.

Cultures of *Methanobacterium thermoautotrophicum* were grown in a medium containing ^{61}Ni enriched to 89.4% of the total nickel present. Since this isotope has a nuclear spin of $I = 3/2$, a single ^{61}Ni nucleus should produce a four-line hyperfine splitting pattern, i.e., $2I + 1 = 4$. As shown in Figure 5-25(C), hydrogenase isolated from the ^{61}Ni-enriched medium shows clear evidence of hyperfine splitting of the signals at $g = 2.01$ and 2.23. From a cursory inspection of Figure 5-25(C), the ^{61}Ni-enriched spectra appear to be anomalous since the signal at $g = 2.23$ is split into three rather than four lines, and the signal at $g = 2.01$ appears to be a quartet with the inner two lines distorted. One must use care in the interpretation of these spectra since the nickel in the medium was less than 90% ^{61}Ni-enriched in the first place. In other words, the spectrum should be a composite produced by normal and ^{61}Ni-enriched species. Figure 5-25(E) shows the simulated spectrum expected for 100% ^{61}Ni-enrichment, while Figure 5-25(D) is a composite of the normal spectrum (20%) and Figure 5-25(E) (80%). The latter spectrum compares closely with the signals from ^{61}Ni-enriched hydrogenase.

The evidence presented here supports the view that the esr spectrum of *Methanobacterium thermoautotrophicum* hydrogenase is due to paramagnetic nickel. The cited authors speculate that the signal is due to Ni(III), which can behave as an effective $S = 1/2$ state at low temperature. They also reason that esr-intractable Ni(II) ($S = 1$ or 0) is produced when the enzyme is reduced by molecular hydrogen. However, one should bear in mind the limitation of the method (and the cited authors were careful in that regard). The esr evidence alone can identify the element responsible for the observed paramagnetism, but the oxidation state of the element is more in question.

5.4.7 Metal Ion Replacements

The ion Co(II) has been used in replacements of other divalent cations. For example, oxyhemoglobin is not amenable to esr ($S = 0$), but if Co(II) replaces Fe(II) the resulting species is not only esr active (low spin, $S = 1/2$) but also functional in the sense that oxygen is bound at the axial position (Hoffman and Petering, 1970). The g-value approaches 2 while cobalt hyperfine coupling is decreased markedly, showing that the unpaired electron is relatively localized on the dioxygen adduct.

High-spin Co(II) ($S = 3/2$) occurs when Co(II) replaces Zn(II) in zinc metalloenzymes. Moreover, the cobalt analogues of carboxypeptidase A, thermolysin, and procarboxypeptidase A are fully functional. High-spin Co(II) is difficult to detect as a consequence of its extremely rapid relaxation rate. As a result, esr measurements are only possible when the sample temperature approaches 4 K. Kennedy *et al.* (1972) conducted an esr study of these three cobalt-substituted enzymes along with a Co_2Zn_2 version of dimeric alkaline phosphatase, and the resulting spectral characteristics are compared with those of model Co(II) complexes in Table 5-3. The various enzyme and model compound spectra exhibit diversity. Dimeric Co_2Zn_2 alkaline phosphatase (from *E. coli*) closely resembles six-coordinate $K_2Co(SO_4)_2 \cdot 6H_2O$. Regarding the other enzymes, the authors state that "these enzyme spectra would be consistent with a structure in which Co(II) forms three strong bonds to the protein and one bond (or indeed none at all) of a different character." When Co(II) carbonic anhydrase was exposed to its inhibitor, *p*-carboxybenzenesulfonamide, the esr spectrum was altered (not shown; $g = 6.1$, 2.2, and 1.6) and found to resemble that of tetrahedral $Co(\gamma\text{-pic})_2Cl_2$ (Table 5-3). The zinc site in carboxypeptidase A, based on the X-ray structure solved by Lipscomb in 1967, is tetrahedral with coordinating groups of histidines 69 and 196, glutamate 72, and water (Blow and Steitz, 1970).

Table 5-3. g- and A-Values of High-Spin Co(II) Complexes and Co(II)-Substituted Metalloenzymes[a]

	g_1 (or $g_\parallel$)	g_2 (or $g\perp$)	g_3	$A_1(\text{cm}^{-1})$ (or $A_\parallel$)	$A_2(\text{cm}^{-1})$	$A_3(\text{cm}^{-1})$ (or $A\perp$)
Six-coordinate						
$K_2Co(SO_4)_2 \cdot 6H_2O$	6.62	3.61	2.50	0.029	0.008	0.0065
$(NH_4)_2Co(SO_4)_2 \cdot 6H_2O$	6.46	3.06		0.025	0.002	
Five-coordinate						
$Co(Et_4dien)Cl_2$	7.08	2.53	1.59	0.030	0.006	0.009
$Co(terpy)Cl_2$	5.93	3.53	1.88	0.014	0.004	0.009
$[Co(Me_6tren)Cl]Cl$	2.29	4.25				
Four-coordinate						
$Copy_2Cl_2$	6.16	2.65	1.79	—	—	—
$Copy_2Br_2$	5.45	3.77	2.10	—	—	—
$Co(\gamma\text{-pic})_2Cl_2$	6.14	2.12	1.50	—	—	—
Enzymes						
[(Thermolysin)Co]	~5.6	~4.4	~2.7	—	—	—
[(CPD)Co]	~5.7	~4.0	~2.2	—	—	—
[ProCPD)Co]	5.6	4.2	2.9	—	—	—
[(AlkPhos)Co_2Zn_2]	6.36	3.4	2.66	0.024	—	—

[a] From Kennedy *et al.*, 1972.

Other substitutions are possible. Cu(II) does not appear to be a good choice because the substituted enzymes are not active and the esr spectrum shows only minimal changes. Haffner *et al.* (1974) reported esr studies of carbonic anhydrase, carboxypeptidase A, thermolysin, superoxide dismutase, and alkaline phosphatase in which Mn(II) replaced Zn(II). These were all tractable by esr, and in all cases the Mn(II) enzyme showed a prominent feature at $g_{eff} = 4.3$ typical of rhombic d^5 systems. The spectra were interpreted to reflect distorted tetrahedral geometry around Mn(II), although it was noted that the method does not easily distinguish between four- and six-coordinate d^5 ions.

5.4.8 ENDOR Applications

The iron/molybdenum nitrogenase of *Azotobacter vinelandii* contains iron, molybdenum, and sulfur in the approximate ratio 1 Mo: 6–8 Fe: 4–6 S. The ^{95}Mo-ENDOR spectrum suggests that molybdenum is an even-spin system of Mo(II) or Mo(VI) integrated into a multi-metal center with a total spin of $S = 3/2$ (Hoffman *et al.*, 1982a). However, the esr spectrum is relatively featureless and is indicative of rhombic symmetry (Figure 5-26). Nitrogenase isolated from a medium enriched in ^{57}Fe ($I = 1/2$) shows line broadening but no resolved hyperfine splitting due to ^{57}Fe.

The ability of ENDOR to further resolve this seemingly inscrutable esr spectrum is demonstrated by a ^{57}Fe-ENDOR study of *A. vinelandii* nitrogenase (Hoffman *et al.*, 1982b). The isotopes ^{57}Fe and ^{1}H are alike in having $I = 1/2$; however, the interpretation of ^{57}Fe-ENDOR spectra will differ from the ^{1}H example presented in the introduction (see Section 5.3.1) since ^{57}Fe/field interactions are weaker than the hyperfine coupling constants. The applicable equation is:

$$\nu_{observed} = (A/2) \pm \nu_{Fe} \qquad (5\text{-}15)$$

Doublets, assuming both members are detectable, will be separated by twice the nuclear resonant frequency ($\nu_{Fe} = 0.440$ MHz at 3200 G), which simplifies the process of interpreting overlapping doublets, and the po-

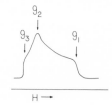

Figure 5-26. Esr spectrum of *A. vinelandii* nitrogenase. The values of g_1, g_2, and g_3 are, respectively, 2.0, 3.65, and 4.32.

sition of the midpoint between doublet members is half the hyperfine coupling constant for that particular chemical environment.

Measurements at the extreme edges of the esr absorption envelope (Figure 5-26, g_1 and g_3) typically yield coupling constants which approach those obtained from single-crystal measurements, in this case the values of A_1 and A_3. However, values obtained from the mid-regions (e.g., g_2) are composite and difficult to interpret. The ^{57}Fe-ENDOR spectrum of the nitrogenase, measured at the g_3 position, is shown in Figure 5-27(A). These lines are clearly due to ^{57}Fe since the natural enzyme shows no such ENDOR spectrum.

Careful measurements of spacings between the ENDOR peaks reveal that none differ by $2\nu_{Fe}$; therefore, each is the prominent member of a doublet pair with the weaker component lost in the baseline noise. Figure 5-27(A) indicates very clearly that the nitrogenase contains a minimum of *six* unique chemical environments. The ENDOR spectrum at g_1 [Figure 5-27(B)] shows some of the $2\nu_{Fe}$ doublet pairs and is also consistent with a minimum of six iron environments. This evidence indicates a complex active site region for *A. vinelandii* nitrogenase. The authors note that if the active site consists of a Fe_6Mo cluster with even-spin Mo, there must be an odd number of both ferric and ferrous iron atoms to obtain a net spin of $S = 3/2$ (assuming the latter assignment is correct). The ordinary esr spectrum of the nitrogenase shows marked changes during catalytic interactions (Wang and Watt, 1984).

ENDOR is applicable to a variety of systems. The interaction of H_2O_2 with catalase and peroxidases results in a probable porphyrin π-cation radical and an oxyferryl $Fe(IV)=O$ center, both being components of compound I (Hager *et al.*, 1972). The esr signal is assigned to the cation radical. A strong ^{17}O-ENDOR signal is detected when horseradish peroxidase reacts with $H_2^{17}O_2$, indicating that ". . . one oxygen atom from the oxidant remains with the intermediate and therefore shows the second oxidizing equivalent to be associated with an oxyferryl center . . ." (Roberts *et al.*, 1981).

ENDOR may be applied to entities more complex than isolated proteins and enzymes. Forman *et al.* (1981) used the method in studies of

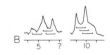

Figure 5-27. ENDOR signals from *A. vinelandii* nitrogenase. Scale is marked in MHz. (*A*) Obtained with the esr spectrometer irradiating at g_3 (Figure 5-26); (*B*) with the esr spectrometer set at g_1. All doublet spacings are identical due to the fixed nmr frequency of ^{57}Fe.

subchloroplast particles enriched in Systems I or II, and found evidence that chlorophyll is the primary reduced product in Photosystem I, while reduced pheophytin is the more likely case in Photosystem II.

It is probably not too strong to state that a multiband (e.g., X and Q) instrument with ENDOR capabilities is a minimum requirement for serious studies of inorganic biochemistry. All too often a simple X-band spectrometer produces relatively featureless spectra which invite ambiguous interpretation. Even with double resonance the limitations of the method must be borne in mind. Also, there have been too many instances where esr spectra due to impurities have been incorrectly assigned to the entity in question (see the last paragraph of Section 5.2.5.4).

5.4.9 Biological Studies with Spin Labels

5.4.9.1 Membrane-Associated Proteins

Spin labels are well-suited for determining the degree of molecular motion (or lack of it) in membranes and membrane-associated proteins, and the literature abounds with examples of this kind of application. The label may be linked to the lipid, the protein, or any other class of substance involved in the membrane system. Moreover, the repertoire of labeled substances is sufficiently diverse that the label may be placed at different positions (Keana, 1978), e.g., the nitroxide group may be located at the surface of the membrane, the middle of the lipid moiety, on the hydrophobic end, etc.

Once introduced, the location of the label relative to the water phase may be confirmed by adding a water-soluble line-broadening agent (such as Ni^{2+}) to the biological preparation. For example, a spin-labeled fatty acid, which had been attached to the erythrocyte band-3 protein in erythrocyte ghosts, presented a spectrum indicative of at least two environments. These spectra were not altered by the addition of aqueous Ni^{2+} (Bittman et al., 1984), agreeing with a label location within the hydrophobic portion of the membrane (or, alternatively, within a hydrophobic region of the protein or a protein aggregate). Spin labels which are presented to the water phase show substantially broadened esr spectra in the presence of Ni^{2+}.

Very often, the spectrum of the spin label is a composite of multiple environments. For example, the spectra of spin-labeled lipids interacting with myelin proteolipid protein show a dependence on the lipid/protein ratio (Brophy et al., 1984). The protein-bound component is motionally restricted while the lipid-associated form is more mobile, resulting in distinctively different esr signals. The ratio of signal intensities, n_f/n_b (fluid/bound), was found to be a function of the lipid/protein ratio (n_t), the number of lipid association sites on the protein (n_1), and the association constant (K_r) (Knowles *et al.*, 1981):

$$n_f/n_b = n_t/(n_1 K_r) - 1/K_r \qquad (5\text{-}16)$$

The spectra of spin-labeled membrane proteins often show modulations associated with function, which is understandable if distinctly different conformational environments impart different degrees of mobility to the label. Such "induced-fit" conformational changes may occur during allosteric regulation or the binding of substrates, cofactors, and coenzymes. As an example, the Ca^{2+}-dependent ATPase of the sarcoplasmic reticulum membrane has been specifically labeled with an iodoacetamide nitroxide (Coan, 1983). Relatively small spectral changes are seen when this protein interacts with its substrate, ATP (or other nucleotides), and inorganic phosphate, but when Ca^{2+} is added, the nucleotide effect is marked while the inorganic phosphate effect is reversed (see Figure 5-28). These properties are consistent with a model of calcium ion-effected allosteric regulation.

5.4.9.2 Soluble Proteins and Structural Proteins

Perhaps the simplest spin label application is the determination of protein rotational correlation time, but in this case the nitroxide group must not rotate relative to the protein tertiary structure. This method has been used to evaluate the correlation time and thus the radius of gyration of ceruloplasmin (Cannistraro and Sacchetti, 1984). The label can be quite sensitive to conformational effects. The glycoprotein fibronectin, which occurs both in plasma and on cell surfaces, is readily labeled at its free sulfhydryl groups using a maleimide nitroxide reagent (Lai and Tooney,

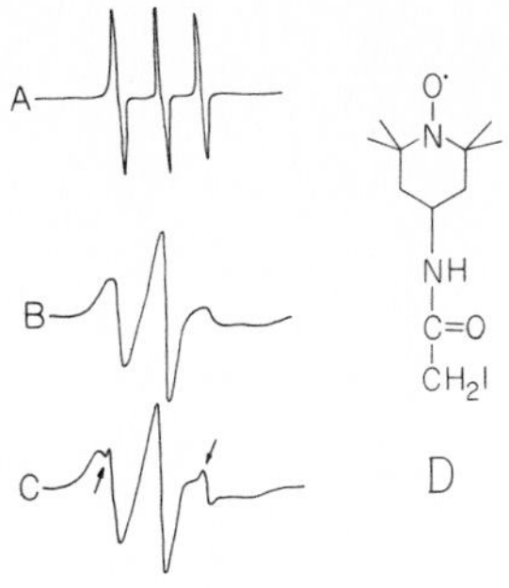

Figure 5-28. Spin label spectra. The spectrum of a free label (i.e., a small molecule alone) is shown in (A). The lines are narrow due to rapid rotation (though somewhat asymmetric). In (B) the spin label is attached to sarcoplasmic reticulum ATPase (the solution contains an AMP derivative), and the lines have become broad and markedly asymmetric due to hindered rotation. Spectrum (C) differs from (B) in having calcium ion added, and it is evident that calcium ion is interacting with this enzyme (note spectral differences marked by the arrows). The spin label structure is shown in (D). Adapted from Coan, 1983.

1984). At low temperatures the nitroxide esr spectra are very broad and reflect highly immobilized sites, but as the temperature is increased from 30 to 60°C, a marked narrowing occurs, presumed to be due to an unfolding of the protein molecule. Also, the correlation time of the low-temperature form differs by a factor of 39 from the value calculated for a globular conformation, suggesting that this protein does not have a globular configuration. Similar conformational changes are observed in spin-labeled tropomyosin (Graceffa and Lehrer, 1984).

If the macromolecule contains *two* paramagnetic centers which are rigid and reorient slowly relative to the electron spin relaxation time of a paramagnetic center, the result is an esr signal amplitude which varies in proportion to the inverse sixth power of the distance between the paramagnetic centers (Niccolai *et al.*, 1982, especially p. 370). Most such experiments involve adding Mn^{2+} to a nitroxide spin-labeled protein, or a Mn^{2+}-containing protein interacting with Cr^{3+} or Co^{2+}. The very strong dependence of signal amplitude on the distance between centers makes the method especially sensitive to fairly small conformational changes. In the other extreme (of slow electron spin relaxation) the nitroxide esr spectrum may show additional splitting due to *electron–electron spin–spin coupling*, and the coupling constant (J) may be sensitive to conformation. In this method, the label must be relatively near to the paramagnetic center. Studies with a series of spin-labeled copper salicylaldimines revealed a strong dependence of J on conformation, with J-values ranging between 3 and 92 Gauss (More *et al.*, 1984).

5.4.9.3 Nucleic Acids

Selective spin labeling of nucleic acids is more difficult as a result of their relatively monotonous structures (compared with proteins), i.e., these molecules are conformationally dynamic, and cross-linking agents may interact with most of the base groups. In the case of tRNAs, more selectivity is possible if the labeling agent can take advantage of the unique chemistry of rare base groups (Dugas, 1977). Spin-labeled DNA has been used to characterize melting transitions between single-strand and double-strand forms (Dugas, 1977). The method is able to detect similar temperature-dependent structural changes in tRNA molecules.

5.4.10 Esr Trends

Future developments in esr are likely to parallel those of nmr, e.g., multidimensional data presentations (2.7.1.1.1) which are more easily interpreted. Hyperfine interactions in simple inorganic systems are readily characterized using spin-echo spectroscopy and two-dimensional data formats (Barkhuijsen *et al.*, 1982). Fourier transform methods may be used to obtain structural information from esr spectra (Narayana and Kevan, 1982), but it should be noted that esr phase memory (i.e., the duration of a free induction decay signal) is much shorter than the analogous nmr phenomenon. There is much room for the development of multidimensional formats for esr double resonance applications.

Image-forming methods based on esr may become biomedically significant (Ohno, 1982). The difficulty in creating esr images is mostly due to the complicating effects of hyperfine splitting. In contrast, nmr images are usually derived from relaxation rate differences in a water-phase signal consisting of a single, featureless line.

References

Albracht, S. P. J., Graf, E. G., and Thauer, R. K., 1982. The epr properties of nickel in hydrogenase from Methanobacterium thermoautotrophicum, *FEBS Lett.* 140:311.

Atherton, N. M., and Weissman, S. I., 1961. Association between sodium and napthalenide ions, *J. Am. Chem. Soc.* 83:1330.

Barkhuijsen, H., deBeer, R., deWild, E. L., and vanOrmondt, D., 1982. Measurement of hyperfine interactions with electron spin-echo spectroscopy. Application to F-centers in KCl, *J. Magn. Reson.* 50:299.

Bartl, A., and Sommer, K.-H., 1984. Entwicklung eines Simulationsprogramms für ESR-Spektren und dessen Anwendung fur die Spektrensimulation ^{33}S-markierter 1,2,3-Dithiazolyle, *J. Prakt. Chem.* 326:165.

Bereman, R. D., and Kosman, D. J., 1977. Stereoelectronic properties of metalloenzymes. 5. Identification and assignment of ligand hyperfine splittings in the electron spin resonance spectrum of galactose oxidase, *J. Am. Chem. Soc.* 99:7322.

Bereman, R. D., Wang, F. T., Najdzionek, J., and Braitsch, D. M., 1976. Stereoelectronic properties of metalloenzymes. 4. Bis(imidotetraphenyldithiophosphino-S,S')copper(II), *J. Am. Chem. Soc.* 98:7266.

Birker, P. J. M. W. L., and Freeman, H. C., 1977. Structure, properties and function of a copper(I)–copper(II) complex of D-penicillamine: pentathallium(I)-μ_8-chloro-dodeca-(D-penicillaminato)-octacuprate(I) hexacupuprate (II) *n*-hydrate, *J. Am. Chem. Soc.* 99:6890.

Bittman, R., Sakaki, T., Tsuji, A., Devaux, P. E., and Ohnishi, S., 1984. Spin-label studies of the oligomeric structure of band-3 protein in erythrocyte membranes and in reconstituted systems, *Biochim. Biophys. Acta* 769:85.

Blow, D. M., and Steitz, T. A., 1970. X-Ray diffraction studies of enzymes, *Annu. Rev. Biochem.* 39:63.

Blumberg, W. E., 1966. Some aspects of models of copper complexes, in *Biochemistry of Copper* (J. Peisach, P. Aisen, and W. E. Blumberg, eds.), Academic Press, New York, p. 49.

Blumberg, W. E., 1981. The study of hemoglobin by electron paramagnetic resonance spectroscopy, *Methods Enzymol.* 76:312.

Bramley, R., and Strach, S. J., 1983. Electron paramagnetic resonance spectroscopy at zero magnetic field, *Chem. Rev.* 83:49.

Brophy, P. L., Horvath, L. I., and Marsh, D., 1984. Stoichiometry and specificity of lipid–protein interaction with myelin proteolipid protein studied by spin-label electron spin resonance, *Biochemistry* 23:860.

Buchachenko, A. L., and Wasserman, A. M., 1982. The structure and dynamics of macromolecules in solutions as studied by esr and nmr techniques, *Pure Appl. Chem.* 54:507.

Cannistraro, S., and Sacchetti, F., 1984. Small angle neutron scattering and spin labeling of human ceruloplasmin, *Phys. Lett. A* 101:175.

Christner, J. A., Munck, E., Janick, P. A., and Siegel, L. M., 1981. Mössbauer spectroscopic studies of *Escherichia coli* sulfite reductase; evidence for coupling between the siroheme and Fe_4S_4 cluster prosthetic groups, *J. Biol. Chem.* 256:2098.

Coan, C., 1983. Sensitivity of spin-labeled sarcoplasmic reticulum to the phosphorylation state of the catalytic site in aqueous media and dimethyl sulfoxide, *Biochemistry* 22:5826.

Coleman, P. M., Freeman, H. C., Guss, J. M., Murata, M., Norris, V. A., Ramshaw, J. A. M., and Venktappa, M. P., 1978. X-ray crystal structure analysis of plastocyanin at 2.7 Å resolution, *Nature* 272:319.

Commoner, B., 1961. Electron spin resonance studies of photosynthetic systems, in *Light and Life* (W. D. McElroy, and B. Glass, eds.), Johns Hopkins Press, Baltimore, Maryland, pp. 356–377.

Commoner, B., Heise, J. J., Lippincott, B. B., Norberg, R. E., Passonneau, J. V., and Townsend, J., 1957. Biological activity of free radicals, *Science* 126:57.

Dorio, M. M., and Freed, J. H., 1979. *Multiple Electron Resonance Spectroscopy*, Plenum Publishing Corp., New York.

Dugas, H., 1977. Spin-labeled nucleic acids, *Acc. Chem. Res.* 10:47.

Eggleton, G. L., Jung, G., and Wright, J. R., 1978. The ^{1}H-nmr spectra of mixed valence complexes of copper derivatives of 1-amino-2,2-dimethyl-2-mercaptoethane, *Bioinorg. Chem.* 8:173.

Forman, A., Davis, M. S., Fujita, I., Hanson, L. K., Smith, K. M., and Fajer, J., 1981. Mechanisms of energy transduction in plant photosynthesis: esr, ENDOR and MO's of the primary acceptors, *Israel J. Chem.* 21:265.

Giordano, R. S., and Bereman, R. D., 1974. Stereoelectronic properties of metalloenzymes. I. A comparison of the coordination of copper(II) in galactose oxidase and a model system, N,N'-ethylenebis(trifluoroacetylacetoniminato)copper(II), *J. Am. Chem. Soc.* 96:1019.

Graceffa, P., and Lehrer, S. S., 1984. Dynamic equilibrium between the two conformational states of spin-labeled tropomyosin, *Biochemistry* 23:2606.

Graf, E. G., and Thauer, R. K., 1981. Hydrogenase from *Methanobacterium thermoautotrophicum*, a nickel-containing enzyme, *FEBS Lett.* 136:165.

Haberditzl, W., 1973. Magnetochemistry: methods of measuring static magnetic susceptibility and their applications in biochemistry, in *Experimental Methods in Biophysical Chemistry* (C. Nicolau, ed.), John Wiley and Sons, New York, Chapter 8, pp. 351–392.

Haffner, P. H., Goodsaid-Zalduondo, F., and Coleman, J. E., 1974. Electron spin resonance of manganese(II)-substituted zinc(II) metalloenzymes, *J. Biol. Chem.* 249:6693.

Hager, L. P., Doubek, D. L., Silverstein, R. M., Hargis, J. H., and Martin, J. C., 1972. Chloroperoxidase. IX. The structure of compound I, *J. Am. Chem. Soc.* 94:4364.

Hansson, O., and Andreasson, L., 1982. Epr-detectable magnetically interacting manganese ions in the photosynthetic oxygen-evolving system after continuous illumination, *Biochim. Biophys. Acta.* 679:261.

Hoffman, B. M., and Petering, D. H., 1970, Coboglobins: Oxygen-carrying cobalt-reconstituted hemoglobin and myoglobin, *Proc. Natl. Acad. Sci. U.S.A.* 67:637.

Hoffman, B. M., Roberts, J. E., and Orme-Johnson, W. H., 1982a. ^{95}Mo and ^{1}H ENDOR spectroscopy of the nitrogenase MoFe protein, *J. Am. Chem. Soc.* 104:860.

Hoffman, B. M., Venters R. A., and Roberts, J. E., 1982b. ^{57}Fe ENDOR of the nitrogenase MoFe protein, *J. Am. Chem. Soc.* 104:4711.

Ingram, D. J. E., 1969. *Biological and Biochemical Applications of Electron Spin Resonance*, Hilger, London.

Janick, P. A., and Siegel, L. M., 1982. Electron paramagnetic resonance and optical spectroscopic evidence for interaction between siroheme and Fe_4S_4 prosthetic groups in *Escherichia coli* sulfite reductase hemoprotein subunit, *Biochemistry* 21:3538.

Keana, J. F. W., 1978. Newer aspects of the synthesis and chemistry of nitroxide spin labels, *Chem. Rev.* 78:37.

Kennedy, F. S., Hill, H. A. O., Kaden, T. A., and Vallee, B. L., 1972. Electron paramagnetic resonance spectra of some active cobalt(II) substituted metalloenzymes and other cobalt(II) complexes, *Biochem. Biophys. Research Commun.* 48:1533.

Khanna, R., Rajan, S., Steinback, K. E., Bose, S., Govindjee, A., and Gutosky, H. S., 1981. Esr and nmr studies on the effects of magnesium ion on chloroplast manganese, *Israel J. Chem.* 21:291.

Knowles, P. F., Watts, A., and Marsh, D., 1981. Spin-label studies of head-group specificity in the interaction of phospholipids with yeast cytochrome oxidase, *Biochemistry* 20:5888.

Kohl, D. H., Wright, J. R., and Weissman, M., 1969. Electron spin resonance studies of free radicals derived from plastoquinone, α- and γ-tocopherol and their relation to free radicals observed in photosynthetic materials, *Biochim. Biophys. Acta* 180:536.

Kroneck, P., Nauman, C., and Hemmerich, P., 1971. Formation and properties of binuclear cupric mercaptide five-membered chelates, *Inorg. Nucl. Chem. Lett.* 7:659.

Krueger, R. J., and Siegel, L. M., 1982. Spinach siroheme enzymes: isolation and characterization of ferredoxin sulfite reductase and comparison of properties with ferredoxin nitrite reductase, *Biochemistry* 21:2892.

Lai, C., and Tooney, N. M., 1984. Electron spin resonance spin label studies of plasma fibronectin: Effect of temperature, *Arch. Biochem. Biophys.* 228:465.

Lai, C., Hopwood, L. E., Hyde, J. S., and Lukiewicz, S., 1982. Esr studies of O_2 uptake by Chinese hamster ovary cells during the cell cycle, *Proc. Natl. Acad. Sci. U.S.A.* 79:1166.

Mims, W. B., and Peisach, J. 1981. Electron spin echo spectroscopy and the study of metalloproteins, *Biol. Magn. Reson.* 3:213.

More, K. M., Eaton, G. R., and Eaton, S. S., 1984. Metal–nitroxyl interactions. 35. Conformational effects on spin–spin interaction in spin-labeled copper salicylaldimines, *Inorg. Chem.* 23:1165.

Narayana, P. A., and Kevan, L., 1982. Fourier transformation of electron spin-echo modulation: Application to solvent shell geometry of O_2^- in dimethyl sulfoxide, *J. Magn. Reson.* 46:84.

Niccolai, N., Tiezzi, E., and Valensin, G., 1982. Manganese(II) as magnetic relaxation probe in the study of biomechanisms and of biomolecules, *Chem. Rev.* 82:359.

Ohno, K., 1982. Esr imaging: A deconvolution method for hyperfine patterns, *J. Magn. Reson.* 50:145.

Pake, G. E., 1962. *Paramagnetic Resonance, an Introductory Monograph*, W. A. Benjamin, Inc., New York.

Poole, C. P., 1982. *Electron Spin Resonance, A Comprehensive Treastise on Experimental Techniques*, 2nd ed., John Wiley and Sons, New York.

Roberts, J. E., Hoffman, B. M., Rutter, R., and Hager, L. P., 1981. ^{17}O-ENDOR of horse-radish peroxidase compound I, *J. Am. Chem. Soc.* 103:7654.

Schwarz, K., and Merz, W., 1961. A physiological role of chromium(III) in glucose utilization (glucose tolerance factor), *Fed. Proc.* 20:111.

Sugiura, Y., and Tanaka, H., 1970. Sulfur-containing chelating agents, XXV. Chelate formation of penicillamine and its related compounds with copper, *Chem. Pharm. Bull.* 18:368.

van Kempen, H., Perenboom, J. A. A. J., and Birker, P. J. M. W. L., 1981. Ferromagnetic exchange coupling in polynuclear copper(I)–copper (II) complexes with penicillamine and related ligands, *Inorg. Chem.* 20:917.

van Vleck, J. H., 1939. The Jahn–Teller effect and crystalline Stark splitting for clusters of the form XY_6, *J. Chem. Phys.* 7:72.

Wang, Z.-C., and Watt, G. D., 1984. H_2-uptake activity of the MoFe protein component of *Azotobacter vinelandii* nitrogenase, *Proc. Natl. Acad. Sci. U.S.A.* 81:376.

Winkler, M. E., and Bereman, R. D., 1980. Stereoelectronic properties of metalloenzymes. 6. Effects of anions and ferricyanide on the copper(II) site of the histidine and the tryptophan modified forms of galactose oxidase., *J. Am. Chem. Soc.* 102:6244.

X-ray Diffraction Methods for the Analysis of Metalloproteins

6.1 Introduction

6.1.1 Perspective

Research in the natural sciences is nearly always carried out in the context of a structural image of the subject material. This image or model, be it founded on experimental fact or imagined from analogous structures, provides a conceptual framework on which to base both theory and experiment. The relevant structural images in chemical research are those of the atomic arrangements in molecules. In the case of biochemistry it is our images of macromolecular structures (including their dynamic behavior) that relate directly to our approach to such basic problems as how enzymes work, how molecular aggregates self-assemble, how biological systems communicate information and transfer energy, how genes replicate, and how membranes transport specific ions. Metal atoms often play important roles in these biochemical processes, and structural images of such metal centers are pertinent to the study of these questions.

Crystallography has long been an indispensable tool for research on organic, inorganic, and metallo-organic compounds. Crystal structures serve as cornerstones in many such studies as is evident in the classic reference volumes by Pauling (1960) and by Cotton and Wilkinson (1972). More recently, macromolecular crystallography has been established as an important physical method in biochemistry. The results of these studies now provide the relevant structural images for much biological research. In a period of just over two decades since the first atomic model of a protein was reported (Kendrew *et al.*, 1960) scores of macromolecular structures have been solved. The holdings of the Protein Data Bank (Bernstein *et al.*, 1977) now exceed 200 coordinate sets from over 125 unique molecules, and the structures of at least another 50 distinct proteins are known.

Metal ions are an integral part of many proteins, and they are often vital to the activity of these molecules. Over one-third of the protein structures determined to date are those of metalloproteins, and this is probably close to the situation at large. In addition, most proteins also bind metal ions at sites other than natural metal centers. In fact, the binding of heavy metals to crystalline proteins is usually necessary for a structure determination. Unnatural binding is also involved in heavy-metal poisoning.

Metalloproteins do fall into functional classes. For example, all oxygen carriers and perhaps the majority of redox proteins require either iron or copper; zinc is often directly involved in the catalytic action of enzymes; and calcium usually plays a role either in structure stabilization or in information transfer. However, there is no apparent association of metal binding with particular structural classes of proteins (Richardson, 1981). For example, hemerythrin is an α-helical bundle; plastocyanin is a barrel or sandwich of antiparallel β-sheets; and carboxypeptidase is an α/β protein. Metalloproteins occur in membranes, in the cytoplasm, and extracellularly. They are also distributed widely with respect to organism or tissue of origin. In short, metalloproteins are typical of proteins generally.

6.1.2 Scope

This chapter is specifically concerned with X-ray crystallographic methods for the study of metalloproteins. Metal complexes with nucleic acids or other macromolecules are not treated, although methods described here may also apply in those cases. The emphasis is on natural metal coordination in proteins and not on ions of crystallization or heavy atoms added for phase determination. Methods for the analysis of small organometallic molecules (such as chorophylls, siderophores, or cobalamin) are outside the scope of this chapter except for their occurrence as cofactors in proteins.

Other diffraction methods that are also applicable to metalloproteins are not described here. For the most part, there are no special advantages with respect to metal centers in either electron microscopy and diffraction or neutron diffraction methods. Scanning electron microscopy with X-ray fluorescence analysis is an exception (see Chapter 11). Recent access to the continuous X-radiation from synchrotron sources has given impetus to other X-ray methods that do relate specifically to metal centers: X-ray microscopy at absorption edges can be used to localize metal deposits in tissues (Kirz and Sayre, 1980). Small-angle solution scattering and membrane diffraction based on anomalous dispersion can give structural information about metal centers in proteins (Stuhrmann and Notbohm, 1981). And perhaps most importantly, X-ray absorption spectroscopy (particularly the extended X-ray absorption fine structure, EXAFS) can give detailed information about the structure of metal centers of proteins in any physical state. Reviews about this important technique (Cramer and Hodgson, 1979; Teo, 1981) can be consulted for details.

Just as the properties of metalloproteins are like those of proteins in general, so too are the methods for diffraction analysis of these structures alike. Thus, other treatments of protein crystallography also apply to metalloproteins, and most of this chapter applies equally well to proteins that do not have metal centers. There is one exception: in many cases the special scattering properties of a native metal atom can be used to special advantage in the structure determination.

Several earlier reviews and books have been written on protein crystallography. Even some of the earliest of these reviews are still quite useful, and in many instances they present a treatment that is complementary to that given here. The reviews by Phillips (1966) and Matthews (1977) and the comprehensive book by Blundell and Johnson (1976) are especially notable. Attention should also be drawn to a particularly lucid account of fundamentals by Cantor and Schimmel (1980). A very recent book by McPherson (1982) emphasizes crystal preparation. Finally, one other review is devoted specifically to metalloproteins. Lipscomb (1980) reviews crystallographic methods and gives a rather comprehensive display of metal centers in proteins as known at that time. This complements the very limited description of results in the present chapter.

6.2 Theoretical Basis

6.2.1 Crystal Properties

Crystals are repetitive, symmetric structures. Many equivalent unit cells are organized on a period lattice, and the matter within each cell is interrelated by a set of self-consistent symmetry operators that pervade the crystal. There are 230 possible symmetric arrangements in a three-dimensional lattice, and these are called space groups. In the case of proteins, which are inherently handed, mirror or inversion operators are not permissible, and only 65 of the space groups are allowed. Details about crystal symmetry are given by Buerger (1963) and in the International Tables (Henry and Lonsdale, 1952).

The X-ray diffraction pattern from a crystal can be described by the convolution of the continuous diffraction from the contents of one unit cell with an interference function due to the period lattice. Thus, as in the case of an optical diffraction grating, diffraction from a crystal is concentrated into discrete diections, and this produces inividual spots on a photographic film, as was first shown by von Laue. The allowed directions can be thought of as arising from reflections off lattice planes in the crystal at angles given by Bragg's law for constructive interference. These reflections can be indexed on a lattice reciprocal to the real crystal lattice. Spacings in this reciprocal lattice are inversely related to the unit cell dimensions, and the diffraction pattern symmetry is indicative of the space group symmetry. Through the use of Ewald's construction, it is possible to predict precisely where the X-ray reflections will fall from a knowledge of the crystal orientation, the unit cell constants, and the X-ray wavelength. Specifics about X-ray optics are given in standard texts, e.g., by Blundell and Johnson (1976) and by James (1962).

6.2.2 Diffraction Theory

The intensities of the coherent diffraction into the various reflections from lattice planes in a crystal depends on the internal structure of the unit cell and on a variety of factors related to the experimental conditions and to the basic physics of the scattering process. The theoretical basis for the description of diffraction intensities in terms of fundamental molecular properties is well established. It is the excellence of this theory that gives X-ray crystallography the potential to be such a definitive science.

The pertinent measurement in an X-ray diffraction experiment is the number of photon pulses recorded as the crystal is rotated completely through the reflecting condition for a particular reciprocal lattice point. This quantity, when suitably corrected for background, is called the total energy or the integrated intensity, I_{hkl}, of the reflection indexed by h, k, l. If a piece of crystal, one so small that absorption of radiation and multiple scattering events may be neglected, is immersed in a beam of unpolarized X-rays that has an incident intensity of I_0 photons per unit time per unit area and is rotated through a reflection with uniform angular velocity ω about an axis normal to the scattering plane, one expects to observe an integrated intensity of:

$$I_{hkl} = \frac{I_0}{\omega} \left(\frac{e^2}{mc^2} \right)^2 \left(\frac{1 + \cos^2 2\theta}{2 \sin 2\theta} \right) \frac{\lambda^3 v}{V^2} \left| F_{hkl} \right|^2 \qquad (6\text{-}1)$$

Most of the parameters in this classic result from Darwin (see James, 1962, or Guinier, 1963, for more detail) will ordinarily be constant in a given experiment. These include I_0, ω, e (the electron charge), m the mass of the electron, c the velocity of light, λ the X-ray wavelength, V the unit cell volume, and v the crystal volume. In a practical case these can be collected into a single undetermined constant K. The factors that involve θ, the Bragg scattering angle, obviously vary from reflection to reflection. These geometrical factors, G, depend on the polarization of the X-ray beam and on the geometry of crystal rotation, but they can readily be calculated from the cell constants and crystal orientation; however, the exact expression may differ depending on the X-ray source and the camera or diffractometer geometry that is employed. Finally, the intensity depends on the magnitude of the structure factor F_{hkl}, which gives the coherent scattering by the contents of the unit cell relative to the scattering by a free electron. The manner in which this factor depends on the molecular structure and dynamics is described below.

Real crystals are not so small that absorption is negligible. The intensity of an individual ray is reduced by a factor $\exp(-\mu t)$ depending on the linear absorption coefficient μ and the pathlength t. This must be integrated over all ray tracings to arrive at the total transmission factor T for a reflection. In principle, there is a further attenuation, or extinction, or the primary beam due to coherent scattering losses and due to destructive interference with doubly reflected rays. This has a dramatic effect on the intensities observed from perfect crystals. However, real crystals are usually not perfect; instead they can be thought of as mosaics of many small independently scattering domains. The reflecting power of protein crystals is also always relatively weak. Consequently, extinction is negligible within the domains of "ideally imperfect" protein crystals, and the intensities of these mosaic blocks are simply additive so that Eq. (6-1) applies even for large crystals. Other factors, notably decay with time due to radiation damage, must also be considered in protein crystallography, but in general Eq. (6-1) is a suitable point of departure for interpreting diffraction observations.

6.2.3 Structure Factor Equation

The intensity formula given above reduces the diffraction problem from that of the crystal to that of the unit cell. That is, it remains to evaluate F_{hkl}. Quite in general, the X-ray scattering into a reciprocal-space direction $\mathbf{S}(|\mathbf{S}| = 2s = 2\sin\theta/\lambda)$ from a volume element dv which is located by a position vector $\mathbf{r}$ and which has an electron density $\rho(\mathbf{r})$

is proportional to $\rho(\mathbf{r}) \exp(2\pi i r \cdot \mathbf{S}) dv$. The total wave scattered from an object, I, is just the sum of the contributions from the individual volume elements. The proportionality factor can be found by relating the total scattering to the Thomson scattering from a free electron, ϕ_e, to obtain the structure factor $F = \phi/\phi_e$ as defined above for Eq. (6-1). Then,

$$F(\mathbf{S}) = \int_{\text{object}} | \int \rho(\mathbf{r}) \exp(2\pi i r \cdot \mathbf{S}) dv \qquad (6\text{-}2)$$

In the case of a crystal, the object of interest is the contents of a unit cell, and only the special directions designated by the Miller indices are allowed. Since the electron density distribution is localized about atomic centers, it is possible to reduce the total integral to the sum of individual atomic integrals. For our purposes it suffices to consider spherically symmetric atoms, and it is possible to evaluate the appropriate integral for each atom type from first principles to produce an atomic scattering factor. Then the structure factor equation can be written as

$$F_{hkl} = \sum_{\substack{\text{atom in} \\ \text{unit cell}}} f_j(s_{hkl}) \exp(-B_j s_{hkl}^2) \exp[2\pi i(hx_j + ky_j + lz_j)] \qquad (6\text{-}3)$$

Here, F_{hkl} is the structure factor of the diffraction spot located by the indices, h, k, and l; f_j is the atomic scattering factor for atom j; s is as defined above; x_j, y_j, and z_j are the positional parameters of the atom in fractions of a unit cell edge; and B_j is an isotropic temperature parameter related by $B = 8\pi^2 \overline{u^2}$ to the mean-square displacement, $\overline{u^2}$, of the atom from its average position. In some cases it is necessary to generalize the structure factor equation to include an occupancy factor for certain atoms (especially for solvent sites), and the treatment of thermal fluctuations must in general be anisotropic.

The diffraction problem is now reduced to the atomic scattering. As mentioned, the atomic scattering factors can be calculated from quantum mechanical descriptions of the atomic electron density distribution. To a first approximation, this scattering factor is independent of the X-ray wavelength. The scattering of X-rays by matter arises from the acceleration of electrons in vibrations excited by the incident X-ray wave. It is to be expected, then, that possible resonance with the natural frequencies of vibration of bound electrons in atoms will modulate the X-ray scattering and make it wavelength dependent. This results in a dispersive property called anomalous scattering, and it is a significant factor for all but the lightest atoms. Consequently, the atomic scattering factor is, in general, complex:

$$f = f_0 + \Delta f'(\lambda) + i\Delta f''(\lambda) \tag{6-4}$$

The normal, wavelength-independent scattering is f_0, and the real and imaginary components of the anomalous scattering are denoted $\Delta f'$ and $\Delta f''$. The property of anomalous scattering has special usefulness in the structure analysis of metal-containing proteins.

6.2.4 Electron Density Equation

The diffraction theory quite naturally produces a description in terms of Fourier transforms. Equation (6-2) gives the structure factor F as the Fourier transform of the electron density distribution ρ. It is a further property of Fourier transform theory that the inverse operation produces the electron density distribution in terms of the structure factors. For a crystal, in which case the structure factors are discrete rather than continuous, this inverse Fourier integral reduces to the Fourier series:

$$\rho(xyz) = \frac{1}{V} \sum_h \sum_k \sum_l F_{hkl} \exp[-2\pi i(hx + ky + lz)] \tag{6-5}$$

The summations are over all space, but in practice useful images can be produced with the restricted set of terms within a sphere $s < s_{max}$ which corresponds to a resolution limit $d_{min} = 0.5/s_{max}$.

The electron density equation in principle permits one to generate an image from the diffraction pattern. However, the structure factor, inasmuch as it represents a wave, is in general a complex number with both an amplitude and a phase, $F = |F| \exp(i\phi)$. As is clearly evident in the intensity equation, only the magnitude of F is observable. The phase information is lost in the observational interactions. This poses the phase problem which is central to any crystal structure determination. Methods for determining these essential phases are discussed below in Section 6.4.

Although an electron density map cannot be calculated until phases have been determined, it is possible to derive some structural information directly from the diffraction intensities. If the squared magnitudes are used in a Fourier series one obtains the Patterson function,

$$P(uvw) = \frac{1}{V} \sum \sum \sum |F_{hkl}|^2 \exp[-2\pi i(hu + kv + lw)] \tag{6-6}$$

It can be shown that the Patterson function corresponds to the autocorrelation function of the electron density and that it has features at the termini of interatomic vectors. For simple structures, this can lead to an interpretation of the structure. For proteins there is generally such enormous overlap in the Patterson function that this mapping of interatomic vectors is hopelessly uninterpretable. However, by use of appropriate difference coefficients, it is often possible to determine the position of heavy metal atoms from modified Patterson functions.

Fourier syntheses are widely used in crystallography. In general, the Fourier series of Eq. (6-5) will have arbitrary coefficients ξ_{hkl}. The F^2 coefficients of a Patterson function are one example. Electron density functions in actual experimental situations are another since the ξ_{hkl} can only be best estimates of the true F_{hkl} values. A kind of Fourier synthesis of particular use in protein crystallography is the difference synthesis. The density for an additive A in a complex PA with protein P is approximated by the Fourier series with coefficients $\xi = (|F_{PA}| - |F_P|) \exp(i\phi_P) \simeq F_A$.

With this brief theoretical background in hand it should be possible to comprehend the process of molecular structure analysis by means of protein crystallography. The flow of this process is indicated in Figure 6-1 for the series of tangible objects that are produced in a typical analysis. Crystals are grown, structure factor amplitudes are measured from the diffraction pattern, phases are evaluated by one means or another, an electron density map is synthesized from these data, this map is inter-

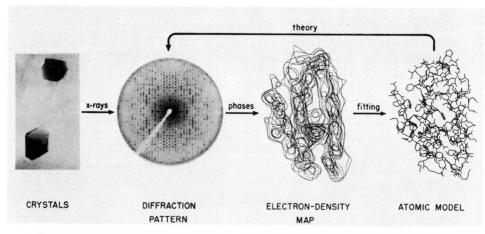

Figure 6-1. The pathway of structure analysis in protein crystallography. The tangible objects used here to illustrate this process are taken from the study of the iron protein myohemerythrin.

preted by fitting an atomic model to the density distribution, and finally this model is refined to achieve an optimal match between the theoretical structure amplitudes calculated from the atomic model with those in the observed diffraction pattern. The various steps in this pathway to the solution of a protein structure problem are described in ensuing sections.

6.3 Experimental Procedures

6.3.1 Crystallization

Probably the least understood aspect of a protein structure analysis is the first and crucial step of crystallization. It is often described as more an art than a science. It is inherently a trial-and-error process, and for the most part it is empirical rules rather than theoretical principles that guide experiments. Nonetheless, hundreds of proteins have been crystallized. A recent but still incomplete compilation of macromolecular crystals has 772 entries of well-characterized crystals (G. L. Gilliland, personal communication).

As with other chemical substances, the solubility of proteins depends on the specific surface characteristics of the molecule and on the properties of the solvent under given experimental conditions. As the composition of the solution is changed so that solubility reaches a saturation point, the protein will precipitate into the solid state. The resulting precipitate might be amorphous, it might be microcrystalline, it might be large single crystals, or it might even be polymorphic. The object of a crystallization experiment, then, is to change the characteristics of a protein solution in a controlled manner that will produce suitable single crystals.

Although the physics of crystal growth is understood for simple systems, protein surfaces are so complex and their properties are dependent on so many variables that protein crystallization would seem to defy theoretical analysis. However, a qualitative understanding of protein solubility properties is given by treating proteins as polyelectrolytes in the Debye–Hückel theory (see Blundell and Johnson, 1976). The crystallization process itself has also been treated theoretically by Kam et al. (1978). Their systematic investigation of the successive phases in crystallization offers considerable insight into the process. The first step is nucleation, during which the total free energy of the aggregate increases with size until a critical size is reached. After this nucleus is formed the crystal grows spontaneously at a rate determined by such factors as diffusion and the sticking coefficient for attachment of a free protein molecule to the new crystal. Finally, the crystal comes to a cessation of growth. The factors governing the terminal size of a protein crystal are poorly understood, but the accumulation of errors and impurities may be important, and certainly the number of nuclei in relation to total available protein is a crucial factor.

Since nucleation is energetically unfavorable it is usually possible to achieve a state of supersaturation in solution with respect to a particular crystalline state. Many crystallization techniques are designed to slowly approach a limited degree of supersaturation. It is important that nucleation be possible but that only a few nuclei be formed if large crystals are to grow. Nucleation often occurs on foreign bodies or substrate surfaces. Alternatively, the introduction of seed crystals can circumvent the nucleation problem.

McPherson (1976) has described many different crystallization techniques, and the details of these procedures will not be repeated here. It is important to note, however, that microtechniques have become increasingly important. Many of the first studied proteins, e.g., hemoglobins, are available in gram quantities, and batch techniques are suitable. However, some of the most interesting protein molecules are in extremely limited supply. Fortunately the actual mass requirements for X-ray quality crystals are quite small. Suitable crystals are usually 0.2 to 0.4 mm on an edge and typically have a volume on the order of 0.02 mm^3. This corresponds to a protein mass of about 15 μg for a protein crystal with a typical solvent content of 45% (Matthews, 1977).

Among the most important of the microtechniques are microdialysis and vapor diffusion. Microdialysis is carried out through a membrane stretched over the end of a capillary tube (Zeppezauer, 1971) or with a small, specially constructed cell. Vapor diffusion is either carried out between a reservoir or drop of precipitant and a drop of protein solution in a concavity slide (Davies and Segal, 1971) or in a hanging drop on a cover slip over a reservoir. Protein and precipitant solutions are deposited with precision micropipettes or microsyringes. In typical microdialysis experiments, 20–50 μl samples of protein solution are used, and vapor diffusion drops usually include 5–10 μl of protein solution at the start. Typically, the stock protein solution is at 10–30 mg/ml. Thus, a one-drop test can be made with about 100 μg of protein, which is enough for a few large crystals, and hundreds of attempts can be made with 10 mg of material.

Protein crystallization is potentially sensitive to a large number of factors. The state of the protein is an important factor as it is affected by pH; the degree of purity; limited proteolysis or other modifications; and the presence of inhibitors, activators, ligands, metal ions, or other additives. Naturally, protein concentration is an influential factor. Temperature and other environmental conditions such as vibration may be important. And, of course, the solvent composition is crucially important. The pH and ionic strength of the solvent and the nature of precipitant are potential variables. Many different precipitating agents have been used, but the widest success is found with ammonium sulfates, phosphates, polyethylene glycol (PEG), and 2-methyl-2,4-pentanediol (MPD).

Membrane proteins present a challenge for crystallization that is quite different from that for soluble proteins. Recently, a few membrane proteins have been crystallized (Garavito and Rosenbusch, 1980; Michel, 1982). Solubilization in a uniquely defined detergent such as β-octyl glucoside seems to be essential. Displacement of the detergent by small amphiphilic molecules like heptane-1,2,3-triol might be an important step. This was necessary for the spectacular success by Michel (1982) in crystallizing a bacterial photosynthetic reaction center. The reaction center, like many other membrane proteins, is a metalloprotein.

6.3.2 Preparation of Heavy-Atom Derivatives

Heavy-atom derivatives of protein crystals are needed for structure analysis by the method of isomorphous replacement. Since this method is of such singular importance in protein crystallography, the experimental preparation of suitable complexes of heavy metals with the crystalline protein is usually an essential step in a structure determination. The search for heavy-atom derivatives in many of the early studies was primarily a blind trial-and-error testing of all compounds on the shelf. Fortunately, the experience from previous work together with a better awareness of the chemistry of heavy metal complexation with proteins now permit a more rational approach to this aspect of protein crystallography. Nonetheless, the preparation of isomorphous derivatives remains a highly empirical process and can pose a most formidable block in a structure analysis.

Theoretical considerations provide some guidance in attempts to prepare heavy-atom derivatives of proteins. Blundell and Johnson (1976) and Blundell and Jenkins (1977) have given excellent summaries of the chemistry of metal complexes. As a general rule, hard metal ions bind to hard ligands, and soft metal ions bind to soft ligands. Hard metal ions are those with filled outer shells so that the electrons are strongly held (generally, class A metals from the left side of the periodic table) whereas soft metal ions have outer shell electrons that are readily polarizable (generally, class B metals from the right side of the periodic table). Hard ligands (e.g., carboxylate, carbonyl, hydroxyl or amide groups; water; fluoride ions) are electronegative and tend to form ionic coordination complexes with metal ions; in contrast, soft ligands (e.g., thiol, thioether, or imidazole groups; chloride or cyanide ions) have delocalizable electrons and tend to form covalent bonds with metal ions.

There are, of course, exceptions to these rules, and metals of intermediate character do exist (Fe, Co, Ni, Cu, and Zn ions from the first transition series are notable examples). Nevertheless, an analysis of heavy-atom sites found in protein crystals (Blundell and Johnson, 1976) fits well with this rationalization. Ions of alkali metals and alkaline earths (Cs, Ba), lanthanides (Sm, Tb, etc.), actinides (Th, U), and certain other metals with "hard" character (Pb, Tl) are generally coordinated by several ligands, usually including aspartate and glutamate side chains and main-chain carbonyls. By comparison, ions of soft metals (Hg, Ag, Au, Pd, Pt, Ir, Os, Re) usually bind specifically to individual cysteines, disulfide bridges, methionines, histidines, or amino termini, although electrostatic binding also occurs.

Since the reactions of hard metal ions with proteins involve complex coordination sites it is usually not possible to ascertain whether such sites might exist from the amino acid sequence or other chemical data. Thus, heavy-atom derivatives of this sort are usually found by a trial-and-error search procedure. One exception is the replacement of natural Ca^{2+} ions by Ba^{2+} or lanthanides. Other metal sites can sometimes be replaced by heavier atoms. The replacement of Zn^{2+} by Cd^{2+} or Hg^{2+} is a notable example. Many of the reactions with soft metal ions are quite specific, and experiments can be targeted to specific amino acid side chains. For example, certain mercurials are sulfhydryl reagents that bind to cysteines; platinum chloride can be quite specific for methionines; and various gold, mercury, and platinum compounds bind to histidines. In addition, specific chemical modifications of proteins can introduce heavy atoms, as in the iodination of tyrosines. Finally, it sometimes is possible to prepare heavy-atom analogues of active-site substrates that will bind to the catalytic center of a crystalline enzyme.

On occasion derivatives have been prepared by forming the complex in solution and then crystallizing this heavy-atom derivative of the protein. Unfortunately, the conditions for crystallization of complexes often differ from those for the native protein, and in any case this is a rather time- and material-consuming process. Instead, one usually finds a stabilizing medium into which native crystals can be transferred without dissolution, and then crystals are soaked in heavy-atom solutions of this medium. The optimal heavy-atom concentration for formation of derivative by soaking is typically 1–5 mM. Limited solubility of a heavy-atom compound in the soaking medium sometimes causes a problem. For example, uranyl phosphate is insoluble and, consequently, the stabilization medium for uranyl ion soaks must exclude phosphate. The pH of the solution affects the reactivity of some heavy atoms. Heavy atoms usually diffuse from the medium into the crystal within a few hours, and the reaction is often complete in a one-day soak. The ultimate test for heavy-atom derivatives formation is given in the X-ray diffraction pattern. However, heavy-atom binding is often indicated by noticeable changes in density or in crystal color.

The notable successes of the isomorphous replacement method to date have led to some relaxation in the search for other useful heavy-atom compounds. However, as increasingly larger structures are attempted, more strongly scattering derivatives are needed. New anomalous scattering methods would also benefit from such derivatives. The availability of soluble, reactive metal cluster compounds would be most desirable. An early effort was made in this direction by the use of $Ta_6Cl_{12}^{2+}$ and $Nb_6Cl_{12}^{2+}$ with triclinic lysozyme (Stanford and Corey, 1968). More recently, Strothkamp $et\ al.$ (1978) have characterized the four-mercury complex, tetrakis(acetoxymercuri)methane (TAMM). TAMM is reactive toward sulfhydryls, but unfortunately it is only sparingly soluble.

6.3.3 Data Measurement

Once a native or derivative protein crystal has been prepared, one wishes to record its X-ray diffraction pattern. Initially it is only the symmetry (space group), cell constants, and diffraction quality that are of interest. These are most conveniently assessed photographically on a precession camera. This instrument designed by Buerger (1964) produces an undistorted picture of the reciprocal lattice on photographic film. Diffraction symmetry is readily apparent in precession photographs, and intensity differences between native and heavy-atom derivative patterns can be seen by superimposing comparable films.

Although much is learned from visual inspection of the sections of the diffraction pattern that are displayed in precession photographs, a structure analysis obviously requires quantitative measurements of the intensities for a full three-dimensional set of reflections. This data collection process can be a considerable task since there are several thousand independent data measurements from even the smallest proteins, and large structures such as viruses may have hundreds of thousands of measurable reflections. The problems presented by such vast data sets are exacerbated by the relative weakness of diffraction from large structures (intensity varies inversely with unit cell volume), and protein crystals are usually quite susceptible to radiation damage. Precession methods have sometimes been used for data collection, but for the most part diffraction data are collected by one of three other methods: automated diffractometry, rotation photography, or electronic area-detector diffractometry.

Diffractometers generally measure the data one reflection at a time. The crystal and detector are moved under computer control from one reflection to the next in a systematic exploration of reciprocal space. The reflected X-ray intensity is usually measured by a scintillation counter as the crystal is rotated through the reflecting position. In order to cope with the weak intensity from protein crystals, peak-top step-scans are often used (Wyckoff *et al.*, 1967) and can be fitted by a Gaussian (Hanson *et al.*, 1979) or a learned peak profile after subtraction of an average background. Automated diffractometers are relatively inefficient for protein crystals but have the advantage of producing digital data that can be of high precision. Diffractometers are discussed in detail by Arndt and Willis (1966) and more recently by Sparks (1982).

Rotation or oscillation photography has become a mainstay of protein crystallography, especially for large molecules and especially at synchrotron sources. The principle of operation is very simple. The crystal is rotated or oscillated through a small angular range ($\Delta\phi = 0.5$–$2°$), and all data are recorded on film. This is an exceptionally efficient method for recording the data. However, the developed film must be read by a densitometer, and the rather complicated pattern must be indexed before useful data are obtained. Perhaps the most difficult aspect of this analysis is the treatment of partially recorded reflections. A useful collection of papers on the rotation method is assembled in Arndt and Wonacott (1977).

Electronic area detectors combine the efficiency of film in simultaneously recording many reflections with the digital recording and potential precision of diffractometers. Two rather different kinds of area detectors have been developed. Xuong *et al.* (1978) have adapted the multiwire proportional counter technology from high-energy physics to the crystallographic problem, and several other multiwire detectors have now been built. Arndt (1977) uses a television detector to scan an X-ray-sensitive phosphor. Solid-state area detectors are being developed. The usual data collection strategy for area detectors is similar to that employed in rotation photography, but in this case the time window for accepting a particular reflection is under computer control. This effectively reduces the background contribution.

The conventional X-ray source is a sealed vacuum tube run at sufficiently high voltage that electrons accelerated from the filament will excite an X-ray emission characteristic of the target material. The target commonly used for protein crystallography is copper; Cu Kα radiation has a wavelength of 1.54 angstroms. A variant on the sealed-tube X-ray generator that has wide application in protein work is the rotating anode generator. By continuously rotating the target it is possible to distribute the heat load over a greater area and achieve greater X-ray intensity. In recent years, synchrotron sources have become accessible. They provide much greater brilliance and offer a continuously tunable choice of wavelengths. The advantage of exposure time reductions by orders of magnitude and the potential for obtaining phase information from anomalous dispersion can outweigh the disadvantages of experimentation at a remote site.

The environment of the crystal is a very important consideration during data measurement. Protein crystals have large expanses of solvent (30–80%) and must be kept in contact with the mother liquor to preserve high crystalline order. This is accomplished by mounting in sealed thin-walled glass capillaries. Some crystals must be kept cool (e.g., 0°C) during data collection, and cooling often reduces radiation damage. Glass and adherent mother liquid around the crystal add to absorption of the X-ray beam and complicate corrections during data reduction.

As was discussed in Section 6.2.2 and shown in Eq. (6-1), the intensity measured in a diffraction experiment depends on a number of factors. Thus:

$$I_{hkl} = K(\text{crystal, zone}) \cdot G(\Omega) \cdot T(\Omega) \cdot R(t, \theta) \cdot |F_{hkl}|^2 \qquad (6\text{-}7)$$

where K is a scale factor, G is a geometrical factor often called the Lorentz-polarization factor, T is the transmission factor, and R is a radiation damage factor. Both G and T depend on the diffraction geometry, Ω, and R depends on an exposure time and scattering angle. These several factors must be evaluated to reduce measured intensities to the desired structure factor amplitudes, $|F_{hkl}|$. Theoretical values are taken for G; T is generally determined experimentally, often from intensity variations due to absorption on rotation about the diffraction vector (ψ scans) for one or a few reflections (North et al., 1968; Huber and Kopfman, 1969), and a parameterized form for R can be determined from repeated measurement of a selection of reflections (Hendrickson, 1976). Scale factors relating data from different zones and different crystals can be determined by least-squares fits relating common data (Hamilton et al., 1965), and the absolute scale can be set by statistical procedures (Wilson, 1942). It is important for later analysis to determine estimated standard deviations for the reduced data. In the case of electronic detectors this can be done by counting statistics with allowance for instrumental instability. The estimation of standard deviations for film data is more difficult, but practical values can be obtained from variation in symmetry mates, film-pack mates, and repeated densitometer readings.

6.4 Structural Analysis

6.4.1 Phase Determination

As discussed in Section 6.2.4 above, both the magnitudes $|F_{hkl}|$ and the phases ϕ_{hkl} of the structure factors are needed to produce an image of the crystalline protein by use of the electron density equation [Eq. (6-5)]. The magnitudes are determined experimentally, as discussed in Section 6.3.3 above, but the phases cannot be measured directly. This presents the phase problem. The methods most commonly used to solve the phase problems for small molecules are of little help for macromolecules. However, a variety of methods have been devised to obtain phase information for protein crystals.

Methods that have been used to help in solving the phase problem for macromolecular crystals can be classified as follows: (1) methods based on heavy atoms, (2) methods based on partially known structure, (3) methods based on geometrical conditions (i.e., sampling redundancy), and (4) methods based on physical conditions (e.g., positivity or atomicity). Some pertinent characteristics of these methods are illustrated in Table 6-1. More details are given in a review (Hendrickson, 1981; see also Section 4.8). Not all of these methods are true solutions to the phase problems in that most are not powerful enough for *ab initio* determination of phases. Moreover, none of the methods is definitive—errors in phases usually contribute much more error to an electron density function than do the errors in measurement of structure factor magnitudes. In many cases it is necessary to combine the phase information from several sources.

Phasing errors vary from reflection to reflection and between sources. This makes it essential that probability considerations enter into the phase determination. Blow and Crick (1959) first treated the problem of errors in connection with the isomorphous replacement methods. They found that the "best" Fourier synthesis, one with minimal mean-square errors in electron density when averaged over the entire unit cell, has as its coefficients the centroids of the structure factor distributions, $P(F,\phi)FdFd\phi$. Since phase errors predominate over amplitude errors, this joint proba-

Table 6-1. Methods for Phase Determination in Protein Crystallography

Method	Type of application		
	Ab initio evaluation	Ambiguity resolution or refinement	Extension
Heavy atom			
Isomorphous replacement	✓		
Anomalous scattering	✓		
Heavy-atom scattering	?	✓	
Partial structure			
Molecular replacement	✓		
Incomplete model		✓	✓
Geometrical conditions			
Molecular averaging	?	✓	?
Solvent constancy		✓	?
Physical conditions			
Direct phase relationships	?	✓	✓
Density modification		✓	✓
Maximum entropy	?	✓	✓

bility distribution can usefully be approximated by the phase probability distribution. Then the "best" Fourier coefficients are:

$$\xi_{hkl}^{best} = |F_{hkl}| \int_0^{2\pi} \exp(i\phi)P(\phi)d\phi \tag{6-8}$$

where $P(\phi)$ is normalized to unit integrated probability. After evaluation, Eq. (6-8) can be expressed as $\xi = m|F| \exp(i\phi_c)$ where m is a figure-of-merit (Dickerson *et al.*, 1961) and ϕ_c is the centroid phase. The figure-of-merit serves as a weighting factor and measures phase reliability since it is approximately related to phase error by $\sigma_\phi \simeq \cos^{-1}(m)$.

The characteristics of a phase probability distribution depend on the source of phasing that it describes. However, they invariably are smoothly varying functions that are at most bimodal. It can be shown (Hendrickson and Lattman, 1970; Hendrickson 1971) that, at least approximately, for all known sources:

$$P(\phi) = N \exp(A \cos \phi + B \sin \phi + C \cos 2\phi + D \sin 2\phi) \tag{6-9}$$

where A, B, C, and D are constant coefficients that encode the phase information, and N is a normalization factor. The phase information from different sources can be combined by multiplying the individual probability distributions, provided that the sources are independent. This is generally done, although it may not always be correct, particularly in the case of heavy-atom derivatives with common sites.

6.4.1.1 Isomorphous Replacement

The method of isomorphous replacement holds a preeminent place in protein crystallography since it can provide *ab initio* phase information for completely new structures. The essence of the method is illustrated in Figure 6-2. The diffraction pattern of a crystal of native molecules has structure factors F_P. Heavy-atom derivative crystals may be prepared as described in Section 6.3.2 by reacting a heavy-atom compound with specific sites on the native molecules. These derivative compounds have structure factors F_{PH}. The array of heavy atoms in the derivative crystal gives a contribution of F_H to F_{PH}. Provided that the native and derivative structures are precisely isomorphous and there are no errors, $F_{PH} = F_P + F_H$ as indicated in the figure, and it follows that:

$$|F_{PH}|^2 = |F_P|^2 + |F_H|^2 + 2|F_P||F_H| \cos(\phi_P - \phi_H) \qquad (6\text{-}10)$$

Thus, in principle, the protein phase ϕ_P must in general have one of two values if $|F_{PH}|$, $|F_P|$, and $F_H = |F_H| \exp(i\phi_H)$ are known. A second perfect derivative could resolve this ambiguity. The magnitudes $|F_{PH}|$ and $|F_P|$ are measurable, and if the heavy-atom coordinates are known, F_H can be calculated

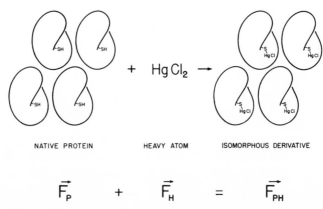

$$\vec{F_P} \quad + \quad \vec{F_H} \quad = \quad \vec{F_{PH}}$$

Figure 6-2. Schematic illustration of the isomorphous replacement reaction of heavy atoms with a protein crystal. Crystallographic structure factors for the corresponding components are related ideally as shown below.

The determination of heavy-atom positions is frequently a major obstacle in the implementation of isomorphous replacement. Positions are usually found by interpreting a Patterson function [Eq. (6-6)] based on difference coefficients

$$(\Delta F)^2 = (|F_{PH}| - |F_P|)^2 \qquad (6\text{-}11)$$

that approximate $|F_H|^2$ (Rossmann, 1960). Alternatively, direct methods from small-molecule crystallography can be used. Inclusion of anomalous scattering data gives improved estimates of $|F_H|$, but substantial errors usually remain. Derivatives with multiple sites are particularly troublesome. However, once one derivative is solved, other—even very complicated ones—can normally be analyzed by difference Fourier methods based on tentative protein phases. Various schemes have been devised to refine the heavy-atom parameters (Blundell and Johnson, 1976; Sygusch, 1977; Bricogne, 1982a), and many of these simultaneously evaluate the phases.

Inaccurate measurements, imperfect isomorphism, and inadequate heavy-atom models create errors that invalidate a straightforward application of Eq. (6-10). Hence, Blow and Crick (1959) described the problem in terms of a lack-of-closure error $\epsilon(\phi)$ for each possible value of ϕ,

$$||F_P| \exp(i\phi) + F_H| = |F_{PH}| + \epsilon(\phi) \qquad (6\text{-}12)$$

and used this to generate the phase probability distribution according to

$$P(\phi) = N \exp(-\epsilon^2(\phi)/2E^2) \qquad (6\text{-}13)$$

where E is an estimated standard deviation at the true phase. The distributions given by Eqs. (6-12) and (6-13) do not exactly yield the representation of Eq. (6-9), but phasing coefficients can be found (Hendrickson, 1971) that give a very good fit. Moreover, with alternative error definitions (Hendrickson and Lattman, 1970; Hendrickson, 1979) an exact representation obtains. The methods of Sygusch (1977) and Bricogne (1982a) circumvent probability distributions and treat experimental errors by variance weighting in the least-squares refinement.

6.4.1.2 Anomalous Scattering

The physical principles of anomalous scattering of X-rays by atoms are discussed briefly in Section 6.2.3. For "light atoms" such as carbon the anomalous scattering is negligible in the useful wavelength range, but "heavy atoms" such as iron, mercury, or uranium may have an anomalous component with Cu $K\alpha$ radiation that is 10 to 30% of the total scattering. At wavelengths near the absorption edge of a lanthanide, such as samarium, the anomalous scattering component can actually be greater than the net total scattering (Templeton *et al.*, 1982). The most readily noticed effect of anomalous scattering is a breakdown of Friedel's law of diffraction symmetry so that $|F_{hkl}| \neq |F_{\bar{h}\bar{k}\bar{l}}|$. In the case of synchrotron experiments, a variation of F_{hkl} with wavelength can be observed, but these measurements are complicated by concomitant variation in absorption.

Until recently, phase information from anomalous scattering has been used almost exclusively as an adjunct to isomorphous replacement. North (1965) and Matthews (1966) pioneered this approach. One can readily show that, in the case of only one kind of anomalous scatterer and in the absence of errors,

$$|F_{PH}(\mathbf{h})|^2 - |F_{PH}(\bar{\mathbf{h}})|^2 = 4|F_P||F_H''| \sin(\phi_P - \phi_H) \qquad (6\text{-}14)$$

where $|F_H''|$ is the contribution from the imaginary component of scattering, $\Delta f''$, for the heavy atom [see Eq. (6-4)]. Since the angle dependence is on $\sin(\phi_P - \phi_H)$, here, whereas it is on $\cos(\phi_P - \phi_H)$ for isomorphous replacement [Eq. (6-10)], the two kinds of information are quite orthogonal and complementary. Equation (6-14) is the basis for a probability formulation similar to that used for isomorphous replacement (Hendrickson, 1979).

Equations analogous to Eq. (6-14) can also be derived for the situation of anomalous scattering without isomorphous replacement. Then, after again simplifying to one kind of anomalous scatterer,

$$|F_{PH}(\mathbf{h})| - |F_{PH}(\bar{\mathbf{h}})| \simeq 2|F_H''| \sin(\phi_{PH} - \phi_H) \qquad (6\text{-}15)$$

to a very good approximation. The observable left-hand side of this equation is known as a Bijvoet difference, and a Patterson function based on the square of these differences (Rossmann, 1961) gives a vector map of the anomalous scatterer structure. If the structure of these heavy-atom tering centers can be interpreted, $|F_H''|$ can be calculated and then phase information can be deduced. However, as in the case of isomorphous replacement, this information is intrinsically ambiguous. Fortunately, in favorable circumstances the real scattering contribution of the heavy atoms, $|F_H'|$, can be used to resolve this ambiguity. More details are given by Smith and Hendrickson (1982). This method of resolved anomalous phasing was first used to determine the structure of the small protein known as crambin directly from the relatively weak anomalous scattering of six sulfur atoms in this molecule (Hendrickson and Teeter, 1981). It has subsequently been used to phase other macromolecules that have native metal ions and also some to which heavy metals have been added. A result from the crambin application is shown in Figure 6-3.

The preceding discussion relates entirely to experiments at a single

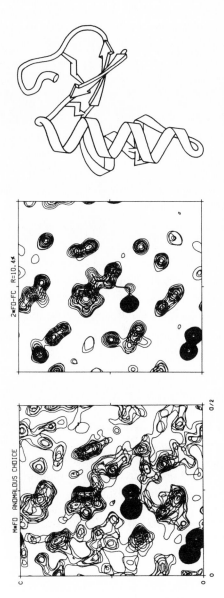

Figure 6-3. Anomalous scattering analysis of crambin. *Left*: A 1.4-Å thick slice of the initial electron density map from which the structure was interpreted. This map is based on resolved anomalous phasing from the six sulfur atoms, four of which appear here as the very dense features. *Center*: Electron density for the same slice after refinement to $R = 0.104$. Bonds in the cysteine-4–proline-5 peptide are drawn superimposed. *Right*: Schematic drawing of the crambin backbone as viewed in the density frames but at a more reduced scale. The disulfide bridges are shown as zigzags. Reproduced from Smith and Hendrickson, 1982, with permission from Oxford University Press.

wavelength. The advent of access to synchrotrons offers the possibility to exploit the dispersive (i.e., wavelength-dependent) nature of anomalous scattering and to tune to maximal effect. It can readily be shown that this leads to an algebraic solution to the phase problem (Karle, 1980). In the case of one kind of anomalous scatterer,

$$
\begin{aligned}
|^{\lambda}F(\mathbf{h})|^2 = {} & |^{0}F(\mathbf{h})|^2 + a(\lambda)|^{0}F_A(\mathbf{h})|^2 \\
& + b(\lambda)|^{0}F(\mathbf{h})||^{0}F_A(\mathbf{h})| \cos[^{0}\phi(\mathbf{h}) - {}^{0}\phi_A(\mathbf{h})] \qquad (6\text{-}16) \\
& + c(\lambda)|^{0}F(\mathbf{h})||^{0}F_A(\mathbf{h})| \sin[^{0}\phi(\mathbf{h}) - {}^{0}\phi_A(\mathbf{h})]
\end{aligned}
$$

Here, $^{\lambda}F$ is the total structure factor at wavelength λ, ^{0}F is the total normal scattering component, and $^{0}F_A$ is the normal scattering contribution from the anomalous scattering centers. The factors a, b, and c, which are functions of the wavelength-dependent scattering factors, embody the only dispersive factors in the equation. Sufficient measurements at multiple wavelengths and for Bijvoet pairs serve to determine all quantities. Calculations indicate that multiple-wavelength methods should be very powerful (Phillips and Hodgson, 1980). Although no protein structure has yet been solved in this way and substantial experimental problems remain to be overcome, the methods show great promise.

Since anomalous scattering is especially pronounced for metal ions, these methods in general hold a special role in the crystallography of metalloproteins.

6.4.1.3 Molecular Replacement

It frequently happens that an unknown crystalline protein is structurally related to another for which the three-dimensional structure is already known. Instances include different chemical states of the same protein, functionally similar proteins from different species, or proteins that show amino acid sequence homology even if not known to be otherwise related. In such cases the prior knowledge of the related structure can be used in the phase determination for the unknown structure. This is done by repositioning the known molecule into the unknown crystal structure, and then using structure factors calculated from this model [Eq. (6-2) or (6-3)] to provide phase estimates for the unknown structure.

Molecular replacement is the term most commonly used to describe this process of structure analysis based on the known structure of a similar molecule or molecular fragment (perhaps molecular repositioning would be more apt). The process is illustrated in Figure 6-4. This method requires that the linear transformation that relates the two structures be deter-

MOLECULAR REPLACEMENT

Figure 6-4. Relationship between similar molecules differently crystallized as exploited in molecular replacement. If the structure of crystal A is known, that of crystal B can be determined by repositioning an isolated molecule (indicated with numerical markers) from A into the B unit cell. A linear transformation involving the rigid body rotation, **R**, and translation, **t**, are used in the repositioning.

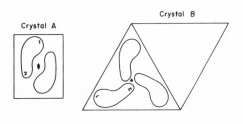

$$\underline{x}_B = \underline{x}'_A = \underline{R}(\theta_1,\theta_2,\theta_3)\cdot\underline{x}_A + \underline{t}(t_x, t_y, t_z)$$

mined. The model or isolated electron density function for the molecule in the known structure, A, must be rotationally oriented and then translated into its place in the unknown crystal structure, B. This transformation

$$\mathbf{x}_B = \mathbf{R}(\theta)\cdot\mathbf{x}_A + \mathbf{t} \qquad (6\text{-}17)$$

can be separated into the rotation matrix, $\tilde{C}$, and the translation vector, **t**.

The rotation function of Rossmann and Blow (1962) can be used to find the orientation of the repositioned molecule. This function compares the integrated point-wise product of the Patterson function of crystal B with that of crystal A after rotation by $\tilde{\mathbf{R}}(\boldsymbol{\theta})$. Lattman and Love (1970) showed that one can conveniently implement molecular replacement by computationally crystallizing the isolated molecule from A in a large asymmetric unit cell, M, and Crowther (1972) developed a fast algorithm for the rotation function. Nevertheless, this remains an uncertain process, particularly when the search molecule is less than the full molecular component of the unknown asymmetric unit.

Having found the orientation, one seeks the translational vector. Several different translational search functions are used. Some, like that of Crowther and Blow (1967), search the Patterson function; others survey for minima in the agreement factor between calculated and observed diffraction data for crystal B. Packing considerations are often helpful (Hendrickson and Ward, 1976).

Phases determined by molecular replacement are of course highly biased toward the model of the structure from crystal A. It is only the observed magnitudes $|F_B|$ that provide new information in an electron density synthesis. Phase information from other sources can help to relieve this bias, but ultimately refinement of the new structure is needed to escape from the input model.

6.4.1.4 Molecular Averaging

Many macromolecules are symmetric aggregates that range from two to sixty or more equivalent subunits. Molecular symmetry is sometimes expressed in the symmetry of the crystal, but oftentimes it is not. A crystal may also have multiple copies in the asymmetric unit even if the molecule is not polymeric. The presence of new crystallographic symmetry creates a larger than minimal structural problem, but the redundancy provides a very powerful source of phase information.

The information content from noncrystallographic symmetry was first developed in reciprocal space (Rossmann and Blow, 1963) and this approach is very instructive. However, practical implementation proves much more effective in real space (Bricogne, 1973). Various procedures have been devised to implement the exploitation of phasing from molecular averaging (Colman, 1974; Argos *et al.*, 1975; Ward *et al.*, 1975; Bricogne, 1976; Nordman, 1980) but the most dramatic successes have come in applications of Bricogne's programs to viral structures, e.g., to influenza virus haemagglutinin (Wilson *et al.*, 1981).

The greatest utility of molecular averaging is in the improvement of initial approximate phases from isomorphous replacement or another source. First, a molecular envelope is determined, and the symmetry operators are defined. Then the initial crude electron density map is averaged and inverted by Fourier transformation [Eq. (6-2)] to produce calculated phases. A probability assessment is made for this information, and the information from molecular averaging is combined with previous information from isomorphous replacement by use of Eq. (6-9). Centroid phases from this joint probability distribution are then used to generate an improved electron density map. Then the process is iterated until convergence. This method has come to be an extraordinarily powerful technique. The result of an application to trimeric hemerythrin (Smith and Hendrickson, 1981) is shown in Figure 6-5.

The molecular averaging procedure described above also imposes a flat solvent continuum in the expanses outside the molecular boundary. Solvent regions in protein crystals are essentially fluid, and little order is detectable beyond the second hydration shell in the best-refined structures. Thus, solvent constancy is a justifiable source of phase information. Procedures to use this information alone for phase improvement have been useful (Hendrickson *et al.*, 1975; Hendrickson, 1981).

The preceding sections describe the four sources of phase information that have played an essential role in the analysis of protein structures determined to date. Without the use of one or a combination of these methods, interpretable electron density maps would not have been produced. Many other phase determination methods have been tried (see Table 6-1), but for the most part these have had marginal utility. One exception

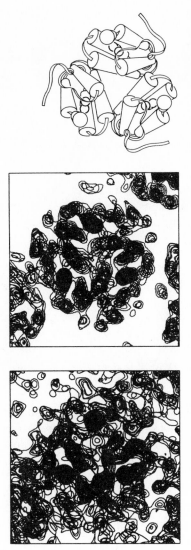

Figure 6-5. Molecular averaging in the structure analysis of trimeric hemerythrin. *Left*: A 12.5-Å thick slice of the original electron density map based on iron-resolved anomalous scattering. The view is along the molecular threefold axis and contains the diiron centers. *Center*: The same density slice after 12 cycles of threefold averaging and solvent leveling to refine the phases. Note the greatly improved clarity of molecular boundaries and definition of density features. *Right*: Schematic drawing of the trimer in the same orientation and scale as in the density slices. Reproduced from Smith and Hendrickson, 1982, with permission from Oxford University Press.

is the use of information from a partial or incomplete model. This crucial kind of information is treated here in the refinement section below. Research in phase determination is very active, and there have been several promising developments recently. Notable among these are efforts to integrate anomalous scattering and isomorphous replacement with direct methods (Hauptman, 1982), several new developments in density modification (Schevitz *et al.,* 1981; Wang, 1981; Bhat and Blow, 1982), and a surge of interest in maximum entropy (Narayan and Nityananda, 1982; Bricogne, 1982b; Collins, 1982).

6.4.2 Map Interpretation and Model Building

Once phases have been determined for the measured set of structure amplitudes from a protein crystal, the electron density distribution can be calculated by Fourier synthesis [Eq. (6-5)]. This resulting three-dimensional function is trivially referred to as an electron density map. The level of detail in an electron density map depends on the extent of data included in the synthesis. Truncation effects obscure the image such that the best point-to-point resolution is $0.715\ d_{min}$ (James, 1948) where d_{min} is the minimal interplanar spacing of the sphere of terms entered into Eq. (6-5). Experimental errors further confuse the image and limit resolution. In practice, protein crystallographers conventionally cite d_{min} as the nominal resolution of a map.

The manner of interpretation of a map and the character of a model depend on the resolution. In the case of small organic molecules where the resolution is typically on the order of 0.8 Å, atoms are well-resolved and can be identified in a list of interpolated peak positions. In contrast, initial maps for proteins may be at 5–6-Å resolution, where adjacent amino acids are not even resolved from one another. Even at 1.5-Å resolution, atoms will not be fully resolved if there is appreciable thermal or static disorder. Consequently, the characteristic features of groups of atoms play a crucial role in the interpretation of electron density maps from proteins.

A map is usually displayed as equi-density contours that are visually examined for recognizable features. Helices appear as dense cylindrical rods at about 6-Å resolution and β-sheets become resolved into individual strands at roughly 4-Å resolution. Thus, much can be learned at rather low resolution; however, it is not until about 3-Å resolution that a detailed tracing of the polypeptide chain can be made. At this level, the polypeptide backbone is clearly isolated but remains continuous, and the distinctive shapes of individual side chains become apparent. The task of interpreting a map in terms of an atomic model is simplified at higher resolution. For example, the carbonyl oxygens are prominent at about 2.5-Å resolution, and this helps greatly in orienting peptide groups. Knowledge of the amino-acid sequence is an invaluable aid in map interpretation.

Various physical presentations are used for contoured maps. Perhaps the most common of these is as a stack of appropriately spaced transparent sheets onto which sections of the map are contoured. Alternatively, portions of a map can be displayed in stereodrawings of a stack of such contours. Sometimes at low resolution a balsa-wood representation is constructed (e.g., of the density within contours above a certain level), but this is usually done only after a molecular boundary has been interpreted. In recent years, the capability to display contoured maps on cathode-ray displays under real-time interactive control by computer graphics has become commonplace. In these cases, a network of contours in three orthogonal directions at a single level is the usual mode of display.

The embodiment of an interpretation of an electron density map is a model. At low resolution this might be an idealized schema of the molecular shape or the location of helical rods. At the chain-tracing level, very often an initial model is a set of approximate positions for the α-carbon atoms. These can often be identified in contour stacks at the side chain branch points, and they usually are evenly spaced since most peptides are transplanar with a C_α—C_α distance of 3.8 Å. Ultimately, a full set of atomic coordinates is needed.

Physical models are constructed in many forms. Earlier, wire models as designed by Kendrew and Watson were fashioned from brass parts connected by set-screw fittings and given a scale of 2 cm/Å. More recently, models based on commercial plastic parts at 1 cm/Å have come into favor. Models of atomic coordinates can be represented numerically in a computer and displayed graphically. The need for detailed physical models has declined as computer graphics systems have become more sophisticated. Moreover, the virtual model in a computer has many advantages for analysis and manipulation.

Methods for constructing atomic models to fit the density in electron density maps of proteins have improved a great deal in recent years. The first such models were built by transferring measurements from a contoured stack to clips on vertical rods in a dense grid. The introduction of Kendrew skeletal models simplified the model-building process, and Richards' optical device (Richards, 1968) superimposed the map images and revolutionized the process. More recently, the half-silvered mirrors of Richards' boxes have been almost universally replaced by computer graphics displays. This has greatly reduced the labor of model building and concomitantly improved the accuracy or results.

The effectiveness of a computer graphics system hinges critically on the software used to drive the displays. Several very good display units are available, and some excellent program systems have now been developed. These include GRIP (Wright, 1982), FRODO (Jones, 1978), and BILDER (Diamond, 1981). Each of these permits the simultaneous display of an electron density map and an atomic model. This image can be rotated under real-time control from a joystick, knobs, or magnetic table, and a fragment of the model can be freed for fitting by independent rotations and translations controlled by other devices. All of these graphics systems include idealization procedures to repair stereochemical irregularities introduced in the building process.

Computer display systems are now commonly used to build protein models into new electron density maps α-carbon backbone models are often used to start such interpretations). However, initial applications were primarily in the refitting of partially refined models into maps based on model phasing. This is still a dominant activity since model revision is an essential ingredient of refinement, as discussed in the next section. Another important aspect of model building concerns the interpretation of difference maps for isomorphous complexes between a native protein and activity effectors, substrates, ligands, cofactors, or other additives. The analysis of such complexes plays an essential role in the study of the chemical activity of proteins. It is also possible to use interactive displays to investigate hypothetical complexes between proteins and their substrates or other macromolecules.

6.4.3 Refinement

The ultimate test of a crystallographic model is its agreement with the observed diffraction pattern. As discussed above, the theory that relates an atomic model to the expected diffraction pattern is well established [Eq. (6-3)]. The most commonly used measure of the agreement of the match between observed, F_{obs}, and calculated, F_{calc}, diffraction data is the ratio R:

$$R = \frac{\sum_{h} \left| |F_{\text{obs}}(\mathbf{h})| - |F_{\text{calc}}(\mathbf{h})| \right|}{\sum_{h} |F_{\text{obs}}(\mathbf{h})|} \tag{6-18}$$

The summation is taken over all data. The R-value expected from a random noncentrosymmetric structure is 0.59 (Wilson, 1950). If a model fits the experimental data to within the accuracy of these observations, then the R-value will be low—perhaps 0.05. On the other hand, a typical starting model has a rather high R-value—perhaps 0.45. It is the goal of a refinement procedure to produce a model that brings the calculated values into optimal agreement with the measurements.

A protein model must meet other criteria in addition to agreement with the diffraction data. The bonding geometry and other stereochemical features of the components from which the protein is built should be in agreement with prior knowledge about these features. In the case of small molecules, the crystallographic data usually suffice to determine atomic parameters within an accuracy that agrees with theoretical and vibrational data. Indeed, these small-molecule structures provide the standard for the geometry of macromolecular components. However, in the case of proteins, the diffraction data alone are insufficient to provide accurate atomic parameters. The intrinsic flexibility and less restrictive lattice contacts in crystalline macromolecules lead to rather large thermal vibrations, dynamic disordering among possible local conformations, and static variations among different unit cells. This causes a rapid fall-off in diffraction intensity with scattering angle and greatly limits the minimal interplanar spacing, d_{min}, from which data are accurately measurable. It is quite usual for large proteins and viruses to have d_{min} of about 3 Å, and protein crystals rarely diffract measurably beyond 1.5 Å. At 3-Å resolution the ratio of diffraction observations to model parameters [four per atom in Eq. (6-3)] is much less than unity.

Various procedures have been used to carry out effective refinement of protein crystal structures in light of the paucity of diffraction data. The first generally useful procedure was a real-space least-squares program by Diamond (1966) that maintained ideal geometry by restricting variable parameters to conformational torsion angles. However, it proves to be essential to permit expression of the real variations in structure if a proper fit is to be reached. Other procedures were devised that alternated free atom movements along gradients in different Fourier syntheses having coefficients of $(|F_{obs}| - |F_{calc}|) \exp(i\phi_{calc})$ with model idealization. However, these programs generally prove to be rather slowly convergent. At present, most protein refinements are carried out by programs that simultaneously minimize the crystallographic residual [Eq. (6-18)] and optimize the stereochemical ideality of potential energy (Konnert, 1976; Hendrickson and Konnert, 1980; Sussman et al., 1977; Jack and Levitt, 1978).

Least-squares programs of this latter kind minimize a composite observational function:

$$\Phi = \phi_{diffraction} + \phi_{bonding} + \phi_{planarity} + \cdots \qquad (6-19)$$

where

$$\phi_{diffraction} = \sum_{reflections} \frac{1}{\sigma_F^2} (|F_{obs}| - |F_{calc}|)^2 \qquad (6-20)$$

and

$$\phi_{\text{bonding}} = \sum_{\text{distances}} \frac{1}{\sigma_D^{\,2}} (d_{\text{ideal}} - d_{\text{model}})^2 \qquad (6\text{-}21)$$

The diffraction observations drive the model to a fit with the F_{obs} data. Simultaneously, the stereochemical observations, such as the ϕ_{bonding} terms, restrain the model to be compatible with prior knowledge regarding the distributions (variance of σ^2) about ideal values for particular features. Restraints related to bonding distances, planarity of groups, chirality at asymmetric centers, nonbonded contacts, and restricted torsion angles may be included. These restraints are directly related to the terms in a typical potential energy description. Variation in thermal parameters and deviations from noncrystallographic symmetry may also be restrained.

Least-squares programs are computationally demanding, and it is important that they be as efficient as possible. Conjugate gradient procedures are helpful in this regard, and Fourier transforms can be used to speed up the most time-consuming steps (Agarwal, 1978; Jack and Levitt, 1978). This is especially important in applications to very large structures.

The structure factor equation [Eq. (6-3)] is highly nonlinear as are the stereochemical factors. This means that the least-squares problem here must be solved iteratively. There is no guarantee that the process will actually converge to the absolute minimum. In fact, it is most likely that the refinement of an initial model will become caught in a local minimum. Manual revision is essential to escape the false minima and to add previously unnoticed elements, such as solvent molecules, to the model. Computer graphics is extremely helpful in this rebuilding process. In the initial stages it may be helpful to combine the phase information from the partially refined model with that from isomorphous replacement or other experiments. Later, maps are usually based exclusively on phases from the current model.

The proceedings of a recent workshop on the refinement of protein structures (Machin *et al.*, 1981) give a good overview of the current status of refinement methods.

6.5 Applications

As explained at the outset, this chapter is intended as an introduction to methods for analyzing crystals of metalloproteins and not as a review of the crystallographic results on metal sites in proteins. Nevertheless, it seems worthwhile to provide here some sense of the character and extent of results obtained by the methods that have been described.

The nature of the resulting information is illustrated by the metal centers from three recent structures: myohemerythrin (Figure 6-6), Cu, Zn superoxide dismutase (Figure 6-7), and satellite tobacco necrosis virus (Figure 6-8). These examples include sites of iron, copper, zinc, and calcium atoms, which are among the most prevalent metal ions in biochemistry. The crystallographic results provide comprehensive and detailed pictures of these centers. Equally definitive models are known for the metal centers in many other proteins.

A reasonably thorough survey of metalloproteins of known structure is given in Table 6-2. The structures are classified by metal center. Although more than seventy distinctive metalloproteins are listed here, this is only a subset of those for which crystal structures have been determined. Each unique chemical sequence is represented only once, but mutants are ignored. Thus, a protein for which structure analyses have been made of several complexes with ligands, substrate analogues, or activators appears as a single entry. For example, human deoxyhemoglobin, carboxyhemoglobin and methemoglobin each form different crystals, and some mutants, such as sickle-cell hemoglobin, form yet other

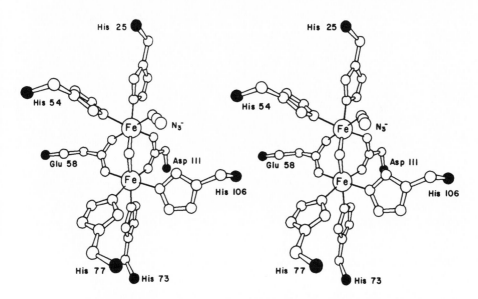

Figure 6-6. The dimeric iron center in azidomet myohemerythrin. This model is based on an incomplete refinement at 1.7-Å resolution that presently has $R = 0.253$. Each Fe is octahedrally coordinated and the two metal atoms are coupled by a μ-oxo bridge. The exogenous ligand (the azide ion, N_3^-, in this case) is ligated at a site on one Fe that also reversibly binds molecular oxygen as the peroxide, O_2^{2-}.

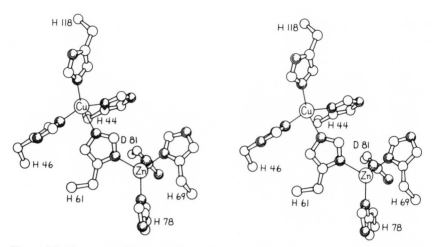

Figure 6-7. The copper-zinc active center in superoxide dismutase. The coordination geometry at the Zn center is tetrahedral with a strong distortion toward a trigonal pyramid with the unique aspartate at the apex. The Zn and Cu atoms are bridged by a bifunctional histidine side chain. The geometry at the Cu center is a tetrahedrally distorted square plane. The axial position toward the viewer is open to the solvent and is thought to be the site of superoxide dismutation. This model is based on a refinement of 2-Å resolution that has R = 0.255 with r.m.s. deviation of 0.03 Å from ideal bond lengths. This figure was provided by Jane Richardson and is reproduced with permission from Tainer *et al.*, 1982.

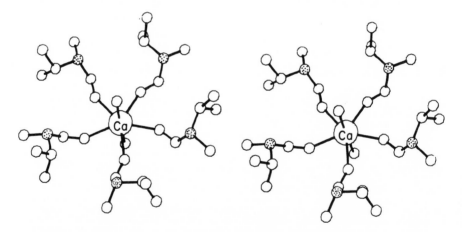

Figure 6-8. A calcium site in the protein coat of a satellite tobacco necrosis virus (STNV). This calcium center lies on an icosohedral fivefold axis of the virus particle and is coordinated equatorially by the carbonyl oxygens of threonines 138 from five symmetry-related subunits and axially by two water molecules. The α-carbon atoms of the coordinating residues ($C_\alpha 138$) are stippled. This is one of three calcium sites in STNV that confer structural stability on the virus. Removal of the calcium causes the particle to swell and is thought to trigger release of the nucleic acid. This figure was provided by Alwyn Jones, Uppsala University, and is based on an unpublished refinement at 2.45-Å resolution.

Table 6-2. Natural Metalloproteins of Known Crystal Structure

Metal	Protein	Source	Molecular mass (kilodaltons)	Sub-units	Molecular symmetry	Unique metal sites	References
	Bacteriochlorophyll	Prosthecochloris aestuarii	127	3	C_3	7	Fenna and Matthews (1975)
	Alkaline phosphatase[a]	Escherichia coli	89	2	C_2	2(?)	Sowadski et al. (1981)
	Staphylococcal nuclease	Staphylococcus aureus	17	1		1	Arnone et al. (1971)
	Phospholipase A₂	Cow pancreas	14	1		1	Dijkstra (1978)
	Phospholipase A₂	Crotalus atrox venum	28	2	C_2	1	Keith et al. (1981)
	Taka amylase	Aspergillus oryzae	50	1		1	Matsuura et al. (1980)
	Parvalbumin	Carp muscle	12	1		2	Kretsinger and Nockolds (1973)
	Ca-binding protein	Cow intestine	8	1		2	Szebenyi et al. (1981)
	Trypsin	Cow pancreas	23	1		1	Bode and Schwager (1976)
	Trypsinogen	Cow pancreas	24	1		2	Kossiakoff et al. (1977)
	Thermolysin[a]	Bacillus thermoproteolyticus	34	1		4	Matthews et al. (1974)
	Concanavalin A[a]	Jack bean	102	4	D_2	1	Hardman et al. (1982); Becker et al. (1975)
	Tomato bushy stunt virus		8,800	180	I	2	Harrison (1980)
	Southern bean mosaic virus		6,600	180	I	2	Abdel-Mequid et al. (1981)
	Satellite tobacco necrosis virus		1,600	60	I	3	Liljas et al. (1982)
	Concanavalin A[a]	Jack bean	102	4	D_2	1	Hardman et al. (1982)
	Myoglobin	Sperm whale	18	1		1	Kendrew et al. (1960)
	Myoglobin	Harbor seal	18	1		1	Scouloudi and Baker (1978)
	Myoglobin	Tuna	17	1		1	Lattman et al. (1971)
	Hemoglobin	Aplysia limacina (mollusc)	16	1		1	Ungaretti et al. (1978)

(continued)

Table 6-2. (continued)

Metal	Protein	Source	Molecular mass (kilodaltons)	Sub-units	Molecular symmetry	Unique metal sites	References
	Hemoglobin	Human	65	$2\alpha, 2\beta$	C_2	2	Perutz et al. (1968)
	Hemoglobin	Horse	65	$2\alpha, 2\beta$	C_2	2	Muirhead and Greer (1970)
	Hemoglobin	Deer	65	$2\alpha, 2\beta$	C_2	2	Schmidt et al. (1977)
	Hemoglobin	Sea lamprey	17	1		1	Hendrickson and Love (1971)
	Hemoglobin	Sea cucumber	17	1		1	Ernst et al. (1982)
	Hemoglobin	Chironomous thummi (insect)	16	1		1	Huber et al. (1969)
	Hemoglobin	Glycera dibranchiata (bloodworm)	16	1		1	Padlan and Love (1974)
	Leghemoglobin	Lupinus luteus (legume)	17	1		1	Vainshtein et al. (1978)
	Leghemoglobin	Soybean	15	1		1	Ollis (1980)
	Myohemerythrin	Themiste zostericola (sipunculan)	14	1		2	Hendrickson et al. (1975)
	Hemerythrin	Themiste dyscrita	108	8	D_4	2	Stenkamp et al. (1981)
	Hemerythrin	Phascolopsis gouldii	108	8	D_4	2	Ward et al. (1975)
	Hemerythrin	Siphonosoma funafuti	42	3	C_3	2	Smith and Hendrickson (1981)
	Cytochrome c	Horse	12	1		1	Dickerson et al. (1971)
	Cytochrome c	Tuna	12	1		1	Takano et al. (1973)
	Cytochrome c	Bonito	12	1		1	Ashida et al. (1973)
	Cytochrome c_2	Rhodospirillum rubrum	13	1		1	Salemme et al. (1973)
	Cytochrome C_{550}	Paracoccus denitrificans	15	1		1	Timkovich and Dickerson (1976)
	Cytochrome C_{551}	Pseudomonas aeruginosa	9	1		1	Almassy and Dickerson (1978)
	Cytochrome C_{555}	Chlorobium thiosulfatophilum	9	1		1	Korszun and Salemme (1977)
	Cytochrome C_{554}	Anacystis nidulans	9	1		1	Ludwig et al. (1982)
	Cytochrome c'	Rhodopseudomonas molischianum	29	2	C_2	1	Weber et al. (1981)
	Cytochrome c_3	Desulfovibrio desulfuricans	13	1		4	Haser et al. (1979)

	Protein	Source					Reference
	Cytochrome c_3	Desulfovibrio vulgaris	13	1		4	Higuchi et al. (1981)
	Cytochrome c_4	Pseudomonas aeruginosa	19	1		2	Sawyer et al. (1981)
	Cytochrome b_5	Calf liver	11	1		1	Mathews et al. (1971)
	Cytochrome b_{562}	Escherichia coli	12	1		1	Mathews et al. (1979)
	Cytochrome c peroxidase	Yeast	34	1		1	Poulos et al. (1980)
	Cytochrome P-450	Pseudomonas putida	40	1		1	Poulos (1982)
	Catalase	Beef liver	232	4	D_2	1	Reid et al. (1981)
	Catalase	Penicillium vitale	280	4	D_2	1	Vainshtein et al. (1981)
	Rubredoxin	Clostridium pasteurianum	6	1		1	Herriott et al. (1970)
	Rubredoxin	Desulfovibrio vulgaris	6	1		1	Adman et al. (1977)
	Ferredoxin	Spirulina platensis	11	1		2	Tsukihara et al. (1978)
	Ferredoxin	Azotobacter vinelandii	13	1		7	Stout et al. (1980)
	Ferredoxin	Peptococcus aerogenes	6	1		8	Adman et al. (1973)
	High-potential iron protein	Chromatium vinosum	9	1		4	Carter et al. (1974)
	Apoferritin	Horse spleen	444	24	O	2(?)	Banyard et al. (1978)
	Superoxide dismutase	Escherichia coli	40	2	C_2	1	Stallings et al. (1982)
94	Superoxide dismutase	Pseudomonas ovalis	40	2	C_2	1	Yamakura et al. (1983)
Cu	Plastocyanin	Poplar leaves	11	1		1	Colman et al. (1978)
	Azurin	Pseudomonas aeruginosa	14	1		1	Adman et al. (1978)
	Azurin	Alcaligenes denitrificans	14	1		1	Baker (1981)
	Superoxide dismutase[a]	Cow erythrocytes	32	2	C_2	1	Tainer et al. (1981)
	Hemocyanin	Spiny lobster	540	6	D_3	2	van Schaik et al. (1982)
101	Hemocyanin	Horseshoe crab	420	6	D_3	2	Magnus and Love (1981)
Zn	Carboxypeptidase A	Cow pancreas	34	1		1	Ludwig et al. (1967)
	Carboxypeptidase B	Cow pancreas	35	1		1	Schmid and Herriott (1976)
	Thermolysin[a]	Bacillus thermoproteolyticus	34	1		1	Matthews et al. (1972)
	DD-carboxypeptidase	Streptomyces albus	19	1		1	Dideberg and Charlier (1981)

(continued)

Table 6-2. (continued)

Metal	Protein	Source	Molecular mass (kilodaltons)	Sub-units	Molecular symmetry	Unique metal sites	References
	Papain	Papaya	23	1		1	Drenth et al. (1968)
	Actinidin	Chinese gooseberry	24	1		1	Baker (1977)
	Carbonic anhydrase C	Human erythrocytes	29	1		1	Liljas et al. (1972)
	Carbonic anhydrase B	Human erythrocytes	29	1		1	Kannan et al. (1975)
	Alkaline phosphatase[a]	Escherichia coli	89	2	C_2	3(?)	Sowadski et al. (1981)
	Alcohol dehydrogenase	Horse liver	80	2	C_2	2	Eklund et al. (1976)
	Superoxide dismutase[a]	Cow erythrocytes	32	2	C_2	1	Tainer et al. (1982)
	Aspartate carbamoyl transferase	Escherichia coli	306	6R, 6C	D_3	1	Honzatko et al. (1982)
	Insulin, 2Zn	Pig pancreas	33	6	C_3	2	Blundell et al. (1971)
	Insulin, 4Zn	Pig pancreas	33	6	C_3	2	Bentley et al. (1976)
	Canavalin	Jack bean	126	6	D_3	2(?)	McPherson (1980)
	Glutathione peroxidase	Cow erythrocytes	84	4	D_2	1	Ladenstein et al. (1979)

[a]These structures contain more than one kind of metal ion and each appears twice in the table.

All of these and many other human hemoglobin structures have been analyzed, but just one representative is cited. Likewise, the complexes of metalloproteins with other macromolecules are omitted. Thus, carboxypeptidase A is included, but its complex with the potato inhibitor is not.

The distinctions made here are somewhat arbitrary, but, for the most part, variations in crystal, quaternary arrangement, or conformational state have rather small effect on the metal centers. On the other hand, for several proteins, it is just these subtle changes at the metal center that cause the variations and control the chemical activity. Thus, each structure in a related series may be highly relevant. Table 6-2 can only provide an entry into the appropriate literature on the structure and function of these metalloproteins.

The sample of known structures is not fully representative of metalloproteins. Several important kinds of metal centers have not yet been characterized structurally. Some of these are presently under study, and results are awaited with great interest. For example, the structures of nitrogenase (Yamane *et al.*, 1982) and sulfite oxidase (Richardson, 1982) will reveal molybdenum centers, D-xylose isomerase (Carrell, 1982) is a cobalt protein that is being analyzed, and the manganese superoxide dismutase structure (Stallings *et al.*, 1981) will add to the sparse data on manganese centers. No nickel protein has been suitably crystallized. Many other metalloprotein structures are under investigation, and several of these will also explore completely new territory. The photosynthetic reaction center (Michel, 1982) is a notable example. In short, the crystallographic analysis of metalloproteins promises to continue to be fruitful and exciting.

References

Abdel-Mequid, S. S., Yamane, T., Fukuyama, K., and Rossmann, M. G., 1981. The location of calcium ions in Southern bean mosaic virus. *Virology* 114:81.

Adman, E. T., Sieker, L. C., and Jensen, L. H., 1973. The structure of a bacterial ferredoxin. *J. Biol. Chem.* 248:3987.

Adman, E. T., Sieker, L. C., Jensen, L. H., Bruschi, M., and LeGall, J., 1977. A structural model of rubredoxin from *Desulfovibrio vulgaris* at 2 Å resolution. *J. Mol. Biol.* 112:113.

Adman, E. T., Stenkamp, R. E., Sieker, L. C., and Jensen, L. H., 1978. A crystallographic model for azurin at 3 Å resolution. *J. Mol. Biol.* 123:35.

Agarwal, R. C., 1978. A new least-squares refinement technique based on the fast Fourier transform algorithm. *Acta Cryst.* A34:791.

Almassy, R. J., and Dickerson, R. E., 1978. *Pseudomonas* cytochrome c_{551} at 2.0 Å resolution: Enlargement of the cytochrome c family. *Proc. Natl. Acad. Sci. U.S.A.* 75:2674.

Argos, P., Ford, G. C., and Rossmann, M. G., 1975. An application of the molecular replacement technique in direct space to a known protein structure. *Acta Cryst.* A31:499.

Arndt, U. W., 1977. Television area detectors. In *The Rotation Method in Crystallography* (U. W. Arndt and A. J. Wonacott, eds.), North-Holland, Amsterdam, pp. 245–262.

Arndt, U. W., and Willis, B. T. M., 1966. *Single Crystal Diffractometry*, Cambridge University Press, Cambridge.

Arndt, U. W., and Wonacott, A. J., 1977. *The Rotation Method in Crystallography*, North-Holland, Amsterdam.

Arnone, A., Bier, C. J., Cotton, F. A., Day, V. W., Hazen, E. E., Richardson, D. C., Richardson, J. S., and Yonath, A., 1971. A high resolution structure of an inhibitor complex of the extracellular nuclease of *Staphylococcus aureus*. *J. Biol. Chem.* 246:2302.

Ashida, T., Tanaka, N., Yamane, T., Tsukihara, T., and Kakudo, M., 1973. The crystal structure of Bonito (Katsuo) ferrocytochrome c at 2.3 Å resolution. *J. Biochem. (Tokyo)* 73:463.

Baker, E. N., 1977. Structure of actinidin: Details of the polypeptide chain conformation and active site from an electron density map at 2.8 Å resolution. *J. Mol. Biol.* 115:263.

Baker, E. N., 1981. Personal communication.

Banyard, S. H., Stammers, D. K., and Harrison, P. M., 1978. Electron density map of apoferritin at 2.8 Å resolution. *Nature* 271:282.

Becker, J. W., Reeke, G. N., Jr., Wang, J. L., Cunningham, B. A., and Edelman, G. M., 1975. The covalent and three-dimensional structure of concanavalin A. *J. Biol. Chem.* 250:1513.

Bentley, G., Dodson, E., Dodson, G., Hodgkin, D., and Mercola, D., 1976. Structure of insulin in 4-zinc insulin. *Nature* 261:166.

Bernstein, F. C., Koetzle, T. F., Williams, G. J. B., Meyer, E. F., Jr., Brice, M. D., Rodgers, J. R., Kennard, O., Shimanouchi, T., and Tasumi, M., 1977. The protein data bank: A computer-based archival file for macromolecular structures. *J. Mol. Biol.* 112:535.

Bhat, T. N. and Blow, D. M., 1982. A density-modification method for the improvement of poorly resolved protein electron-density maps. *Acta Cryst.* A38:21.

Blow, D. M., and Crick, F. H. C., 1959. The treatment of errors in the isomorphous replacement method. *Acta Cryst.* 12:794.

Blundell, T. L., and Jenkins, J., 1977. Binding of heavy metals to proteins. *Chem. Soc. Rev.* 6:139.

Blundell, T. L., and Johnson, L. N., 1976. *Protein Crystallography,* Academic Press, London.

Blundell, T. L., Cutfield, J. F., Cutfield, S. M., Dodson, E. J., Dodson, G. G., Hodgkin, D. C., Mercola, D. A., and Vijayan, M., 1971. Atomic positions in rhombohedral 2-zinc insulin crystals. *Nature* 231:506.

Bode, W., and Schwager, P., 1976. The refined crystal structure of bovine β-trypsin at 1.8 Å resolution. *J. Mol. Biol.* 98:693.

Bricogne, G., 1974. Geometric sources of redundancy in intensity data and their use for phase determination. *Acta Cryst.* A30:395.

Bricogne, G., 1976. Methods and programs for direct-space exploitation of geometric redundancies. *Acta Cryst.* A32:832.

Bricogne, G., 1982a. Multiple isomorphous relacement. In *Computational Crystallography* (D. Sayre, ed.), Oxford University Press, Oxford, pp. 223–230.

Bricogne, G., 1982b. Generalized density-modification methods. In *Computational Crystallography* (D. Sayre, ed.), Oxford University Press, Oxford, pp. 258–264.

Buerger, M. J., 1963. *Elementary Crystallography,* Wiley, New York.

Buerger, M. J., 1964. *The Precession Method,* Wiley, New York.

Cantor, C. R., and Schimmel, P. R., 1980. *Biophysical Chemistry,* Part II, Freeman, San Francisco, Chapter 13.

Carrell, H. L., 1982. Structural studies of D-xylose isomerase. Abstracts of the American Crystallographic Association, La Jolla Meeting, Ser. 2, Vol. 10, No. 2, p. 35.

Carter, C. W., Jr., Kraut, J., Freer, S. T., Xuong, Ng.-H., Alden, R. A., and Bartsch, R. G., 1974. Two-angstrom crystal structure of oxidized *Chromatium* high-potential iron protein. *J. Biol. Chem.* 249:4212.

Collins, D. M., 1982. Electron density images from imperfect data by iterative entropy maximization. *Nature* 298:49.

Colman, P. M., 1974. Noncrystallographic symmetry and the sampling theorem. *Z. Kristollogr.* 140:344.

Colman, P. M., Freeman, H. C., Guss, J. M., Murata, M., Norris, V. A., Ramshaw, J. A. M., and Venkatappa, M. P., 1978. X-ray crystal structure analysis of plastocyanin at 2.7 Å resolution. *Nature* 272:319.

Cotton, F. A., and Wilkinson, G., 1972. *Advanced Inorganic Chemistry,* 3rd ed., Interscience, New York.

Cramer, S. P., and Hodgson, K. O., 1978. X-ray absorption spectroscopy: A new structural method and its applications to bioinorganic chemistry. *Progr. Inorg. Chem.* 25:1.

Crowther, R., 1972. The fast rotation function. In *The Molecular Replacement Method* (M. G. Rossmann, ed.), Gordon and Breach, New York, pp. 174–177.

Crowther, R. A., and Blow, B. D. M., 1967. A method of positioning a known molecule in an unknown crystal structure. *Acta Cryst.* 23:544.

Davies, D. R., and Segal, D. M., 1971. Protein crystallization: Micro techniques involving vapor diffusion. *Methods Enzymol.* 22:266.

Diamond, R., 1966. A mathematical model-building procedure for proteins. *Acta Cryst.* 21:253.

Diamond, R., 1981. Bilder: A computer graphics program for biopolymers and its application to the interpretation of the structure of tobacco mosaic virus discs at 2.8 Å resolution. In *Biomolecular Structure, Conformation, Function and Evolution,* Vol. 1 (R. Srinivasan, ed.), Pergamon, Oxford, pp. 567–588.

Dickerson, R. E., Kendrew, J. C., and Strandberg, B. E., 1961. The crystal structure of myoglobin: Phase determination to a resolution of 2 Å by the method of isomorphous replacement. *Acta Cryst.* 14:1188.

Dickerson, R. E., Takano, T., Eisenberg, D., Kallai, O. B., Samson, L., Cooper, A., and Margoliash, E., 1971. Ferricytochrome c—I. General features of the horse and bonito proteins at 2.8 Å resolution. *J. Biol. Chem.* 246:1511.

Dideberg, O., and Charlier, P., 1981. Crystal structure determination of a DD carboxypeptidase at 2.5 Å resolution. *Acta Cryst.* A37:C-30.

Dijkstra, B. W., Drenth, J., Kalk, K. H., and Vandermaelen, P. J., 1978. Three-dimensional structure and disulfide bond connections in bovine pancreatic phospholipase A_2. *J. Mol. Biol.* 124:53.

Drenth, J., Jansonius, J. N., Koekoek, R., Swen, H. M., and Wolthers, B. G., 1968. Structure of papain. *Nature* 218:929.

Eklund, H., Nordstrom, B., Zeppezauer, E., Soderlund, G., Ohlsson, I., Boiwe, T., Soderbert, B.-O., Tapia, O., Branden, C.-I., and Akeson, A., 1976. Three-dimensional structure of horse liver alcohol dehydrogenase at 2.4 Å resolution. *J. Mol. Biol.* 102:27.

Ernst, S. R., Franke, R. R., Kitto, G. B., Pattridge, K., and Hackert, M. L., 1982. Structure of hemoglobin from the sea cucumber *Molpadia arenicola*: Rotation and translation function tribulations. Abstracts of the American Crystallographic Association, La Jolla Meeting, Ser. 2, Vol. 10, No. 2, p. 37.

Fenna, R. L., and Matthews, B. W., 1975. Chlorophyll arrangement in a bacteriochlorophyll protein from *Chlorobium limicola*. *Nature* 258:573.

Garavito, M., and Rosenbusch, J. P., 1980. Three-dimensional crystals of an integral membrane protein: An initial X-ray analysis. *J. Cell. Biol.* 86:327.

Guinier, A., 1963. *X-ray Diffraction in Crystals, Imperfect Crystals and Amorphous Bodies*, Freeman, San Francisco.

Hanson, J. C., Watenpaugh, K. D., Sieker, L., and Jensen, L. H., 1979. A limited-range step-scan method for collecting X-ray diffraction data. *Acta Cryst.* A35:616.

Hardman, K. D., Agarwal, R. C., and Freiser, M. J., 1982. Manganese and calcium binding sites of concanavalin A. *J. Mol. Biol.* 157:69.

Harrison, S. C., 1980. Protein interfaces and intersubunit bonding—The case of tomato bushy stunt virus. *Biophys. J.* 32:139.

Hamilton, W. C., Rollett, J. S., and Sparks, R. A., 1965. On the relative scaling of x-ray photographs. *Acta Cryst.* 18:129.

Haser, R., Pierrot, M., Frey, M., Payan, F., Astier, J. P., Bruschi, M., and LeGall, J., 1979. Structure and sequence of the multihaem cytochrome C_3. *Nature* 282:806.

Hauptman, H., 1982. On integrating the techniques of direct methods and isomorphous replacement—I. The theoretical basis. *Acta Cryst.* A38:289.

Hendrickson, W. A., 1971. A procedure for representing arbitrary phase probability distributions in a simplified form. *Acta Cryst.* B27:1472.

Hendrickson, W. A., 1976. Radiation damage in protein crystallography. *J. Mol. Biol.* 106:889.

Hendrickson, W. A., 1979. Phase information from anomalous scattering measurements. *Acta Cryst.* A35:245.

Hendrickson, W. A., 1981. Phase evaluation in macromolecular crystallography. In *Structural Aspects of Biomolecules* (R. Srinivasan and V. Pattabhi, eds.), Macmillan India, New Delhi, pp. 31–80.

Hendrickson, W. A., and Konnert, J. H., 1980. Incorporation of stereochemical information into crystallographic refinement. In *Computing in Crystallography* (R. Diamond, S. Ramaseshan and K. Venkatesan, eds.), Indian Institute of Science, Bangalore, pp. 13.01–13.23.

Hendrickson, W. A., and Lattman, E. E., 1970. Representation of phase probability distributions for simplified combination of independent phase information. *Acta Cryst.* B26:136.

Hendrickson, W. A., and Love, W. E., 1971. Structure of lamprey haemoglobin. *Nature New Biol.* 232:197.

Hendrickson, W. A., and Teeter, M. M., 1981. Structure of the hydrophobic protein crambin determined directly from the anomalous scattering of sulphur. *Nature* 290:107.

Hendrickson, W. A., and Ward, K. B., 1976. A packing function for delimiting the allowable locations of crystallized macromolecules. *Acta Cryst.* A32:778.

Hendrickson, W. A., Klippenstein, G. L., and Ward, K. B., 1975. Tertiary structure of myohemerythrin at low resolution. *Proc. Natl. Acad. Sci. U.S.A.* 72:2160.

Henry, N. F. M., and Lonsdale, K., 1952. *International Tables for X-ray Crystallography*, Volume 1, Kynoch Press, Birmingham.

Herriott, J. R., Sicker, L. C., Jensen, L. H., and Lovenberg, W., 1970. Structure of rubredoxin: An X-ray study to 2.5 Å resolution. *J. Mol. Biol.* 50:391.

Higuchi, Y., Bando, S., Kusunoki, M., Matsuura, Y., Yasuoka, N., Kakudo, M., Yamanaka, T., Yogi, T., and Iinokuchi, N., 1981. The structure of cytochrome C3 from *Desulfovibrio vulgaris*, Miyazaki at 2.5 Å resolution. *Acta Cryst.* A37:C-29.

Honzatko, R. B., Crawford, J. L., Monaco, H. L., Ladner, J. E., Edwards, B. F. P., Evans,

D. R., Warren, S. G., Wiley, D. C., Ladner, R. C., and Lipscomb, W. N., 1982. Crystal and molecular structures of native and CTP-liganded aspartate carbamoyltransferase from *Escherichia coli. J. Mol. Biol.* 160:219.

Huber, R., and Kopfman, G., 1969. Experimental absorption correction: Results. *Acta Cryst.* A25:143.

Huber, R., Epp, O., and Formanek, H., 1969. Ausklärung der molekülstruktur des insektenhämoglobins. *Naturwissenschaften* 56:362.

Jack, A., and Levitt, M., 1978. Refinement of large structures by simultaneous minimization of energy and R factor. *Acta Cryst.* A34:931.

James, R. W., 1948. False detail in three-dimensional Fourier representations of crystal structures. *Acta Cryst.* 1:132.

James, R. W., 1962. *The Optical Principles of the Diffraction of X-rays,* Bell, London.

Jones, T. A., 1978. A graphics model building and refinement system for macromolecules. *J. Appl. Cryst.* 11:268.

Kam, Z., Shore, H. B., and Feher, G., 1978. On the crystallization of proteins. *J. Mol. Biol.* 123:539.

Kannan, K. K., Nostrand, B., Fridborg, K., Lovgren, S., Ohlsson, A., and Petef, M., 1975. Crystal structure of human erythrocyte carbonic anhydrase B. Three-dimensional structure at a nominal 2.2 Å resolution. *Proc. Natl. Acad. Sci. U.S.A.* 72:51.

Karle, J., 1980. Some developments in anomalous dispersion for the structural investigation of macromolecular systems in biology. *Int. J. Quantum Chem.* 7:357.

Keith, C., Feldman, D. S., Deganello, S., Glick, J., Ward, K. B., Jones, E. O., and Sigler, P. B., 1981. The 2.5 Å crystal structure of a dimeric phospholipase A$_2$ from the venom of *Crotalus atrox. J. Biol. Chem.* 256:8602.

Kendrew, J. C., Dickerson, R. E., Strandberg, B. E., Hart, R. G., Davies, D. R., Phillips, D. C., and Shore, V. C., 1960. Structure of Myoglobin—A three-dimensional Fourier synthesis at 2 Å resolution. *Nature* 185:442.

Kirz, J., and Sayre, D., 1980. Soft X-ray microscopy of biological specimens. In *Synchrotron Radiation Research* (H. Winick and S. Doniach, eds.), Plenum Press, New York, pp. 277–322.

Konnert, J. H., 1976. A restrained-parameter structure-factor least-squares refinement procedure for large asymmetric units. *Acta Cryst.* A32:614.

Korszun, Z. R., and Salemme, F. R., 1977. Structure of cytochrome c$_{555}$ of *Chlorobium thiosulfatophilum:* Primitive low-potential cytochrome c. *Proc. Natl. Acad. Sci. U.S.A.* 74:5244.

Kossiakoff, A. A., Chambers, J. L., Kay, L. M., and Stroud, R. M., 1977. Structure of bovine trypsinogen at 1.9 Å resolution. *Biochemistry* 16:654.

Kretsinger, R. H., and Nockolds, C. E., 1973. Carp muscle calcium-binding protein. *J. Biol. Chem.* 248:3313.

Ladenstein, R., Epp, O., Bartels, K., Jones, A., Huber, R., and Wendell, A., 1979. Structure analysis and molecular model of the solenoenzyme glutathione peroxidase at 2.8 Å resolution. *J. Mol. Biol.* 134:199.

Lattman, E. E., and Love, W. E., 1970. A rotational search procedure for detecting a known molecule in a crystal. *Acta Cryst.* B26:1854.

Lattman, E. E., Nockold, C. E., Krestsinger, R. H., and Love, W. E., 1971. Structure of yellow fin tuna metmyoglobin at 6 Å resolution. *J. Mol. Biol.* 60:271.

Liljas, A., Kannan, K. K., Bergsten, P.-C., Waara, I., Fridborg, K., Strandberg, B., Carlbom, U., Jarup, L., Lovgren, S., and Petef, M., 1972. Crystal structure of human carbonic anhydrase-c. *Nature New Biol.* 235:131.

Liljas, L., Unge, T., Jones, T. A., Fridborg, K., Lovgren, S., Skoglund, U., and Strandberg,

B., 1982. Structure of satellite tobacco necrosis virus at 3.0 Å resolution. *J. Mol. Biol.* 159:93.

Lipscomb, W. N., 1980. Metal ion environments: The X-ray diffraction method and results. In *Advances in Inorganic Biochemistry,* Vol. 2, *Structure and Function of Metalloproteins* (D. W. Darnall and R. G. Wilkins, eds.), Elsevier North-Holland, New York, pp. 265–302.

Ludwig, M. L., Hartsuck, J. A., Steitz, T. A., Muirhead, H., Coppola, J. C., Reeke, G. N., and Lipscomb, W. N., 1967. The structure of carboxypeptidase A, IV. Preliminary results at 2.8 Å resolution, and a substrate complex at 6 Å resolution. *Proc. Natl. Acad. Sci. U.S.A.* 57:511.

Ludwig, M. L., Pattridge, K. A., Powers, T. B., Dickerson, R. E., and Takano, T., 1982. Structure analysis of a ferricytochrome c from the cyanobacterium, *Anacystis nidutans. Electron Transport and Oxygen Utilization* (Chien Ho, ed.), Elsevier North-Holland, New York, pp. 27–32.

Machin, P. A., Campbell, J. W., and Elder, M., 1981. *Refinement of Protein Structures,* Science and Engineering Research Council, Daresbury Laboratory, Warrenton, England.

McPherson, A., Jr., 1976. The growth and preliminary investigation of protein and nucleic acid crystals for X-ray diffraction analysis. *Methods Biochem. Anal.* 23:249.

McPherson, A., 1980. The three-dimensional structure of canavalin at 3.0 Å resolution by X-ray diffraction analysis. *J. Biol. Chem.* 255:10472.

McPherson, A., 1982. *Preparation and Analysis of Protein Crystals.* Wiley, New York.

Magnus, K. A., and Love, W. E., 1981. Crystals of a 70,000 dalton subunit of *Limulus polyphemus* hemocyanin. In *Invertebrate Oxygen-Binding Proteins* (J. Lamy and J. Lamy, eds.), Marcel Dekker, New York, pp. 363–368.

Mathews, F. S., Bethge, P. H., and Czerwinski, E. W., 1979. The structure of cytochrome b_{562} from *Escherichia coli* at 2.5 Å resolution. *J. Biol. Chem.* 254:1699.

Mathews, F. S., Levine, M., and Argos, P., 1971. The structure of calf liver cytochrome b_5 at 2.8 Å resolution. *Nature New Biol.* 233:15.

Matsuura, Y., Kusunoki, M., Harada, W., Tanaka, N., Iga, Y., Yasuoka, N., Toda, H., Narita, K., and Kakudo, M., 1980. Molecular structure of Taka-amylase A.—I. Backbone chain folding at 3 Å resolution. *J. Biochem.* 87:1555.

Matthews, B. W., 1966. The extension of the isomorphous replacement method to include anomalous scattering measurements. *Acta Cryst.* 20:82.

Matthews, B. W., 1977. X-ray structure of proteins. In *The Proteins,* 3rd ed., Vol. 3 (H. Neurath and R. L. Hill, eds.), Academic Press, New York, pp. 403–590.

Matthews, B. W., Jansonius, J. N., Colman, P. M., Schoenborn, B. P., and Dupourque, D., 1972. Three-dimensional structure of thermolysin. *Nature New Biol.* 238:37.

Matthews, B. W., Weaver, L. H., and Kester, W. R., 1974. The conformation of thermolysin. *J. Biol. Chem.* 249:8030.

Michel, H., 1982. Three-dimensional crystals of a membrane protein complex—the photosynthetic reaction center from *Rhodopseudomonas viridis. J. Mol. Biol.* 158:567.

Muirhead, H., and Greer, J., 1970. Three-dimensional Fourier synthesis of human deoxyhaemoglobin at 3.5 Å resolution. *Nature* 228:516.

Narayan, R., and Nityananda, R., 1982. The maximum determinant method and the maximum entropy method. *Acta Cryst.* A38:122.

Nordman, C. E., 1980. Procedures for the detection and idealization of non-crystallographic symmetry with application to phase refinement of the satellite tobacco necrosis virus structure. *Acta Cryst.* A36:747.

North, A. C. T., 1965. The combination of isomorphous replacement and anomalous scattering data in phase determination of non-centrosymmetric reflexions. *Acta Cryst.* 18:212.

North, A. C. T., Phillips, D. C., and Mathews, F. S., 1968. A semi-empirical method of absorption correction. *Acta Cryst.* A24:351.

Ollis, D. L., 1980. Dissertation, Department of Inorganic Chemistry, University of Sydney, Australia.

Padlan, E. A., and Love, W. E., 1974. Three-dimensional structure of hemoglobin from the polychaete annelid, *Glycera dibranchiata*, at 2.5 Å resolution. *J. Biol. Chem.* 249:4067.

Pauling, L., 1960. *The Nature of the Chemical Bond*, 3rd ed., Cornell University Press, Ithaca, New York.

Perutz, M. F., Muirhead, H., Cox, J. M., and Goaman, L. C. G., 1968. Three-dimensional Fourier synthesis of horse oxyhaemoglobin at 2.8 Å resolution: The atomic model. *Nature* 219:131.

Phillips, D. C., 1966. Advances in protein crystallography. *Adv. Res. Diff. Meth.* 2:75.

Phillips, J. C., and Hodgson, K. O., 1980. The use of anomalous scattering effects to phase diffraction patterns from macromolecules. *Acta Cryst.* A36:856.

Poulos, T. L., 1982. A 4 Å electron density map of cytochrome P450. Abstracts of the American Crystallographic Association, La Jolla Meeting, Ser. 2, Vol. 10, No. 2, p. 37.

Poulos, T. L., Freer, S. T., Alden, R. A., Edwards, S. L., Skogland, J., Takio, K., Eriksson, B., Xuong, Ng. h., Yonetani, T., and Kraut, J., 1980. The crystal structure of cytochrome c peroxidase. *J. Biol. Chem.* 255:575.

Reid, T. J., III, Murthy, M. R. N., Sicignano, A., Tanaka, N., Musick, W. D. L., and Rossmann, M. G., 1981. Structure and heme environment of beef liver catalase at 2.5 Å resolution. *Proc. Natl. Acad. Sci. U.S.A.* 78:4767.

Richards, F. M., 1968. The matching of physical models to three-dimensional electron-density maps: A simple optical device. *J. Mol. Biol.* 37:225.

Richardson, D. C., 1982. Personal communication.

Richardson, J. S., 1981. The anatomy and taxonomy of protein structure. *Adv. Prot. Chem.* 34:167.

Ringe Ponzi, D., Yamakura, F., Suzuki, K., Petsko, G. A., and Ohmori, D., 1983. Structure of iron superoxide-dismutase from *Pseudomonas ovalis* at 2.9 Å resolution. *Proc. Natl. Acad. Sci. U.S.A.* 80:3879.

Rossmann, M. G., 1960. An accurate determination of the position and shape of heavy-atom replacement groups in proteins. *Acta Cryst.* 13:221.

Rossmann, M. G., 1961. The position of anomalous scatterers in protein crystals. *Acta Cryst.* 14:383.

Rossmann, M. G., and Blow, D. M., 1962. The detection of subunits within the crystallographic asymmetric unit. *Acta Cryst.* 15:24.

Rossmann, M. G., and Blow, D. M., 1963. Determination of phases by the conditions of non-crystallographic symmetry. *Acta Cryst.* 16:39.

Salemme, F. R., Kraut, J., and Kamen, M. D., 1973. Structural bases for function in cytochromes c—An interpretation of comparative X-ray and biochemical data. *J. Biol. Chem.* 248:7701.

Sawyer, L., Jones, C. L., Damas, A. M., Harding, M. M., Gould, R. O., and Ambler, R. P., 1981. Cytochrome c$_4$ from *Pseudomonas aeruginosa. J. Mol. Biol.* 155:831.

Schevitz, R. W., Podjarny, A. D., Zwick, M., Hughes, J. J., and Sigler, P. B., 1981. Improving and extending the phases of medium- and low-resolution macromolecular structure factors by density modification. *Acta Cryst.* A37:669.

Schmid, M. F., and Herriott, J. R., 1976. Structure of carboxypeptidase B at 2.8 Å resolution. *J. Mol. Biol.* 103:175.

Schmidt, W. C., Jr., Girling, R. L., Houston, T. E., Sproul, G. D., Amma, E. L., and Huisman, T. H. J., 1977. The structure of sickling deer Type III hemoglobin by molecular replacement. *Acta Cryst.* B33:335.

Scouloudi, H., and Baker, E. N., 1978. X-ray crystallographic studies of seal myoglobin. *J. Mol. Biol.* 126:637.

Smith, J. L., and Hendrickson, W. A., 1981. Iron-resolved anomalous phasing and local symmetry averaging in the structure solution of trimeric hemerythrin. *Acta Cryst.* A37:C-11.

Smith, J. L., and Hendrickson, W. A., 1982. Resolved anomalous phase determination in macromolecular crystallography. In *Computational Crystallography* (D. Sayre, ed.), Oxford University Press, Oxford, pp. 209–222.

Sowadski, J. M., Foster, B. A., and Wyckoff, H. W., 1981. Structure of alkaline phosphatase with zinc/magnesium cobalt or cadmium in the functional metal sites. *J. Mol. Biol.* 150:245.

Sparks, R. A., 1982. Data collection with diffractometers. In *Comutational Crystallography* (D. Sayre, ed.), Oxford University Press, Oxford, pp. 1–18.

Stallings, W. C., Pattridge, K. A., Powers, T. B., Fee, J. A., and Ludwig, M. L., 1981. Characterization of crystals of tetrameric manganese superoxide dismutase from *Thermus thermophilus* HB. *J. Biol. Chem.* 256:5857.

Stallings, W. C., Powers, T. B., Pattridge, K. A., Fee, J. A., and Ludwig, M. L., 1983. Superoxide dismutase from *Escherichia coli* at 3.1 Å resolution—A structure unlike that of copper-zinc protein at both monomer and dimer levels. *Proc. Natl. Acad. Sci. U.S.A.* 80:3884.

Stanford, R. H., Jr., and Corey, R. B., 1968. Determination of the structure of proteins by X-ray diffraction: Possible use of large heavy ions in phase determination. In *Structural Chemistry and Molecular Biology* (A. Rich and N. Davidson, eds.), Freeman, San Francisco, pp. 47–54.

Stenkamp, R. E., Sieker, L. C., Jensen, L. H., and Sanders-Loehr, J., 1981. Structure of the binuclear iron complex in metazidohaemerythrin from *Themiste Dyscritum* at 2.2 Å resolution. *Nature* 291:263.

Stout, C. D., Ghosh, D., Pattabhi, V., and Robbins, A. H., 1980. Iron-sulfur clusters in *Azotobacter* ferredoxin at 2.5 Å resolution. *J. Biol. Chem.* 255:1797.

Strothkamp, K. G., Lehmann, J., and Lippard, S. J., 1978. Tetrakis (acetoxymercuri) methane: A polymetallic reagent for labelling sulfur in nucleic acids. *Proc. Natl. Acad. Sci. U.S.A.* 75:1181.

Stuhrmann, H. B., and Notbohm, H., 1981. Configuration of the four iron atoms in dissolved human hemoglobin as studied by anomalous dispersion. *Proc. Natl. Acad. Sci. U.S.A.* 78:6216.

Sussman, J. L., Holbrook, S. R., Church, G. M., and Kim, S.-H., 1977. A structure-factor least-squares refinement procedure for macromolecular structures using constrained *and* restrained parameters. *Acta Cryst.* A33:800.

Sygusch, J., 1977. Minimum-variance Fourier coefficients from the isomorphous replacement method by least-squares analysis. *Acta Cryst.* A33:512.

Szebenyi, D. M. E., Obendorf, S. K., and Moffat, K., 1981. Structure of vitamin D-dependent calcium-binding protein from bovine intestine. *Nature* 294:327.

Tanier, J. A., Getzoff, E. D., Beem, K. M., Richardson, J. S., and Richardson, D. C., 1982. Determination and analysis of the 2 Å structure of copper, zinc superoxide dismutase. *J. Mol. Biol.* 160:181.

Takano, T., Kallai, O. B., Swanson, R., and Dickerson, R. E., 1973. The structure of ferrocytochrome c at 2.45 Å resolution. *J. Biol. Chem.* 248:5234.

Templeton, L. K., Templeton, D. H., Phizackerley, R. P., and Hodgson, K. O., 1982. L_3-edge anomalous scattering by gadolinium and samarium measured at high resolution with synchrotron radiation. *Acta Cryst.* A38:74.

Teo, B. K., 1980. Chemical applications of extended X-ray absorption fine structure (EXAFS) spectroscopy. *Acc. Chem. Res.* 13:412.

Timkovich, R., and Dickerson, R. E., 1976. The structure of *Paracoccus denitrificans* cytochrome c_{550}. *J. Biol. Chem.* 251:4033.

Tsukihara, T., Fukuyama, K., Tahara, H., Katsube, Y., Matsuura, Y., Tanaka, N., Kakudo, M., Wada, K., and Matsubara, H., 1978. X-ray analysis of ferredoxin from *Spirulina platensis*. *J. Biochem.* 84:1645.

Ungaretti, L., Bolognesi, M., Cannillo, E., Oberti, R., and Rossi, G., 1978. The crystal structure of met-myoglobin from *Aplysia limacina* at 5 Å resolution. *Acta Cryst.* B34:3658.

Vainshtein, B., Harutyunyan, E., Kuranova, I. P., Borisov, V. V., Sosfenov, N. I., Pavlovsky, A. G., Grebenko, A. I., and Nebrasov, Y. B., 1978. X-ray structural investigation of leg-hemoglobin. 4. Determination of structure at 2.8 Å resolution. *Kristallografiya* 23:517.

Vainshtein, B. K., Melik-Adamyan, W. R., Barynin, V. V., Vagin, A. A., and Grebenko, A. I., 1981. Three-dimensional structure of the enzyme catalase. *Nature* 293:411.

van Schaick, E. J. M., Schutter, W. G., Gaykema, W. P. J., Schepman, A. M. H., and Hol. W. G. J., 1982. Structure of *Panulirus interruptus* hemocyanin at 5 Å resolution. *J. Mol. Biol.* 158:457.

Wang, B.-C., 1981. Protein structure determination by the single isomorphous replacement method with a phase selection and refinement process. *Acta Cryst.* A37:C-11.

Ward, K. B., Hendrickson, W. A., and Klippenstein, G. L., 1975. Quaternary and tertiary structure of haemerythrin. *Nature* 257:818.

Weber, P. C., Howard, A., Xuong, Ng. h., and Salemme, F. R., 1981. Crystallographic structure of *Rhodospirillum molischianum* ferricytochrome c' at 2.5 Å resolution. *J. Mol. Biol.* 153:399.

Wilson, A. J. C., 1942. Determination of absolute from relative X-ray intensity data. *Nature* 150:152.

Wilson, A. J. C., 1950. Largest likely values for the reliability index. *Acta Cryst.* 3:397.

Wilson, I. A., Skehel, J. J., and Wiley, D. C., 1981. Structure of the haemagglutinin membrane glycoprotein of influenza virus at 3 Å resolution. *Nature* 289:366.

Wright, W. V., 1982. Grip—An Interactive Computer Graphics System for Molecular Studies. In *Computational Crystallography* (D. Sayre, ed.), Oxford University Press, Oxford, pp. 294–302.

Wyckoff, H. W., Doscher, M., Tsernoglou, D., Inagami, T., Johnson, L. N., Hardman, K. D., Allewell, N. M., Kelly, D. M., and Richards, F. M., 1967. Design of a diffractometer and flow cell system for X-ray analysis of crystalline proteins with applications to the crystal chemistry of ribonuclease-S. *J. Mol. Biol.* 27:563.

Xuong, Ng.-h., Freer, S. T., Hamlin, R., Nielsen, C., and Vernon, W., 1978. The electron stationary picture method for high-speed measurement of reflection intensities from crystals with large unit cells. *Acta Cryst.* A34:289.

Yamane, T., Weininger, M. S., Mortenson, L. E., and Rossmann, M. G., 1982. Molecular symmetry of the MoFe protein of nitrogenase. *J. Biol. Chem.* 257:1221.

Zeppezauer, M., 1971. Formation of large crystals. *Methods Enzymol.* 22:253.

Electron Energy Levels: Electron Spectroscopy and Related Methods

Methods for probing the electronic energy levels within atoms and molecules include visible and ultraviolet spectroscopy (both absorption and emission). In these cases the observed transitions between the ground state and excited state atomic or molecular orbitals are internal events which are often difficult to interpret. As a general rule the optical spectroscopic methods lack element specificity for the purpose of nondestructive biomolecular studies unless it has been rigorously established from other evidence that a given chromophore is localized at a particular set of atoms. Since this volume is concerned with approaches which either offer absolute atom specificity, as in nmr or Mössbauer spectroscopy, or sufficient self-contained information to allow a reasonable conclusion as to the type of atoms involved, the older optical methods have been excluded. The reader is referred to a reference on that subject (Harris and Bertolucci, 1978). The author does not mean to diminish the importance of ultraviolet and visible spectroscopy, e.g., as in studies of bonding environments at transition metal ions (Sutton, 1968), and Chapter 9 examines magnetic circular dichroism.

This chapter is concerned with a group of electron-oriented physical methods of high significance to chemistry which have developed markedly during the past decade (Brundle and Baker, 1984). They are closely related, often being available as different functions within the same multipurpose instrument. The group includes electron spectroscopy for chemical analysis (ESCA), ultraviolet photoelectron spectroscopy (UPS), secondary electron or Auger electron spectroscopy (AES; Auger is pronounced "Oh-zhay"), low-energy electron diffraction (LEED), electron-stimulated desorption (ESD), and secondary ion mass spectroscopy (SIMS). In addition, the Auger mode may be integrated with focused beam techniques in a method known as scanning Auger microscopy (SAM).

Of these, ESCA, Auger spectroscopy, SAM, and to some extent UPS are potentially applicable to the problems of bioinorganic chemistry since they possess a degree of element specificity (but not absolute specificity) and supply direct evidence of an element's valence state. All have the common feature of surface sensitivity, i.e., they probe the outermost layers of a sample rather than its bulk.

7.1 Electron Spectroscopy for Chemical Analysis (ESCA) and Ultraviolet Photoelectron Spectroscopy (UPS)

Photoelectric events are defined as those in which an atomic or molecular electron acquires all the energy of an impinging photon and is removed completely from the atom or molecule to create an electron–cation pair, as shown in Figure 7-1. In obeying energy conservation, the kinetic energy of the emerging electron will be exactly equal to the difference between the original photon energy, $h\nu$, and the ionization energy, E_i, of the orbital from which the electron came, as required in Eq. 7-1. If the photon energy is constant, electrons coming from deeper orbitals will be identifiable by their lower residual kinetic energies while electrons from the outermost (valence) orbitals will have relatively more energy.

$$\tfrac{1}{2}m_e v_e^2 = h\nu - E_i \tag{7-1}$$

where m_e and v_e are, respectively, the mass and the velocity of the electron. Photons energetic enough to create photoelectrons are those with wavelengths beginning in the near infrared and extending to bluer regions of the spectrum. The actual energy involved depends, of course, on the nature of the irradiated sample. At still shorter wavelengths, e.g., the γ-ray portion of the spectrum, photoelectric events become less probable while pair production and Compton scattering begin to dominate the photo–electron interactions.

ESCA is based on a constant energy photon source operating in the soft X-ray region of the spectrum. This method is largely due to the early

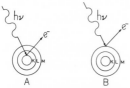

Figure 7-1. Photoelectron production. In the ESCA method an energetic photon, typically in the X-ray region of the spectrum, is able to remove inner shell electrons. In case A the K-shell electron departs with an energy equal to the difference between the X-ray photon energy and the minimum energy needed to remove the electron from the atom or molecule. Ultraviolet photoelectron spectrometry (UPS) makes use of less energetic ultraviolet photons. As shown in case B, these probe only the outermost (valence) shells.

innovations of K. Siegbahn (Siegbahn *et al.*, 1970). Similarly, D.W. Turner originated UPS (Turner *et al.*, 1970), using instead a vacuum ultraviolet source based on the 58.4-nm line of helium. UPS instruments cannot probe the coulomb field much deeper than the outermost filled orbitals of atomic and molecular species while ESCA, owing to its more energetic source, is able to remove electrons from inner atomic cores. UPS is synonymous with the term "*photo electron spectroscopy of outer shells*" (PESOS), while ESCA is "*photo electron spectroscopy of inner shells*" (PESIS). Inner shell energy levels are more element specifc simply because the inner electrons are dominated by the nuclear coulomb field, which is in turn directly related to the element's atomic number. PESOS is more profitably applied to smaller molecules, which may be observed in unusual states (e.g., neutral rather than zwitterionic glycine; see Cannington and Ham, 1983).

All of the electron instrumentation, including Auger and SAM, requires well-designed vacuum systems as a result of the weakly penetrating photons and electrons involved. This property determines that the methods are surface oriented, as stated earlier.

7.1.1 Instrumentation for Producing Photoelectron Spectra

The basic components of a typical photoelectron spectrometer are shown in Figure 7-2. Electrons which are driven from the sample pass first through a set of collimating slits before entering the analyzing electric field (alternatively, a magnetic field) where dispersion according to elec-

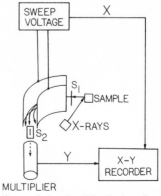

Figure 7-2. Simplified schematic of an ESCA Spectrometer. The dispersive element shown here consists of two coaxial cylindrical sections. Monochromatic X-rays impinge on the sample, where photoelectrons originating in numerous energy levels are generated. The first slit (S_1) forms a collimated beam of these "polychromatic" electrons, and these are deflected into curved paths when they enter the electric field between the cylindrical plates. The less energetic electrons (i.e., those originating in the innermost atomic shells) follow the more curved paths. A spectrum is produced when a linear sweep voltage ramp is applied to the cylindrical plates, which causes the range of electron energies to be scanned across the second slit (S_2). Electrons passing through this slit are detected in the cascaded multiplier, and the signal is displayed as "*y*" on the recorder. A more efficient electron dispersive element is shown in Figure 7-9.

tron velocities and thus residual kinetic energy occurs. A second slit arrangement admits the weak currents emerging from the analyzer to the first dynode of a cascaded multiplier, i.e., the latter is essentially a photomultiplier without the photocathode. The whole assembly must be hermetically shielded to keep out stray fields.

Photoelectron spectra are generated by applying a voltage ramp to the analyzer while recording the current output of the multiplier on a time base. The recorder could be a simple X-time plotter or a computer-operated signal digitizer. It is extremely important that the photon source be constant. In this fashion the electrons from deeper orbitals enter the analyzer at lower velocities and are deflected into the multiplier slit when the applied field is low, while valence electrons are not recorded until the analyzer voltage is high. In practice the ramp voltage is calibrated in ionization energy units, e.g., electron volts (eV). Instrument calibration is not a trivial matter, and if nonlinearity is not taken into account, the systematic energy error between two different instruments may actually *exceed* the core electron chemical shift range of the element in question (Lee, 1982). An example photoelectron spectrum of the valence levels of N_2 is shown in Figure 7-3.

7.1.2 Koopmans' Theorem

To a good first approximation the ionization energies measured by ESCA and UPS have the same identity as the negative value of orbital energies (or eigenvalues) calculated for self-consistent field molecular orbitals. This is known as Koopmans' Theorem, after its originator (Koopmans, 1933). It should be emphasized that the theorem is an approximation

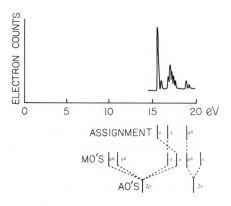

Figure 7-3. A portion of the outer shell photoelectron spectrum of N_2 with assignments shown below. The ordinate (electron counts) corresponds to y in Figure 7-2, and the abscissa (electron volts) is x in Figure 7-2. Energy levels expected from simple molecular orbital (MO) theory are not in complete agreement with the observed spectrum, e.g., the energy of the σ bonding orbital differs considerably. Vibrational fine structure permits classification of an energy level as bonding or antibonding.

in polyelectronic molecules, although the agreement appears to be close (Brundle and Robin, 1971). ESCA and UPS are fortuitous developments for theoretical chemistry since these methods permit quantitative measurements of the energy levels sought after in wave-mechanical calculations. As an example, Figure 7-3 contains a correlation diagram showing the relationship between the theory for N_2 and its photoelectron spectrum. Similarly, self-consistent charge molecular orbital calculations (SCC-MO) for uracil and uracil derivatives were found to be in good agreement with the respective X-ray photoelectron spectra, while a different theoretical method (CNDO/2) produced less favorable results (Maksic *et al.*, 1983).

7.1.3 Empirical Correlations between Atomic Type and Inner Shell Ionization Energy

A consideration of the range of energies involved in inner shell photoelectron spectra will reveal that the dispersion is large in these techniques. More sophisticated molecules, e.g., biomolecules, will obviously lead to a more "cluttered" result than Figure 7-3. Nevertheless, reliable empirical correlations between atoms (or atom group types) and ionization energies not only allow one to discriminate between the various elements but also to make distinctions between oxidation states and possibly the bond types of a particular element. Table 7-1 illustrates the shift ranges

Table 7-1. Chemical Effects on Core Binding Energies: Approximate Shift Ranges of the Biological Elements

Element	Shift range (eV)[a]	Element	Shift range (eV)[a]
B	10	Ca	4
C	12	V	5
N	11	Cr	4
O	6	Mn	3
F	8	Fe	11
Na	4	Co	8
Mg	3	Ni	8
Si	8	Cu	8
P	9	Zn	4
S	16	Se	5
Cl	12	Mo	6
K	4	I	8

[a]Shift ranges have been rounded off to the nearest integer.

of a collection of biologically interesting elements. The biological elements have not been fully explored, which is not surprising in view of the relatively brief history of photoelectron spectroscopy.

7.1.4 The "Direction" of Binding Energy Shifts in ESCA

The inner shell electron binding energies depend both on the nuclear coulomb field and pairwise electron–electron repulsion terms. It is through the latter effect that chemical differences are able to significantly alter the core electron energies, e.g., an inductive decrease in electron density due to neighboring atoms lowers the negative charge density and thus enhances the electron-nucleus interaction. The result is an *increase* in binding energy. Conversely, environments which increase atomic electron density *decrease* the binding energy. Chemical oxidation state has by far the largest effect on binding energies, consistent with wave-mechanical calculations (Brundle and Robin, 1971). ESCA information is thus redundant with Mössbauer, the advantage being that ESCA has access to more elements than the latter (see Chapter 4). Theoretical calculations of core energies may be applied to organic and biological structures with varying degrees of success. Table 7-2 presents a comparison between calculations for morphine and the empirically determined binding energies.

7.1.5 Correlations between Photoelectron, Nmr, and Mössbauer Chemical Shifts

There are correlations between the line shifts observed in electron spectroscopy and nmr chemical shifts (Carlson, 1975; Lindberg, 1974).

Table 7-2. Koopmans' Theorem Ionization Potentials and PES (eV) for Morphine

MO	Eigenvalue[a]	Eigenvalue (corrected by -1.85 eV)	PES (Exptl.)
76	-6.64	-8.49	8.3
75	-7.35	-9.20	
74	-7.62	-9.47	9.4
73	-8.24	-10.09	10.2
72	-9.80	-11.65	10.8
71	-10.31	-12.16	12.0
70	-10.34	-12.19	

[a]*Ab initio* LCAO-MO-SCF calculation. Adapted from Popkie *et al.*, 1976.

This is not surprising since the net electronic charge associated with an atom determines both the diamagnetic shift component of the nmr line and the coulomb field from which the electron energy levels derive. It is also predictable that the relationship between the two types of shifts is not simple. Line positions in the electron spectrum result from the above cited coulomb forces, while additional phenomena, e.g., paramagnetic and circulating ring current effects, further complicate the nmr spectrum.

ESCA chemical shifts also correlate with Mössbauer isomer shifts (Carlson, 1975), as shown in Figure 7-4. In general, the Mössbauer shift depends on innermost shell electron density while ESCA modulations involve shells further out.

7.1.6 Lineshapes of Photoelectron Energy Levels

The photoelectron spectrum of N_2 (Figure 7-3) shows more features than its accompanying correlation diagram warrants. For example, the π bonding level is split into several equally spaced levels of fine structure corresponding to different vibrational energy states. When resolvable, the vibrational structure leads to a direct classification of the energy level as a bonding, nonbonding, or antibonding state. This can be seen as follows: the removal of an antibonding electron by photoionization increases the force constant between the two atoms and therefore increases the vibrational frequency. Bonding electrons behave in the opposite sense while the removal of nonbonding electrons has little or no effect on the force constant. Since the ground state vibrational frequency of N_2 is 2345 cm^{-1} (Raman) the lower-frequency spacing at π (1850 cm^{-1}) identifies the level as a bonding orbital, i.e., photoionization weakens the bonding force constant. One will note that this result is based on the combined information from two different physical methods.

More often the electron originating in a bonding or antibonding orbital will give rise to a broadened, unresolved peak which is more or less symmetrical. Emissions from nonbonding orbitals are usually more narrow and skewed.

Figure 7-4. Correlation of Mössbauer and ESCA chemical shifts. The ordinate is the ESCA binding energy of ruthenium $3d_{5/2}$ electrons, and the abscissa is the Mössbauer shift of ^{99}Ru, relative to ruthenium metal. The correlation of these two variables in widely different chemical environments is good. Adapted from the data of Prather and Zatko (see Carlson, 1975, and references cited therein).

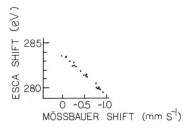

7.1.7 Sample Requirements for ESCA and UPS

The implications of a method which probes a sample to a linear depth of only about 3 nm are truly profound. This means that the sample need only be a thin film on some substrate, and this could be as little as a monomolecular layer for some biopolymers. Particulate residues collected on microporous filters effectively shield the filter material itself from the spectrometer (Blaisdell and Grieger, 1979) and ultrafiltration is thus one possible method for collecting the small amounts of biopolymers needed for photoelectron spectroscopy. One must be especially careful to avoid surface charging, which can shift the observed energy levels, and a conductive backing is essential.

The sample is usually exposed to a high vacuum. In the case of polymers, evaporation is not a problem. Liquid samples are also possible (Lindberg *et al.*, 1976), a development of considerable importance to biological studies. Most current applications of ESCA appear to be limited to the level of complexity of proteins (or lower).

The radiations used in photoelectron spectroscopy are of the "hard" variety, which are able to induce radiation chemical changes. In the strict sense, the photoelectron methods should not be classified as nondestructive. This will be particularly true in samples subject to chemical chain reactions. However, as is the case with Mössbauer spectroscopy, the total dose on a sample may be kept low. Also, flowing systems may circumvent spurious reactions (Lindberg *et al.*, 1976).

7.1.8 Biological Applications of Photoelectron Spectroscopy

Approximate core electron chemical shift ranges of the different biological elements are collected in Table 7-1. From Table 7-1, it will be seen that the shifts of the inner shell electrons of B, C, N, O, and F are moderately sensitive to oxidation state, as are also those of Si, P, Cl, and I. It will be noted that sulfur is the element most influenced by chemical differences, which in part explains the frequency of papers involving sulfur.

Of the metals, only Fe, Co, Ni, and Cu have significant shift ranges. Na, K, Mg, and Ca exist (biologically) in fixed valence states and thus experience small chemical effects, i.e., the ranges given in the table are essentially those between the free metal and the corresponding metallic cation.

At least two approaches may be followed in applying the photoelectron methods to biological systems. In one version, the electron spectra are obtained as a function of depth into the specimen. Ratios of line intensities may then be used to establish a profile of element concentrations at different layers. In this method the necessary removal of successive layers is inherently destructive, e.g., as in the case of plasma etching. Elemental composition is thus obtained at the sacrifice of the specimen's original chemical structure. Profiles at cell surfaces and organelle membranes may be expected to depend on function, including altered function in disease (Millard, 1978).

In a second approach, an attempt is made to preserve the sample's chemical integrity, and attention is focused on the redox and bonding status of a particular element. Chemical status is reflected in linewidth, satellites, and shift effects. The sample is usually a purified compound (protein, coenzyme, lipid, etc.), and the degree to which the element participates in the biochemical function will be reflected in its electron spectrum. If significant spectral changes occur in different functional states it is reasonable to assume that the element participates in the biological mechanism, while the absence of effects suggests (but may not prove) nonparticipation. The potential of electron spectroscopy for probing metal ions in enzymes was recognized by the originator of the method (Siegbahn, 1974).

It is far more difficult to attain the goal of the second approach since all forms of electron spectroscopy subject the sample to a source of ionizing radiation. Deleterious radition-induced chemical reactions are certain to build up with increasing dose, particularly if there is an opportunity for a chain mechanism. Justifiable concerns have arisen over questions of sample integrity and purity (Wurzbach *et al.*, 1977; Solomon *et al.*, 1975; Weser *et al.*, 1979). It is prudent to evaluate total dose effects in any photoelectron measurements on biological substances. Moving sample techniques, i.e., those in which the beam continually impinges on fresh material, appear to be particularly warranted for biochemical samples (Siegbahn, 1974).

7.1.9 Core Electron Energies and Shifts in Biomolecules

ESCA lines due to abundant biological elements such as carbon and oxygen become severely cluttered with increasing molecular weight, much more so than is the case for nmr resonances (see Section 2.4.1). The smaller molecules are more tractable, e.g., photoelectrons originating in nitrogen atoms have been used in studies of pyrroles, purines, pyrimidines, smaller peptides, and amino acids (Sherwood, 1976).

ESCA is applicable to much larger molecules if the element in question occurs in 1:1 equivalence (or at least in a small molar ratio) with the biomolecule. An example is glutathione peroxidase, an enzyme with four peptide subunits. Each subunit contains a single atom of selenium (Flohe *et al.*, 1973). One question regarding this enzyme centers on the nature of the ligand atoms to which the selenium is bound. X-ray photoelectron measurements on the model selenomethionine, selenocystine, Na_2SeO_3 and SeO_2 characterized selenium $3d$ energy levels at 55.2, 55.5, 58.1, and 59.1 eV, respectively (Rupp and Weser, 1975). The assigned selenium $3d$ line in glutathione peroxidase has been measured at 55.3 ± 0.3 eV, as shown in Figure 7-5 (Chiu *et al.*, 1977). On the basis of this information, one may reach a tentative conclusion that selenium is not bound to oxygen in the enzyme. The evidence is more consistent with an environment such as C—Se—C, C—Se—H, or perhaps S—Se—S.

The selenium $3d$ binding energies noted above fall within a range of about 4 eV, which is a relatively pronounced chemical effect. Core levels of sulfur also experience large shifts. In both elements the principal factor is oxidation state. Chemical effects on core electrons are largest when the radius of the valence shell is small. In other elements, notably copper, one also observes satellite peaks, which may introduce an additional source of information (Weser and Rupp, 1978), e.g., on paramagnetism and oxidation state (see Section 7.1.10).

It should not be assumed that the coordination of a metal ion to a ligand atom will necessarily alter the core levels of the ligand atom, even though its bonding environment is altered. In the case of sulfur, the $2p$ binding energies span about 7 eV for oxidation states ranging between S^{2-} and SO_4^{2-}. Studies with the model compound shown in Figure 7-6 indicate that sulfur core energies are not necessarily sensitive to metal coordination (Thompson *et al.*, 1979), e.g., the S_{2p} line is found at 164 eV with or without the cupric ion present. However, an exposure of the copper chelate to a 10^{-5}-ampere argon-ion beam led to rapid radiolytic oxidation at sulfur, resulting in a line shift to 168 eV. The latter reaction

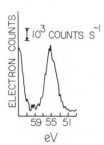

Figure 7-5. A 55-eV photoelectron emission from glutathione peroxidase, assigned as originating from a $3d$ selenium energy level. Based on comparisons with photoelectron spectra of simple selenium compounds, the line at 55 eV is consistent with a selenium atom that is *not* oxygen bound.

Figure 7-6. Sulfur $2p$ X-ray photoelectron spectra obtained from an N,N,S,S-coordinated copper complex. Conditions of the spectra are as follows: (A) Ligand without copper; (B) the complex (structure shown at the right); (C) the complex after a 10-s exposure to a 10-μA argon ion beam; (D) the complex after a 30-s exposure to the argon ion beam. It is seen that coordination with copper has little effect on the sulfur core level, but radiolytic damage leads to a rapid production of spectrometric artifacts. In view of these effects, one must use caution in the interpretation of ESCA spectra. Adapted from Thompson et al., 1979.

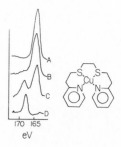

mechanism definitely involved copper since no shift was observed when the metal free ligand was similarly treated (Thompson et al., 1979).

The example of Figure 7-6 may explain a metal ion-dependent shift observed in several copper proteins. Sulfur $2p$ lines were observed both at 164 eV and 169 eV in French bean plastocyanin (Solomon et al., 1975) and in Limulus oxyhemocyanin, Rhus vernicifera stellacyanin, and Pseudomonas aeruginosa azurin (Wurzbach et al., 1977). The evidence concerning plastocyanin has been interpreted as an impurity effect (Peeling et al., 1977); however, the apoprotein (prepared by dialysis against CN^-) presented only the 164-eV peak, and addition of Co(II) or Cu(II) restored the 169-eV line (Solomon et al., 1975). Cobalt is known to replace the copper atom in some proteins (McMillin et al., 1974). One possible conclusion is that the metal ion is adjacent to the sulfur atom and thus sensitizes the latter to radiolytic oxidation. Alternatively, the 169-eV line could be due to charge transfer between sulfur and the metal (Larsson, 1977). A question of this type appears to be difficult to resolve on the basis of ESCA evidence alone.

In other cases the chemical shift of the coordinating ligand is altered in the presence of the metal ion. Nitrogen $1s$ levels of octaethylporphyrin were found to shift in the presence of metal ions, and linewidths could be correlated with the number of unpaired electrons in the ion (Karweik and Winograd, 1976). Other special applications are possible, e.g., the determination of sulfoxide groups in oligopeptides (Jones et al., 1978).

7.1.10 Satellite Peaks

The so-called "shake-up" satellite lines are observed among transition elements (Carlson, 1975). Shake-up satellite peak intensities depend on the spin multiplicity state (i.e., high or low spin) of paramagnetic Fe(III) complexes which undergo reversible, temperature-dependent spin multiplicity transitions. In contrast, binuclear complexes with anomalous magnetic behavior (e.g., cupric acetate) show no such temperature effects, even though their magnetic susceptibilities vary with temperature. This is of considerable significance since paramagnetic atoms which are anti-ferromagnetically coupled in polynuclear complexes [e.g., binuclear Cu(II)] are often difficult to characterize by methods such as esr. In contrast, the ESCA spectrum may yield direct evidence of the atom's oxidation state. Thus, ESCA spectroscopy should be viewed as complementary to those methods.

An example of satellite structure present in a Cu(II) complex is shown in Figure 7-7. The spectrum of a Cu(I) complex, also shown in Figure 7-7, presents a single narrow line which is almost free of nearby structure. The shoulder to the left of this line may be due to partial radiolytic decomposition of the sample. The oxidation state of copper is also reflected in the chemical shift of the Cu $2p$ binding energy. Cu(I) and Cu(II) differ by about 2.5 eV, while Cu(II) and the less commonly encountered Cu(III) differ by about 2 eV (Gagne $et\ al.$, 1980).

The evidence is consistent with shake-up excitations originating in an interaction of the departing photoelectron with unpaired d-orbital valence electrons (Ioffe $et\ al.$, 1978). For example, satellite peaks are observed for Cu(II), which is paramagnetic, but not Cu(I), which is diamagnetic (Carlson, 1975; Weser and Rupp, 1978; Rupp and Weser, 1976). Some examples are given in Table 7-3. Shake-up satellites are also observed in the nitro $1s$ spectra of very polar nitro-amines, in which charge transfer is possible (Nakagaki $et\ al.$, 1982).

The ability to detect and characterize paramagnetic properties by means of ESCA increases the value of the method as a probe of metalloprotein structure, i.e., knowledge of the kind of paramagnetism associated with a transition metal ion may be related to its coordinating ligands and their geometrical arrangement. Evidence of this type is also available through Mössbauer spectroscopy, which is not applicable to many iso-

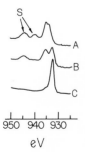

Figure 7-7. Satellite structure in paramagnetic copper complexes. Copper is coordinated to macrocyclic ligands in structures of the type Cu_2L. (A) Cu(II)Cu(II)L; (B) Cu(II)Cu(I)L; (C) Cu(I)Cu(I)L. The diamagnetic example (C) shows only a single photoline for the copper $2p_{3/2}$ core level while paramagnetic forms (A and B) show satellites (S) to the left. Satellites are produced when other (internal) electron transitions are associated with the photoionization event. One will note that the satellite lines are toward the more energetic (i.e., corresponding to electrons that are harder to remove) region of the ESCA spectrum. Adapted from Gagne $et\ al.$, 1980.

Table 7-3. Electron Binding Energies and Satellite Splitting[a]

Compound	Cu $2p_{1/2}$ (eV)		Cu $2p_{3/2}$ (eV)		N $1s$ (eV)	O $1s$ (eV)	S $2p$ (eV)
	Satellite	Main peak	Satellite	Main peak			
CuO (Cu^{2+})	963.7	955.9	943.8	935.7		532.0	
CuCN (Cu^+)	—	954.1	—	934.1			
Glycine					399.2	530.5	
Alanine					400.3	530.4	
Glycine, sodium salt					400.4	531.4	
Alanine, sodium salt					399.3	531.1	
Cu-(glycine)$_2$ (Cu^{2+})	963.2	954.8	942.8	934.8	399.9	531.6	
Cu-(alanine)$_2$ (Cu^{2+})	963.1	955.1	942.9	934.9	400.2	532.2	
Cystine					400.2	530.4	163.0
Cu-cystine (Cu^{2+})	963.1	954.3	943.6	934.3	399.8	531.2	163.7
Cysteine					400.4	530.5	162.9
Cu-cysteine (mixed)	—	952.0	—	932.3	400.7	531.0	167.8
Penicillamine					400.7	530.6	163.2
Penicillamine disulfide					400.4	530.5	163.1
Cu-penicillamine (mixed)	—	953.1	—	933.3	399.9	531.5	163.2

[a]Adapted from Rupp and Weser, 1976.

topes (see Chapter 4). ESCA offers the additional possibility of direct observations of the coordinating ligands in the same spectrum, since all of the elements present are recorded.

7.1.11 Line Intensities

The radius of a protein molecule may be sufficient to cause partial obscuration of deeply buried atoms. Thus, photoelectron intensities may reflect the peripheral or internal location of a particular atom. For example, in superoxide dismutase the $2p$ and $3p$ lines of copper were found to be fully twice the intensity of the corresponding core levels of zinc, suggesting that copper is located closer to the surface than zinc (Van der Deen et al., 1977). The latter conclusion may be examined in light of the known crystallographic structure of the enzyme (Richardson et al., 1975). Superoxide dismutase is composed of two identical subunits, each containing single atoms of copper and zinc. The copper and zinc sites are in close proximity ($\sim$6 Angstroms),and the copper atom is in fact nearer the surface, consistent with its proposed redox role in an electron shuttle mechanism (Fielden et al., 1974). One should nevertheless regard depth information with care as artifactual aggregate structures may arise during specimen preparation.

Other measurements of line intensities are more definitive. The oxygen $1s$ region of chlorophyll a presents a shoulder identifiable as water in a 1:1 H_2O:chlorophyll a equivalence (Winograd et al., 1976). On dehydration ($\sim$250°C) this emission line is lost. Various measurements have led to indirect inference of the mechanistically important hydration water, but the ESCA method produced the first spectroscopic detection of its existence.

7.1.12 Calibration Standards

In the early literature, reported values of inner core energies varied unacceptably, largely due to the choice of inappropriate reference standards. In studies on biopolymers one may elect to calibrate internally, e.g., using $1s$ levels of C and N, but in doing so, care must be taken to avoid detection of the sample backing material.

7.1.13 Signal Improvement Methods

All forms of electron spectroscopy, including ESCA and Auger (see Section 7.2), are amenable to various signal enhancement techniques.

Albert and Joyner (1979) have described deconvolution methods which are appropriate in the often encountered situation of overlapping signals. Digital Fourier filtering is also applicable (Betty and Horlick, 1976a). In the latter method a computer acquires the spectrum digitized at N points. The spectrum is then Fourier transformed to the time domain, where it may be multiplied by a trapezoidal envelope function, in similar fashion to the convolution and deconvolution methods used in pulsed FT nmr (see Section 2.4.4.4). The inverse transform yields the original spectrum with an improved signal-to-noise ratio. Autocorrelation analysis is applicable to most forms of instrumentation (Betty and Horlick, 1976b).

7.2 Auger (Secondary Electron) Spectroscopy

The photoelectric methods of ionization described in the preceding sections are complemented by other techniques which excite surface atoms and molecules with charged particles. Typically, the excitation source is a beam of electrons, focused or diffused, in the 0.3–3-keV energy range. The spectrometer arrangement is similar to that of Figure 7-2, but the light source is replaced by an electron beam. As in photoelectron spectroscopy, the dispersed energy spectrum of electrons emanating from the sample surface is found to contain numerous lines of discrete energy.

Electron spectra produced in this fashion will always show a prominent peak identical in energy to the source, which is simply elastic backscattering or "reflection" of the impinging electron beam. Other lines appearing at lower energies include a broad emission of secondary electrons and much narrower but weak lines which may be classified into one of two categories: (1) loss peaks or (2) Auger emissions. The loss peaks are due to moderated source electrons while Auger electrons are created when primary ionized atoms or molecules dissipate excess energy.

7.2.1 Loss Peaks

Inelastic collisions between source electrons and the sample involve several different mechanisms. In some encounters, primary energy is lost as a lattice excitation (e.g., a phonon) while in others atomic core electrons are ejected, creating ions. The primary electron is scattered with a decrement in energy equal to the amount consumed in the collision mechanism. Thus, loss peaks are inelastically backscattered source electrons, and their identity becomes apparent when the source energy is changed, i.e., a given loss peak also varies but maintains a constant energy difference from the reflection peak. In contrast, Auger electrons are source energy independent.

Ionization loss spectroscopy (Gerlach, 1971) is useful in determining the elemental composition of surfaces, and is adaptable for element-specific imaging of surfaces (Joy and Maher, 1978). Each element has a unique set of core ionization energies, and these determine the loss peak–source energy differences. The information is similar to that of ESCA.

7.2.2 Auger Electrons

Events leading to the emission of an Auger electron are shown in Figure 7-8. The bombarding primary electron has transferred part of its energy to an L-shell electron, which leaves the atom as a secondary electron. Excess energy may now be dissipated by one of two processes: (1) an outer electron may fill the lower vacancy with the emission of a photon or (2) the same transition will occur but the photon's equivalent in energy is absorbed by another outer shell electron. About 90% of the transitions which stabilize primary cations proceed by the latter mechanism, and the ejected outer shell electron is the basis of Auger phenomena.

One will note that the Auger electron leaves behind a *doubly charged* cation. In Figure 7-8 the L-shell vacancy is filled by an M-electron, and the transition energy is absorbed by a second M-shell electron (i.e., an *LMM* Auger emission). The kinetic energy with which the Auger electron escapes is given by Eq. (7-2), where $\Delta E_{M,L}(+1)$ is the L–M transition

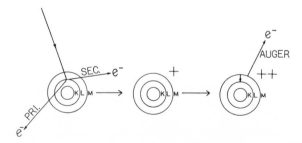

Figure 7-8. Events leading to the production of an Auger electron. The energetic primary (from the electron gun, Figure 7-9) ejects a core electron, e.g., one from the L-shell in the example shown here. The result is a secondary electron/positive atomic ion (or molecular ion) pair. The vacancy in the L-shell is quickly filled by an upper level electron, which may dissipate its energy by (1) radiating a photon or (2) transferring all of the transition energy to another upper shell electron, which leaves the atom. The latter process is the source of Auger electrons, and results in a doubly ionized atom. In the example shown here an M-shell electron filled the L-vacancy and an M-shell electron was ejected, i.e., an *LMM* Auger event.

energy in the singly charged primary cation and $E_{i,M}(+2)$ is the M-shell ionization energy leading to a doubly charged positive ion.

$$\tfrac{1}{2}m_e v_A{}^2 = \Delta E_{M,L}(+1) - E_{i,M}(+2) \tag{7-2}$$

A whole series of Auger lines is possible, e.g., *KLL*, etc., and the relative proportions of these emissions will depend on a series of transition probabilities and primary ionization cross sections.

7.2.3 Auger Instrumentation

Auger spectrometers place a heavy demand on vacuum systems and dispersive elements. The Auger lines occur in the low-energy spectral region where they are overshadowed by a broad secondary electron emission band (i.e., electrons created by primary ionizations). Use of a cylindrical mirror analyzer decreases noise by limiting the current input at the cascade multiplier (Palmberg *et al.*, 1969), and good signal-to-noise ratios are possible even when working with the typical picoampere range of Auger and loss peak emissions. The detection of weak signals is hampered primarily by the vacuum system since surface contamination builds up in less than an hour even under an ultrahigh vacuum ($\sim 10^{-11}$ torr).

The simplified schematic of an Auger spectrometer is shown in Figure 7-9. A point of difference between this arrangement and the photoelectron spectrometer discussed earlier is the use of synchronous detection. A small sinusoidal voltage is superimposed on the sweep ramp input applied to the electrostatic dispersive element (the cylindrical mirror). The modulating signal is also mixed with the output of the electron multiplier in

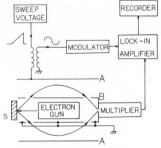

Figure 7-9. An Auger electron spectrometer. The sweep voltage and multiplier components have the same function as in Figure 7-2, and the electron gun replaces the X-ray source. The dispersive element consists of outer (*A*) and inner (*B*) concentric metal cylinders. By varying the voltage between these, Auger electrons originating in the sample (S) (with a range of discrete energies) may be brought successively into focus at the multiplier aperture, i.e., the electric field between the cylinders determines the electron energy which will refocus at the aperture. A spectrum is thus created by recording ramp voltage versus multiplier output. A point of difference between ESCA and Auger spectrometers is the AC modulation signal applied to the ramp voltage in the Auger design, which allows slope detection and thus discrimination against baseline undulations present in the direct signal. The electron gun may be used to produce scanning images of the sample, as in scanning electron microscopy.

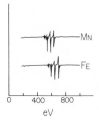

Figure 7-10. Analogous Auger spectra of iron and manganese. The iron peaks are located at higher energies (i.e., more tenaciously bound electrons) simply because iron has a large atomic number and therefore a stronger nuclear coulomb field than manganese.

a phase-sensitive detector. This causes the Auger spectrometer to operate in a slope detection mode, and spectra are recorded as dI/dE versus E, where I is the detector current and E the Auger electron kinetic energy. Auger linewidths, typically about 5 eV, are much narrower than the broad emission due to ionization electrons, and slope detection thus strongly emphasizes the Auger spectrum. The Auger effects almost vanish if the spectrum is recorded as I versus E.

A set of Auger spectra are presented in Figure 7-10. On first inspection the line groups due to iron and manganese appear quite similar. However, the difference in coulomb fields leads to a relative shift of tens of electron volts with only a unit increment in atomic number. Auger spectroscopy is thus element specific, and meets an important criterion for a physical method which might be used to study the role of the elements in biology. Line positions also depend on chemical form (Hiraki *et al.*, 1979), and line shape may also be related to chemical characteristics (Figure 7-11).

7.2.4 Potential Biological Applications of Auger Spectroscopy

Since the information contained in Auger spectra is qualitatively similar to that of ESCA, Auger spectroscopy should also be applicable to biological materials. However, relatively fragile biological structures may be rapidly damaged in the electron beam, and the total electron dose should be kept as low as possible.

Numerous interactions which occur at surfaces are appropriate subjects for Auger measurements. For example, Hoglund and Odeblad (1977) studied the association between Na^+ and keratin in NaCl/keratin combinations. Using the carbon $1s$ line (keratin methylenes) as an internal

Figure 7-11. *LMM* Auger spectra of titanium in different chemical environments. The lineshapes are distinctly different.

reference, the shift of the sodium $1D$ Auger line was recorded as a function of the NaCl:keratin molar ratio. The outcome of an experiment of this type depends on whether or not the surface has binding sites for the ion in question. If so, the line position will shift as the Na^+:keratin molar ratio is increased since two Na^+ environments will exist at molar ratios greater than that needed to saturate the sites. In the absence of saturable sites, the surface ions will average a constant environment, and the shift will not depend on the molar ratio. In the case of Na^+ and keratin the line was observed to shift, indicating a probable bidentate (two ligand) bonding of Na^+ to keratin, most likely at carbonyl positions.

The ability of Auger spectroscopy to characterize detailed surface composition far exceeds many earlier methods. For example, the X-ray powder diffraction patterns obtained from urinary calculi lead to an identification of the material as calcium oxalate. More subtle qualities appear in the Auger spectrum (Rabinowitz and Elliot, 1976); e.g., in addition to the expected lines of Ca, O, and C, one also observes N, P, S, and Si. On etching away the outer 200-Angstrom layer, lines from Fe, K, and Cu are also found.

The linear energy transfer (i.e., deposition of energy per unit length) of electrons is much greater than that of X-rays. The Auger source is thus efficient in activating the outermost sample layer. Spectra may be obtained rapidly (seconds). However, where biological specimens are concerned, total dose is the primary consideration, as noted above, since radiolytic reactions easily degrade fragile structures. From this perspective Auger has no distinct advantages over ESCA or PES. These methods should be viewed as complementary (Spicer et al., 1977).

The detection of spectra in the derivative mode does not preclude recovery of an integrated signal. Also, several signal enhancement methods are available (see Section 7.1.13).

A major disadvantage associated with Auger spectroscopy is surface charging due to accumulation of trapped electrons, especially in nonconducting biological materials. Coulomb repulsion can introduce apparent energy shifts. In view of the relatively small range of chemical effects on core electrons, one should be careful in Auger studies with dielectric materials, i.e., most biological specimens. Perhaps the best biological application of Auger spectroscopy is found in its ability to locate specific elements on a surface, which is the subject of the following section.

7.2.5 Auger Imaging Methods (Scanning Auger Microscopy, SAM)

Auger surface images may be constructed by scanning a sample with an electron beam while monitoring the scattered electrons. In doing this the analyzer is set to record the Auger line of a specific element. Thus, one obtains image contrast based on the surface distribution of the element in question. This is a version of scanning electron microscopy, and it meets one of the criteria of this book in being element specific. Scanning Auger microscopy is suitable for biological applications (Woodruff, 1977; Janssen and Venables, 1979), and the method produces good image contrast (Carlson, 1975). Element-mapped images are comparable to those obtained by electron microprobe analyzers which use X-ray detectors (see Figure 11-2 and Section 11.1.1).

The Auger method is commonly described as a three-dimensional probe for surface chemical analysis. Two of the dimensions are those just described, i.e., the literal image of element distribution on the surface. The third dimension is element composition versus depth, which necessitates successive ablations of the surface with an ion beam, etc. The latter method is also used with ESCA and was discussed earlier (see Section 7.1.8). Figure 11-3 (Chapter 11) was not obtained by the SAM method, but indicates the kind of results one may expect to obtain in element-specific mapping.

References

Albert, C., and Joiner, R. W., 1979. The application of deconvolution methods in electron spectroscopy—A review, *J. Electron Spectrosc. Relat. Phenom.* 16:1.

Betty, K. R., and Horlick, G., 1976a. A simple and versatile Fourier domain digital filter, *Appl. Spectrosc.* 30:23.

Betty, K. R., and Horlick, G. 1976b. Autocorrelation analysis of noisy periodic signals utilizing a serial analog memory, *Anal. Chem.* 48:1979.

Blaisdell, J. M., and Grieger, G. R., 1979. ESCA study of air filters, *Am. Lab.* 11:85.

Brundle, C. R., and Baker, A. D. (eds.), 1984. *Electron Spectroscopy: Theory, Techniques and Applications,* Vols. 1–5, Academic Press, New York.

Brundle, C. R., and Robin, M. B., 1971. Photoelectron Spectroscopy, in *Determination of Organic Structures by Physical Methods.* Vol. 3 (F. Nachod and J. J. Zuckerman, eds.), Academic Press, New York, pp. 1–71.

Cannington, P. H., and Ham, N. S., 1983. He(I) and He(II) Photoelectron spectra of glycine and related molecules, *J. Electron Spectrosc. Relat. Phenom.* 32:139.

Carlson, T. A., 1975. *Photoelectron and Auger Spectroscopy,* Plenum Press, New York.

Chiu, D., Tappel, A. L., and Millard, M. M., 1977. Improved procedure for X-ray photoelectron spectroscopy of selenium-glutathione peroxidase and application to the rat liver enzyme, *Arch. Biochem. Biophys.* 184:209.

Fielden, E. M., Roberts, P. B., Bray, R. C., Lowe, D. J., Mautner, G. N., Rotilio, G., and Calabrese, L., 1974. The mechanism of action of superoxide dismutase from pulse radiolysis and electron paramagnetic resonance. Evidence that only half the active sites function in catalysis, *Biochem. J.* 139:49.

Flohe, L., Guenzler, W. A., and Schock, H. H., 1973. Glutathione peroxidase. Selenoenzyme, *FEBS Lett.* 32:132.

Gagne, R. R., Allison, J. L., Koval, C. A., Mialki, W. S., Smith, T. J., and Walton, R. A., 1980. X-ray photoelectron spectra of copper (I) and copper (II) complexes derived from macrocyclic ligands, *J. Am. Chem. Soc.* 102:1905.

Gerlach, R. L., 1971. Ionization spectroscopy of contaminated metal surfaces, *J. Vac. Sci. Technol.* 8:599.

Harris, D. C., and Bertolucci, M. D., 1978. *Symmetry and Spectroscopy, An Introduction to Vibrational and Electronic Spectroscopy,* Oxford University Press, New York.

Hiraki, A., Kim, S., Kammura, W., and Iwami, M., 1979. Chemical effect in (*LVV*) Auger spectra of third-period elements (aluminum, silicon, phosphorous and sulfur) dissolved in copper, *Appl. Phys. Lett.* 34:194.

Hoglund, A., and Odeblad, E., 1977. Auger electron spectra of sodium bound to keratin and other biological materials, *Phys. Scr.* 16:370.

Ioffe, M. S., Ivleva, I. N., and Borod'kd, Yu G., 1978. Temperature effects in X-ray photoelectron spectra of paramagnetic cupric and iron complexes with anomalous magnetic properties, *Chem. Phys. Lett.* 59:549.

Janssen, A. P., and Venables, J. A., 1979. Scanning Auger microscopy—An introduction for biologists, *Scanning Electron Microsc.* 1979(2):259.

Jones, D., Distefano, G., Toniolo, C., and Bonora, G. M., 1978. Linear oligopeptides. Part 47. A new method for determining sulfoxides in peptide molecules using X-ray photoelectron spectroscopy, *Biopolymers* 17:2703.

Joy, D. C., and Maher, D. M., 1978. A practical electron spectrometer for chemical analysis, *J. Microsc. (Oxford)* 114:117.

Karweik, D., and Winograd, N., 1976. Nitrogen charge distributions in free-base porphyrins, metalloporphyrins, and their reduced analogs observed by X-ray photoelectron spectroscopy, *Inorg. Chem.* 15:2336.

Koopmans, T., 1933. The distribution of wave function and characteristic value among the individual electrons of an atom, *Physica* 1:104.

Larsson, S., 1977. Sulfur 2p photoelectron spectrum of blue copper proteins. Comment on papers by Solomon *et al.,* and Peeling *et al., J. Am. Chem. Soc.* 99:7708.

Lee, R. N., 1982. The effect of nonlinearities on XPS calibration, *J. Electron Spectrosc. Relat. Phenom.* 28:195.

Lindberg, B. J., 1974. Can we expect any meaningful correlations between nmr and ESCA-Shifts?, *J. Electron Spectrosc. Relat. Phenom.* 5:149.

Lindberg, B., Asplund, L., Fellner-Feldegg, H., Kelfve, D., Siegbahn, H., and Siegbahn, K., 1976. ESCA applied to liquids. ESCA spectra from molecular ions in solution, *Chem. Phys. Lett.* 39:8.

McMillin, D. R., Rosenberg, R. C., and Gray, H. B., 1974. Preparation and spectroscopic studies of cobat (II) derivatives of blue copper proteins, *Proc. Natl. Acad. Sci. U.S.A.* 71:4760.

Maksic, Z. B., Rupnik, K., and Veseli, A., 1983. Semiempirical studies of core electron binding energies Part II. SCC-MO calculations for uracil and its derivatives, *J. Electron Spectrosc. Relat. Phenom.* 32:163.

Millard, M. M., 1978. Surface characterization of biological materials by X-ray photoelectron spectroscopy, *Contemp. Top. Anal. Clin. Chem.* 3:1.

Nakagaki, R., Frost, D. C., and McDowell, C. A., 1982. Shake-up satellites in the nitro N 1s spectra of highly polar nitro-aromatic amines, *J. Electron Spectrosc. Relat. Phenom.* 27:69.

Palmberg, P. W., Bohn, G. K., and Trach, J. C., 1969. High sensitivity Auger electron spectrometer, *Appl. Phys. Lett.* 15:254.

Peeling, J., Haslett, B. G., Evans, I. M., Clark, D. T., and Boulter, D., 1977. Some observations on the ESCA spectra of plastocyanins, *J. Am. Chem. Soc.* 99:1025.

Popkie, H., Koski, W., and Kaufman, J., 1976. *Ab-initio* LCAO-MO-SCF calculations of morphine and nalorphine and measurements of their photoelectron spectra, *J. Am. Chem. Soc.* 98:1342.

Rabinowitz, I. N., and Elliot, J. S., 1976. Auger analysis: Its application to urinary calculi, in *Proceedings of the International Colloquium of Renal Lithiasis (1975)*, (B. Finlayson and W. C. Thomas, eds.), University Presses of Florida, Gainesville, Florida, p. 149.

Richardson, J. S., Thomas, K. A., Rubin, B. H., and Richardson, D. C., 1975. Crystal structure of bovine Cu, Zn superoxide dismutase at 3 angstrom resolution: Chain tracing and metal ligands, *Proc. Natl. Acad. Sci. U.S.A.* 72:1349.

Rupp, H., and Weser, U., 1975. X-ray photoelectron spectroscopy of some selenium containing amino acids, *Bioinorg. Chem.* 5:21.

Rupp, H., and Weser, U., 1976. Copper(I) and copper(II) in complexes of biochemical significance studied by X-ray photoelectron spectroscopy, *Biochem. Biophys. Acta.* 446:151.

Sherwood, P. M. A., 1976. Photoelectron spectroscopy, *New Tech. Biophys. Cell. Biol.* 3:93.

Siegbahn, K., 1974. Electron spectroscopy—An outlook, *J. Electron Spectrosc. Related Phenom.* 5:3.

Siegbahn, K., Nordling, C., Johansson, G., Hedman, J., Heden, P. F., Hamrin, K., Gelius, U., Bergmark, T., Woerme, L. O., Manne, R., and Baer, Y., 1970. *ESCA Applied to Free Molecules*, American Elsevier, New York.

Solomon, E. I., Clendening, P. J., Gray, H. B., and Grunthaner, F. J., 1975. Direct observation of sulfur coordination in bean plastocyanin by X-ray photoelectron spectroscopy, *J. Amer. Chem. Soc.* 97:3878.

Spicer, W. E., Linda V.I., and Helms, C. R., 1977. Frontiers in surface and interface analysis, *Research Development* 28, pp. 20–31.

Sutton, D., 1968. *Electrical Spectra of Transition Metal Complexes,* McGraw-Hill, New York.

Thompson, M., Whelan, J., Zemon, D. J., Bosnich, B., Soloman, E. I., and Gray, H. B., 1979. Sulpher 2p photoelectron spectra of 1,8–Bis (2′–pyridyl)–3, 6–dithiaoctane and its copper II complex. Possible interpretation of the S2p 168 eV peak in blue copper proteins, *J. Amer. Chem. Soc.* 101:2482.

Turner, D. W., Baker, C., Baker, A. D., and Brundle, C. R., 1970. *Molecular Photoelectron Spectroscopy: A Handbook of He 584 Å Spectra,* Wiley-Interscience, New York.

Van der Deen, H., Van Driel, R., Jonkman-Beuker, A. H., Sawatzky, G. A., and Wever, R., 1977. X-ray photoelectron spectroscopic studies of hemocyanin and superoxide dismutase, in *Structure and Function of Haemocyanin, Proceedings of the European Molecular Biology, 5th International Workshop,* (J. V. Bannister ed.), Springer-Verlag, Berlin, pp. 172–179.

Weser, U., and Rupp, H., 1978. Assignment of the correct oxidation state of biochemically active copper, *Proceedings of the Third International Symposium on Trace Element–Metabolism in Man and Animals,* (M. Kirchgessner, ed.), Weihenstepan Institute, Weihenstepan, Germany, pp. 40–43.

Weser, U., Younes, M., Hartmann, H. J., and Zienau, S., 1979. X-ray photoelectron spectrometric aspects of the copper chromophore in plastocyanin, *FEBS Lett.* 97:311.

Winograd, N., Shepard, A., Karweik, D., Koester, V., and Fong, F., 1976. X-ray photoelectron spectroscopic studies of the thermal stability of chlorophyll-*a* monohydrate, *J. Am. Chem. Soc.* 98:2369.

Woodruff, D. P., 1977. Auger electron spectroscopy: Principles, developments and applications, in *Developments in Electron Microscopy and Analysis,* Proceedings of the Electron Microscopy Analysis Group, Institute of Physics, University of Bristol, (J. A. Venable, ed.), Academic Press, New York, pp. 337–342.

Wurzbach, J. A., Grunthaner, P. J., Dooley, D. M., Gray, H. B., Grunthaner, F. J., Gay, R. R., and Solomon, E. I., 1977. Sulfur 2p photoelectron spectrum of limulus oxyhemocyanin. Reply to observations on the ESCA spectra of plastocyanins, *J. Am. Chem. Soc.* 99:1257.

Laser Applications: Resonance Raman (RR) Spectroscopy and Related Methods

8.1 Perspective

An application of laser technology to the detection of quadrupole resonance phenomena was described in Chapter 3 (see Section 3.4.2). The present chapter explores resonance Raman (RR) spectroscopy. It should be noted that RR spectroscopy is but one of many potential applications involving lasers (Brewer and Mooradian, 1974; Bradersen, 1979; Morris and Wallan, 1979; Butler *et al.*, 1978; Omenotto, 1979; Steinfeld, 1978; Horrocks *et al.*, 1980; Parker, 1983); yet it does have the distinction of providing information of a relatively selective nature. Also, it is the one biological application of lasers which has been used sufficiently to justify its inclusion as a whole chapter. Raman and RR spectroscopy are relatively old methods with biological applications extending back to the 1930s. The laser improvements are most recent.

The basic phenomenon underlying all modern forms of Raman spectroscopy is relatively straightforward (Thomas and Kyogoku, 1977). In Figure 8-1 the laser generates an intense beam of monochromatic light. Nearly all of the light passes unhindered through the sample cell, but a small amount of it is scattered. A spectrophotometric scan of the scattered light reveals both elastic and inelastic encounters. Most photons are scattered elastically, i.e., their wavelength is not altered. However, some are scattered inelastically through interactions with the vibrational energy levels within the sample molecules. Photons leaving with longer wavelengths (Stokes lines) are the result of encounters which cause vibrational excitation within the molecule while those departing with shorter wavelengths (anti-Stokes lines) are due to the reverse process, i.e., a vibrational quantum is, in effect, combined with the source photon. The latter events are ordinarily the least frequent. Inelastic interactions are known as

Raman scattering, and they obey the energy conservation rule of Eq. (8-1), where the subscripts *s, i,* and *v* identify, respectively, scattered, incident, and vibrational photons.

$$h\nu_s = h\nu_i \pm h\nu_v \qquad (8\text{-}1)$$

or simply:

$$\nu s = \nu_i \pm \nu_v \qquad (8\text{-}2)$$

The information contained within Raman spectra closely parallels the direct vibrational absorptions observed in infrared spectra, although the selection rules are quite different (Horak and Horak, 1979) in these two methods. Scale calibration (cm^{-1}) in Raman spectrometers permits a direct comparison with infrared spectra. The sample Raman spectrum shown in Figure 8-2 is compared with the corresponding infrared spectrum.

A fortuitous consequence of Raman selection rules is the relatively weak scattering from water (which contrasts with water's intense infrared absorptions). Raman studies of biological materials (Carey, 1978; Thomas and Kyogoku, 1977) may thus be conducted under conditions which closely resemble the living state, e.g., in aqueous solutions. The main limitation in conventional Raman investigations of biological entities is simply one of excessive spectral cluttering due to molecular complexity. This is easily appreciated since each chemical bond may give rise to more than one vibrational mode, and there are thousands of bonds in a relatively small protein molecule. Solvent effects are to be expected in any biological study, both in the form of interfering lines from the solvent and solvent-induced changes in the spectrum of the molecule under investigation. The

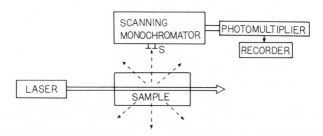

Figure 8-1. Essential components of the laser Raman spectrometer. An intense beam from the laser is scattered by the sample and the scattered light includes the original wavelength plus weak components at the Raman frequencies. Some of the scattered light enters the entrance slits (S) and is dispersed in the scanning monochromator, preferably a double-grating design. The very weak beam emerging from the monochromator is detected by the photomultiplier. As a general rule the spectrum is recorded as amplitude (ordinate) versus wavenumbers in cm^{-1} (abscissa). The scale thus agrees with the common method for presenting infrared spectra.

Figure 8-2. A comparison of the infrared and Raman spectra of *o*-xylene. Vibrational transitions are found at the same wavenumbers, but line amplitudes vary considerably as a result of differing selection rules in the two phenomena. Raman line intensities (*I*) are brighter than the background, as in emission spectra, while infrared radiation is absorbed, resulting in decreased transmittance (*T*) at the resonant frequencies. Thus the line amplitudes have opposite polarity, as shown.

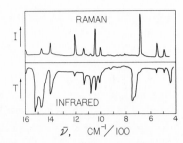

solvent is usually water, though others (DMSO, DMF) are possible. Many biological Raman studies have been carried out and are useful (Spiro and Gaber, 1977; Frushour and Koening, 1975), but in general the selectivity is relatively low. The distinguishable groups generally provide conformational information (see Section 8.6.1). Improvements in conventional Raman have been along the line of eliminating fluorescence problems.

8.2 Resonance Raman (RR) Spectroscopy

The distinction between Raman and RR spectroscopy is illustrated in Figure 8-3. In ordinary Raman spectroscopy the source wavelength lies well away from any absorption bands (wavelength NR, Figure 8-3), but in RR spectroscopy the wavelength of the incident beam is deliberately selected to fall within an absorption band of the substance under investigation (wavelength RR, Figure 8-3). Under the latter conditions the irradiating source is resonant with an electronic transition of the subject molecule (thus the name "resonant Raman"), and the result is a marked enhancement of line intensity, often as much as 300-fold, in the Raman effect produced by chromophore-associated bonding groups. This creates a large degree of selectivity since the information contained in the scattered spectrum reflects only a small portion of a macromolecule. At wave-

Figure 8-3. Raman excitation in relation to visible light absorption band with $\lambda_{max} \sim 450$ nm. A normal Raman spectrum is produced when the monochromatic source is at NR, well away from the absorption band. Excitation at RR leads to resonance-enhanced Raman spectra while excitation at PR causes an intermediate or pre-resonance condition.

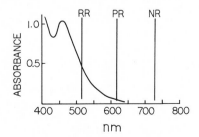

length PR (Figure 8-3), the scattering may show partial enhancement and a "pre-resonance" condition is said to exist. For example, pre-resonance enhancement apparently causes the 2,4′-bithiazole-associated modes to dominate the Raman spectrum of bleomycin (Freedman *et al.*, 1983).

The mechanism of RR scattering is shown in Figure 8-4. A photon promotes an electron from the ground state to the electronic excited state. RR photons are then scattered when the electron returns to one of the bonding vibrational states. The incident photon may also promote the ground state electron to an excited vibrational level with similar results. The RR spectrum thus reveals bonding state vibrational levels and is comparable to ordinary infrared, the distinction being in symmetry-determined selection rules.

RR enhancement is confined to the region of space occupied by the ground and excited state molecular orbitals of the chromophore. In some cases the chromophore is delocalized over several atoms, which causes the involvement of relatively more bonding groups in the RR effect. This degree of selectivity is comparable to esr and MCD (Chapters 5 and 9, respectively). Examples of the latter type are porphyrins and carotenoid pigments. In other cases the orbitals may be more localized. Intense visible and ultraviolet absorptions due to ligand–metal charge transfer effects are common among metalloproteins, and RR thus provides a direct assessment of the bonding environment at metal ion coordination sites. The resonance enhancement in RR is large and depends on proximity to the absorption maximum, as indicated by Eq. (8-3), which is true only for the pre-resonance condition.

$$I = K(\nu_0 - \nu_{\text{vib}})^4 (\nu_{\text{abs}}^2 + \nu_0^2)^2/(\nu_{\text{abs}}^2 - \nu_0^2)^2 \qquad (8\text{-}3)$$

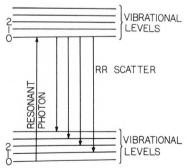

Figure 8-4. Energy levels involved in resonance Raman scattering. The scattering process is distinct from fluorescence and phosphorescence, where there may be a substantial time delay between absorption of the exciting photon and the re-radiation of a second photon at a longer wavelength. In RR scattering the mechanism may be described as the promotion of an electron to an excited state, followed by immediate decay to a vibrational level of the ground state other than the one from which it originated. In the example shown here, the vibrational quantum number (0) does not change on going from the ground state to the electronic excited state (a Franck–Condon transition). Resonance Raman scattering may also involve electronic excitations between *different* vibrational levels.

The vibrational energy levels observed in RR scattering are those of the ground state (Figure 8-4). Vibrational frequencies depend on atomic masses and bond strengths, and functional group correlations have been established for numerous structural types. Uncertain assignments may be resolved by isotope replacement (*vide infra*). If the atomic identities and vibrational modes are known, then RR is able to provide direct measurements of *bond orders,* permitting the user to decide whether the bonding is weak or strong. This is the principal sort of information contained in both Raman and RR spectra. The resonant Raman effect also conveys a picture of chromophore delocalization by accentuating the vibrational resonances of the chromophore's structural framework. If the light-absorbing moiety is separate from the biopolymer (e.g., a prosthetic group), a combination of RR with high-resolution methods such as nmr and X-ray diffraction permits an intensive scrutiny of the component.

The method is not without limitations. If irradiation into the absorption band causes fluorescence, the relatively weak Raman scattering may be overwhelmed by the fluorescence background. Also, intensely absorbing substances may simply attenuate the scattered light. In some situations the symmetry (or pseudosymmetry) of the chromophore may prevent the detection of certain vibrational modes (Horak and Horak, 1979).

8.3 Sample Considerations

Water's relatively weak Raman scattering permits measurements in aqueous solutions. Resonance Raman is one of the few physical methods suitable for *in vivo* experimentation. If an organism's pigmentation is simple, i.e., due mostly to a single chromophore, then the RR spectrum may be no more complex than that of a simple *in vitro* system. Obviously, organisms which contain several overlapping pigments produce more ambiguous results. Suitable life forms are bacteria and unicellular plants and animals. These may be grown in culture media enriched in specific isotopes. As noted above and discussed elsewhere (Section 8.6.2.5), isotope replacement leads to frequency shifts which clearly identify vibrational modes involving the element in question. Although enriched isotope media are expensive, the method is preferable to chemical replacement, which only works for labile atoms (e.g., some metal ions), is more or less restricted to the level of purified proteins and other biopolymers, and may lead to artifact formation.

Typically, concentrations in RR measurements are in the 10^{-4}–10^{-7} M range (Tsai and Morris, 1975). In addition to liquid water, the sample may also be liquid crystalline or solid. Laser microprobe instrumentation may be combined with a Raman scattering spectrometer (Dhamelincourt *et al.*, 1979) for gathering detailed information through a sample imaging system. Clearly, RR is one of the *least* restrictive methods with respect to sample conditions.

8.4 Symmetry and the Intensity of Vibrational Absorptions

There are $3n - 3$ internal degrees of freedom in a polyatomic molecule containing n atoms. If the molecule is nonlinear, three of the degrees of freedom are rotational (of the whole molecular framework). The typical biological molecule thus has $3n - 6$ vibrational degrees of freedom. As a test of this equation, monatomic argon is found to have zero degrees of freedom while water has three. The three fundamental vibrational modes of water are shown in Figure 8-5(A). (Higher harmonics of these fundamentals are possible.) The number of modes increases rapidly with molecular weight, as shown in Figure 8-5(B).

The total number of vibrational degrees of freedom does not necessarily correspond, line for line, with vibrational spectra as symmetry (or pseudosymmetry) eliminates or greatly attenuates some of the modes (Horak and Horak, 1979). This is a consequence of the mechanism by which vibrational energy is absorbed or emitted. Infrared absorption occurs when the vibrational mode changes the molecular electric dipole moment. Mode ν_2 of water [Figure 8-5(A)], for example, is expected to be infrared active. In contrast, the symmetrical vibration of CO_2 shown in Figure 8-5(C) cannot change the dipole moment (zero in this case) and thus does not absorb infrared radiation.

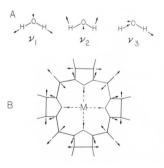

A

ν_1 ν_2 ν_3

B

C $\leftarrow O{=}C{=}O \rightarrow$

Figure 8-5. Vibrational modes of simple and complex molecular structures. The mode shown for the substituted porphyrin structure (*B*) is one of its 35 permissible in-plane, RR-active vibrations.

Raman interaction is based on molecular polarizability rather than directly upon dipole moment since the photon–molecule interaction is based on induced dipole moments. If the vibrational mode does not change molecular polarizability, then there is no mechanism for energy transfer in the scattering process. The induced dipole, m, involved in the Raman effect depends on polarizability, α, and the electric field, ϵ, created by the photon:

$$m = \alpha\epsilon$$

(8-4)

A full account of the applicable principles of symmetry and group theory is beyond the scope of this volume, and the reader should consult an excellent reference on this subject (Cotton, 1967). In using group theory to determine the vibrational modes responsible for Raman spectra one applies the principle that "a fundamental transition will be Raman active . . . if the normal mode involved belongs to the same (symmetry) representation as one or more of the components of the polarizability tensor of the molecule" (p. 265 of Cotton, 1967). The tensor components are quadratic functions of the Cartesian coordinates (for example, x^2, xy, or a combination such as $x^2 - y^2$), and character tables include these with the representations which they generate. As an example, in the case of the carbonate ion or any other molecule with D_{3h} symmetry vibrational modes belonging to the A'_1, E', and E'' representations should be Raman active (the reader should examine the complete set of character tables in Cotton, 1967).

Symmetry effects are not confined to the realm of simple molecules. Although biopolymers are inherently chiral, some structures may have pseudosymmetry, e.g., as in porphyrins and other metal ion coordination sites. In these cases the disturbance created by the bulk of the macromolecule may be, to a first approximation, somewhat isolated by intervening chemical bonds. Under such conditions a mode which would be symmetry forbidden in a simple structure might show up weakly in the biomolecule.

Raman intensities may be related to the Raman tensor, the latter usually being defined as the derivative of molecular polarizability with respect to the mass-weighted normal coordinate system. It is not a trivial matter to relate theory to the observed Raman intensities, but examples of such calculations are found in theoretical treatments of planar and pyramidal MX_3 molecules (Bote and Montero, 1984). If certain conditions of pseudosymmetry are satisfied, then it may be possible to relate intensities to the polarizability properties of biological molecules. This avenue is open for development.

8.5 Vibrational Energy Levels

A simple model for a vibrating structure is a potential energy well like that of the two-atom harmonic oscillator shown in Figure 8-6(A). Since the potential energy of a parabolic function is $kx^2/2$, where k is the force constant and x the linear displacement from equilibrium, the wave equation is:

$$\frac{d^2\psi}{dx^2} + \frac{8\pi^2\mu}{h^2} (E - \tfrac{1}{2}kx^2)\psi = 0 \tag{8-5}$$

In Eq. (8-5), ψ is the wave amplitude and μ the reduced mass:

$$\mu = m_1m_2/(m_1 + m_2) \tag{8-6}$$

The solution of Eq. (8-5) is exact, leading to quantized vibrational energy levels determined by a quantum number v and the spacing of the levels, $h\nu$:

$$E = (v + \tfrac{1}{2})h\nu \tag{8-7}$$

From Eq. (8-5), E is seen to be a function of the atomic masses (through m) and the force constant, k. The latter quantity is closely related to bond order, or more correctly, bond strength. In isotopic replacement studies k remains constant while μ is varied. The effect is easily seen in the following equation for a simplistic classical two-body harmonic oscillator:

$$\nu = [1/(2\pi)] (k/\mu)^{1/2} \tag{8-8}$$

If μ is increased by replacing one of the atoms with a heavier isotope, ν will *decrease* predictably.

In a real, polyatomic molecule the situation is much more complex, and an evaluation of the kinetic and potential energies of vibration is not trivial (Barrow, 1962). To some extent isotope replacement must affect *all* vibrational modes, though some will be frequency-shifted more than

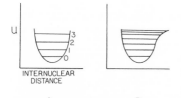

INTERNUCLEAR
DISTANCE

A   B

Figure 8-6. Potential energy diagrams for harmonic and anharmonic oscillators. The ordinate is potential energy (u) and the abscissa, internuclear distance. A classical harmonic oscillator (e.g., two masses connected by a spring) is described by the parabolic potential energy well shown in A. Molecular vibrational levels are quantized, and the potential energy of each level has a discrete value, shown as the horizontal lines in A (vibrational quantum numbers are presented at the right). The anharmonic oscillator (B) has a potential energy well of the type characterizing chemical bonds, i.e., u approaches a finite limit at large internuclear distances, where dissociation occurs.

others. Also, the "anharmonic oscillator" potential energy well of Figure 8-6(B) is a better description of a real molecule since it correctly predicts dissociation at higher vibrational energies. Quantum mechanical treatment of anharmonic oscillators is not exact, but one result is a closer spacing of energy levels at higher vibrational amplitudes, as shown in Figure 8-6(B).

8.6 Applications

8.6.1 Conventional Raman Spectroscopy (Excitation Away from Absorption Bands)

Conventional Raman spectra contain information from all portions of the molecule. Consequently, the spectrum will be dominated by abundant bonding groups or strong resonances which occur in relatively uncluttered regions. Examples are collected in Table 8-1, and on inspection it will be seen that much of the information conveyed by ordinary Raman spectra may be related to molecular conformation.

The amide linkages of peptide backbones exhibit shifts (Table 8-1) which depend on the kind of secondary structure present (or absent).

Table 8-1. Protein Modes[a] Which Reflect Secondary and Tertiary Structure by Shifts or Their Presence or Absence

Group, Mode	Range, $\bar{\nu}$	Remarks
C—N stretch and N—H bend, peptide backbone	$1275-1235$ cm^{-1}	α-Helix 1275, weak; random 1245, broad; β-pleated sheet 1235, sharp.
C—O stretch, peptide backbone	$1670-1655$ cm^{-1}	α-Helix 1655; random 1665; β-pleated sheet 1670.
—S—S—, peptide	510 cm^{-1}	Strong, amplitude reflects the number of intrapeptide and quaternary disulfide bonds.
Tyrosine, aromatic	830 cm^{-1} & 850 cm^{-1} (Fermi doublet)	Doublet shape depends on whether the tyrosine is internal or solvent-exposed. If it is solvent-exposed and hydrogen-bonded to water, the line at 850 cm^{-1} is stronger. The opposite is true for internal locations.
C—S stretch, methionine or cysteine	$700-630$ cm^{-1}	Line position depends on conformation at the C—C bond adjacent (700 trans, 630 gauche).

[a]These are strong modes.

Naturally occurring proteins differ in the relative amounts of secondary and tertiary structure present, e.g., some have large amounts of helical coiling (hemoglobin) while others are more random, consisting mostly of tertiary structure. As a result, the Raman spectrum of a real protein will present composite profiles in the 1275–1235 cm^{-1} and 1670–1655 cm^{-1} regions. Deconvolution techniques may be applied to the observed profile for an estimate of the relative contributions from helical, pleated, and random chain arrangements. The agreement between Raman characterizations and crystallographically established structures appears to be quite good with differences amounting to a few percent (Lippert *et al.*, 1976; Kanna *et al.*, 1975).

Characteristics of tertiary structure are also recognizable in the Raman spectrum, e.g., disulfide bonds, average conformation at C—C—S groupings, solvated tyrosyl residues (Table 8-1), etc. Conformational studies offer a noncrystallographic approach to space-filling structure. If the peptide primary sequence is known along with the features of secondary and tertiary interactions, a conformational analysis by computer (see Section 2.7.1.2.2) may be conducted intelligently since many possible structures will be inconsistent with the available information. Theoretical methods which minimize thermodynamic interactions are applicable to small peptides. Extension to larger entities necessitates a combination of spectroscopies (nmr, Raman, etc.), in addition to the sequence data.

The examples cited in the previous paragraph involved peptides, but one may approach other biological materials, e.g., membranes (Gaber and Peticolas, 1977) and nucleic acids (Luoma and Marshall, 1978). Time-resolved and coherent anti-Stokes spectrometers (see Sections 8.1.1 and 8.7.1) produce similar information with the advantage that fluorescence is avoided.

8.6.2 Resonance Raman Applications

8.6.2.1 Porphyrin Chromophores

Most existing biological applications of RR have been oriented toward porphyrins and related structures. Vibrational modes associated with the porphyrin macrocycle are affected by several factors, e.g., the spin and oxidation state of the central metal atom influences both wavelengths and intensities. Similar effects may derive from interactions of the metal with axially coordinating ligands (Kitagawa *et al.*, 1978; Ozaki *et al.*, 1978). The latter type of effects are of obvious interest with regard to protein function.

The assignment of RR lines to specific vibrational modes depends upon studies of symmetric porphyrins (Kitagawa *et al.*, 1978). One such reference substance is octaethylporphyrin, which has D_{4h} point symmetry in its Ni^{2+}, Co^{2+}, Cu^{2+}, and Zn^{2+} derivatives. The polarizations of RR lines relate to vibrational symmetry, and each may be characterized by an intensity ratio $p = I_\perp/I_\parallel$, where $I_\perp$ is the intensity of scattering with the electric vector perpendicular to the exciting laser beam and $I_\parallel$ is scattering with the vector parallel. In the D_{4h} symmetry group the A_{1g} ($p = 1/8$ or 1/3), B_{1g} and B_{2g} ($p = 3/4$), and A_{2g} ($p = \infty$) modes are readily distinguished (Kitagawa *et al.*, 1975). As shown in Figure 8-5(B) the analogous vibrational modes of the series of metalloporphyrins are clearly indicated.

Figure 8-5(B) illustrates an important feature of the porphyrin macrocycle. Owing to its central location the metal ion has *gerade* symmetry and thus may not enter vibrational motion [see the mode depicted in Figure 8-5(B)]. As a consequence, the mass of the metal ion *per se* does not affect in-plane vibration frequencies; instead, the observed differences are due to alteration in the force constants (i.e., bond strengths) accompanying metal coordination (isotope replacement is compromised). If metal orbitals conjugate with π-bonding orbitals the result will be (1) stronger π-bonding and (2) an increase in the optical absorption frequency (in this case a $\pi-\pi^*$ Q-band). As shown in Figure 8-7, an increase in the optical absorption band frequency is paralleled by increases in RR frequencies.

Metal conjugation with π-electrons also appears to be the dominant source of shifts induced by axial ligands. The iron in Fe^{3+} (high-spin) octaethylporphyrin is out-of-plane in the axial chloro derivative but in-plane in the imidazole derivative. Better conjugation is probably responsible for higher Raman frequencies in the latter. In contrast, RR makes

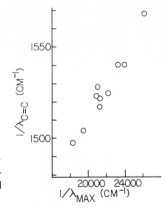

Figure 8-7. Correlation between the C—C vibration frequency and the electronic absorption maximum in conjugated hydrocarbons. Adapted from Carey, 1978, and references cited therein.

Table 8-2. Resonance Raman Redox Markers of Ferric and Ferrous
Cytochrome b_{562} Compared with Those Expected for Various Spin Forms and
Oxidation States of Heme Iron[a]

Band	Valence state	Spin state	Expected position (cm^{-1})	Observed position Ferric (cm^{-1})	Observed position Ferrous (cm^{-1})
Oxidation state marker	Ferric	na	1374–1373	1370	
band A (polarized band)	Ferrous	na	1362–1358		1362
Band B (depolarized)	Ferric	na	1565–1562	1563	
	Ferrous	na	1548–1546		1548
Spin state marker band	Ferric	Low	1588–1582	1588	
C (anomalously	Ferric	High	1555	none	
polarized)	Ferrous	Low	1586–1584		1586
	Ferrous	High	1552		none
Oxidation and spin state	Ferric	Low	1508–1502	1506	
marker band E	Ferric	High	1482	none	
(polarized)	Ferrous	Low	1493		1493
	Ferrous	High	1973		none
Band F (depolarized)	Ferric	Low	1642–1636	1639	
	Ferric	High	1608	none	
	Ferrous	Low	1620		
	Ferrous	High	1607		

[a]Adapted from Bullock and Myer, 1978.

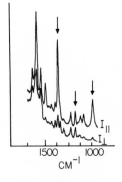

Figure 8-8. Polarized RR scattering from *E. coli* cytochrome b_{562}. The polarization analyzer is parallel to the incident beam in the upper recording and perpendicular to it in the lower recording. It is seen that the intensity ratios vary widely. The center arrow marks a feature with comparable parallel and perpendicular intensities while the outer arrows label lines with large polarization differences. Adapted from Bullock and Myer, 1978.

no distinction between axial halides, i.e., electronegativity is not a determining factor.

Other effects include pH-induced shifts and changes in line intensities. The importance of model compound studies lies in providing insight into the type of behavior one may expect of the structure under investigation. The inclusion of the framework within a protein matrix will introduce asymmetry, resulting in altered scattering polarizations, splitting of resonances, and modes which involve movements of the metal. Nevertheless, in many cases a degree of pseudosymmetry may be retained.

Empirical studies with heme iron have revealed marker bands which characterize oxidation and spin states (Spiro and Strekas, 1974). An application to cytochrome b_{562} isolated from $E.$ $coli$ is illustrated in Table 8-2. Figure 8-8 presents the scattering spectra for reduced b_{562}, and it will be noted that the two polarization traces ($I_\perp$ and $I_\parallel$) provide the ratio information, described above, which confirms band identity. From Table 8-2 one may infer that cytochrome b_{562} exists as a low-spin hemoprotein in both valence states (Bullock and Myer, 1978). Cytochrome b_{562} probably contains two histidines and three methionines (Bullock and Myer, 1978), and the low-spin nature of the heme iron could be due to histidine/histidine/methionine coordination at the axial positions.

Anomalous oxidation state markers may occur in some instances. In reduced cytochromes P-450 the band IV marker is found shifted to unusually low frequencies. This has been interpreted as an effect of strong π-basicity on the part of the axial ligand at heme iron, causing delocalization of d-orbital electrons from iron into the porphyrin π-antibonds, thus weakening macrocycle force constants (Ozaki et $al.$, 1978). The RR evidence thus complements evidence from nmr (Keller et $al.$, 1972), esr (Collman et $al.$, 1975), and Mössbauer (Koch et $al.$, 1975) which places an R—S$^-$ ligand at the axial position. In agreement, the carbonyl complexes with these proteins show similar shifts in line position (Ozaki et $al.$, 1978).

8.6.2.2 Chlorophylls

RR methods are applicable to chlorophylls due to the strong absorption of these pigments and ample evidence that the RR scattering bands are sensitive to environment (Lutz et $al.$, 1979). Much current interest centers on the structural characteristics of the so-called "quantum lens," i.e., domains of closely associated chlorophyll a. The latter entities absorb visible light photons, the energy of which is then communicated throughout the domain as an exciton until the excitation energy is dissipated irreversibly at a primary redox reaction center associated with the domain.

Two basic postulates of chloroplast pigment structure have been proposed. In one (Katz, 1973), the domain aggregation is envisioned as a
spontaneous association of the lipid-like chlorophyll molecules in the manner of stacked waffles. Katz has produced clear evidence that chlorophyll
a exhibits a marked tendency to form aggregates of this type. In the second
postulate, Lutz *et al.* (1979) proposed a somewhat different situation involving a closer association of the chlorophyll with protein (a view consistent with the probable role of proteins in most biological manifestations
of specificity). In either case there must be a means of conveying absorbed
energy to the reactive photocenter.

Lutz and coworkers (1979) have compared RR spectra obtained from
the living alga *B. alpina* with various model systems. *Marked* differences
were observed in the 1700–1550 cm^{-1} spectral region, not only between
the various model systems but also between these and the *in vivo* chlorophyll. The models included monomeric chlorophyll *a* (acetone solvent),
chlorophyll *a* oligomers, and hydrated chlorophyll *a* polymers. These
results raise questions about the presence of self-associated chlorophyll
aggregates in living chloroplasts (postulate one). Model systems and reconstituted mixtures are not inappropriate subjects for investigation; rather,
their validity may be judged by the degree to which the *in vivo* condition
is mimicked, not only in the sense of function but also in physical characteristics. RR and other methods may be used in this fashion, best described as fingerprinting. The advantage of RR lies in its ability to characterize pigments in the *in vivo* condition.

8.6.2.3 Carotenoid Pigments

Carotenoid substances appear throughout the plant and animal king-
doms, occurring in photosynthetic membranes and in association with
protein, as in human visual pigments (Mathies *et al.*, 1977). These are
convenient RR subjects due to their pigmentation, and function-related
changes may be observed in their RR spectra (Marcus, 1978).

The binding of carotenoids to proteins is often accompanied by shifts
in the visible light absorption spectrum. It was shown in Figure 8-7 that
RR frequencies correlate with the absorption maximum of the chromophore, consistent with changes in delocalization of the chromophore-
associated π-electrons since the latter affect both the force constants and
the energy gap between the ground and excited states. The RR spectrum
may thus present evidence of the molecular features of pigment/protein
interactions, i.e., the various RR frequencies may not be equally influenced on going from the unbound to the bound state.

Examples are found in lobster carotenoproteins, ovoverdin and the crustacyanins, which bind astaxanthin (Salares *et al.*, 1978). In examining these proteins it was possible to establish that the RR spectra obtained *in situ* were identical to those of the isolated carotenoproteins, illustrating again the selectivity of RR in probing complex mixture and providing some assurance of the validity of measurements obtained on isolated components. The RR effects and the red shift in λ_{max} on binding were consistent with structural models which envision charged groups at the protein's binding site. Interaction with the charge would polarize the carotenoid molecule.

Astaxanthin itself aggregates in an aqueous environment, and its λ_{max} shifts toward the red. However, the RR frequencies for C—C stretching vibrations are not altered, indicating that the process of aggregation alters the excited state rather than the ground state.

8.6.2.4 Charge Transfer Bands in Metalloproteins and Metalloenzymes

RR studies of metalloprotein and metalloenzyme charge transfer bands are particularly appropriate to the subject of this reference series. Charge transfer absorption spectra are associated with many of the known metalloenzymes and metalloproteins (Larrabee, 1978). As a general rule, the bonding electrons involved in light absorption are fairly localized, involving only the metal ion and its immediate ligands. Also, charge transfer chromophores usually absorb strongly and produce RR effects at low concentrations.

Due to the direct involvement of a metal atom one may expect to observe metal–ligand modes. These will generally occur in the 500 to 50 cm^{-1} range owing to the kind of bond strengths and masses involved in metal ion–ligand vibrations (Desbois *et al.*, 1978; Vergoten *et al.*, 1978). In many cases this is the only active region of the spectrum, as shown for the copper enzyme laccase in Figure 8-9 (Siiman *et al.*, 1974). Absence of higher-frequency modes typical of organic structures is consistent with the aforementioned localization of the chromophore. Assignments are often difficult. Representative examples are shown in Table 8-3, and it should be understood that some of these are still in question. Comparisons

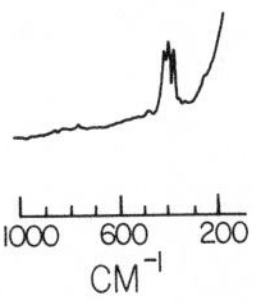

Figure 8-9. Metal–ligand modes in the 600–200 cm^{-1} range in the RR spectrum of laccase. The spectral features near 400 cm^{-1} are presumed to be metal–ligand vibrations. The sharp upturn in the baseline below 200 cm^{-1} is a result of source interference, i.e., the monochromator is approaching the excitation wavelength. Adapted from Siiman *et al.*, 1974.

Table 8-3. Metal–Ligand Vibrational Frequency Ranges

Bond	Substance		Reference
Fe—S	Rubredoxin	3H, 368 cm^{-1}	Long and Loehr, 1970
Fe—S	Fe$_4$S$_4$ (Pseudo-cubane cluster)	400–100 cm^{-1}	Tang et al., 1975
Cu—O	—	500–200 cm^{-1}	—
Cu—N	—	500–200 cm^{-1}	—
Cu—S	—	250 cm^{-1}	Siiman et al., 1976

with model compounds may also aid in the assignment of metal–ligand modes, but there are cases, notably the blue copper proteins, where low molecular weight counterparts have been hard to find (Solomon et al., 1976). The few possibilities include fourteen-copper atom clusters which have X-ray crystallographically proven Cu(II)—sulfur binding interactions (Lucia and Garnier, 1979). Similarly, a related class of Fe(III) thiolate complexes may be models for ferredoxins (Siiman and Carey, 1980).

The information provided by RR, granting a certain assignment, should reflect bond strengths at the metal ion site. Also, functional changes may be detected if the metal ion is at a catalytic location with saturating amounts of the substrate present. The most successful applications of RR to metal-containing proteins have involved labeling the biomolecule with specific isotopes. Isotope replacement is discussed further in the following section.

Iron–ligand stretching bands may be used as a fingerprint of quaternary structure in iron porphyrin-containing proteins (Hori and Kitagawa, 1980), where a 212 cm^{-1} line (^{54}Fe) shifts during conformational changes.

8.6.2.5 Isotope Replacement Methods

It was shown earlier (see Section 8.5) that the element selectivity inherent in the Raman spectroscopies lies in isotope replacement. The vibrational modes which are strongly coupled to a particular nucleus will be found to experience a frequency shift when the nucleus is replaced with a lighter or heavier isotope [Kitagawa et al., 1979; see Eq. (8-8)]. For example, the strongest N$_4$Mg(II) framework-associated vibrations of chlorophyll could be assigned by replacing naturally occurring ^{14}N and ^{24}Mg and ^{15}N and ^{26}Mg, respectively (Lutz et al., 1976).

Modes associated with hydrocarbon backbones may be identified in biological molecules isolated from organisms grown in heavy water and thus presenting C—D bonds in place of C—H. The production of D$_2$O-cultured microorganisms is longstanding (Moses et al., 1958), but the potential of these sources has only been slightly exploited. The use of organisms grown in isotope-enriched media is generally necessary when the element in question is covalently attached to the biological molecule. It is difficult, if not impossible in most cases, to effect a direct chemical replacement of covalent entities without destroying the compound in question.

In some circumstances the isotope is easily replaced, examples being the oxygen transport proteins. The oxygen of oxyhemocyanin is associated with two moles of copper (Lontie and Witters, 1973; Ghiretti, 1978), and RR excitation at the oxygen-dependent visible absorption band reveals a vibrational mode at 742 cm^{-1} (Loehr *et al.*, 1974). This vibrational frequency is characteristic of oxygen in a peroxide (—O—O—) linkage. In support of the assignment, the mode shifts to 704 cm^{-1} when naturally occurring oxygen is replaced by $^{18}O_2$.

A similar case is found for oxyhemerythrin (Kurtz *et al.*, 1976, 1977). A possible peroxide mode occurs at 844 cm^{-1} when the visible absorption is excited at 514.5 nm, and an oxygen–metal mode is observed near 500 cm^{-1} under these conditions. Based on the results shown in Figure 8-10, partial replacement of ^{16}O with ^{18}O reveals some of the characteristics of the oxygen binding site. As in the case of hemocyanin, the substitution of $^{18}O_2$ for $^{16}O_2$ leads to a shift to lower frequency ($\sim$800 cm^{-1}), clearly identifying the RR line as an oxygen vibrational mode. Furthermore, a broader line corresponding to $^{16,18}O_2$ is found between the RR resonances due to $^{16}O_2$- and $^{18}O_2$-substituted hemerythrin. This may be interpreted as being a result of the isotopic asymmetry of $^{16,18}O_2$ and the nonequivalence of the oxygen atoms in the protein-bound state. Figure 8-10(A) shows two possibilities, which lead to two distinct frequencies since the isotopic masses must interact with unique force constants. The broad line is thus due to overlapping lines, as shown beneath the spectrum in Figure 8-10.

An alternative interpretation favors end-on binding of O_2 to Fe (Kurtz

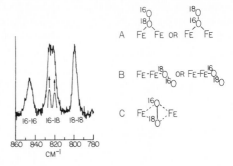

Figure 8-10. Resonance Raman spectra of hemerythrin bound to O_2 which contains equal amounts of the isotopes ^{16}O and ^{18}O. Possible modes of hemerythrin–O_2 bonding are depicted in (A)–(C). Adapted from Kurtz *et al.*, 1976.

et al., 1976). An end-on bent structure of the type shown in Figure 8-10(B) is indicated for oxyhemoglobin (Duff *et al.*, 1979), based on RR studies using isotopically unsymmetrical O_2. Evidence of this type clearly rules out some possibilities such as the broadside structure of Figure 8-10(C). One must realize that the RR evidence alone merely shows that the two oxygens are in different environments, leaving some questions about the detailed structure.

8.6.2.6 Complex Biological Structures

The much higher selectivity of resonance Raman scattering permits its application to unmodified biological materials such as cytochrome heme groups in functionally active mitochondria (Adar and Erecinska, 1978). In most investigations of this type the object is to extend measurements on purified biological substances into the native environment. It may then be possible, under favorable circumstances, to characterize the structural peculiarities which accompany function. Non-resonant Raman techniques have also been applied to fairly complex systems, e.g., to study phospholipid/protein interactions (Bertoluzza *et al.*, 1983).

Biological preparations (e.g., organelles, cells) often contain several major pigments which may be distinguished if their optical absorption bands do not coincide. For example, mitochondrial cytochrome oxidase may be observed by exciting at ~442 nm while cytochromes *b* and *c* are revealed in the excitation range 520–568 nm (Adar and Erecinska, 1978; Salmeen *et al.*, 1978).

The RR spectra of whole mitochondria may be compared with those of the isolated cytochrome *b*-c_1 complex in Figure 8-11. Relatively subtle differences are observed (e.g., near 1300 cm^{-1} with 568.2-nm excitation and at 1568 and 1505 cm^{-1} with 530.9-nm excitation; also, a shift from 1538 cm^{-1} to 1530 cm^{-1}). It is not possible to determine the origin of these changes on the basis of the evidence presented in Figure 8-11, and the observed properties could be due to competing mitochondrial pigments, heme modulations resulting from protein conformation, or functional heme–heme interactions (Adar and Erecinska, 1978). The important

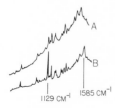

Figure 8-11. Resonance Raman scattering from the cytochrome *b*–c_1 complex with excitation at 568.2 nm. A. Whole mitochondria; B. The isolated *b*–c_1 complex. Adapted from Adar and Erecinska, 1978.

conclusion to be drawn from Figure 8-11 is that RR allows an access to specific substances in rather complex biological mixtures while ordinary Raman spectra are usually too cluttered to be of any use at this level of complexity.

8.6.2.7 Excitation at Ultraviolet Wavelengths

Proteins contain aromatic amino acids (e.g., phenylalanine) while nucleic acids contain similar conjugated structures (pyrimidines and purines). As a result, these materials are strong ultraviolet absorbers. Some of these substances are also fluorescent. Irradiation of the UV absorption bands will produce RR enhanced modes; however, since these groups are widely present in nucleic acids and proteins, the specificity is lost if one examines complex biological preparations. *Most* biological preparations are, therefore, poor subjects for RR probes of the ultraviolet portion of the spectrum, and while there have been attempts to study DNA in living cells (Nocentini and Chinsky, 1983), it is difficult to be certain of the origin of the observed resonances, e.g., some of the bands might originate either in DNA or RNA. Only the relatively near-ultraviolet wavelengths may be used easily (Eickman *et al.*, 1978). Lasers which span the whole spectrum including the ultraviolet region are available, and the limitation is not due to a lack of appropriate excitation sources. Better results are obtained with purified substances. For example, UV resonance Raman spectra of ^{18}O-substituted uracil could be compared with those of natural uracil to obtain a clear assignment of carbonyl-associated vibrational modes (Chinsky *et al.*, 1983). It should be noted that conventional Raman studies of DNA (e.g., at 514.5-nm excitation) have been numerous and offer a means for characterizing nucleic acid conformation and secondary structure (Prescott *et al.*, 1984). The Raman spectra of such complex substances rarely yield to *a priori* interpretations and generally require a source of empirical correlations based on compounds of known structure and conformation.

8.6.2.8 Time-Ordered Spectroscopy in the Picosecond Regime

Using combinations of mode-locked argon-ion lasers and ring dye lasers, it is possible to generate low-power pulses with time durations at or *less than* one picosecond. These pulses may be amplified using Nd/YAG-pumped dye lasers and used to disturb the equilibrium state of a specimen, as in other forms of kinetic spectroscopy (see Section 10.2). Transients thus created are detected by various means, and the possibilities include absorption and fluorescence, in addition to the Raman mode. The extremely rapid time-ordered processes which appear in these spectra may be recorded using a streak camera/vidicon combination. The most rapid chemical events (defined here as being equal to or greater than the time scale of a bond vibration) are accessible by these methods, which are, in fact, able to resolve purely physical processes. The methods are well-suited for characterizing complex events associated with biological pigments (Hilinski and Rentzepis, 1983).

As examples, strong pulses of visible (532 nm) or ultraviolet (355 nm) light cause oxymyoglobin and carboxymyoglobin to release, respectively, O_2 and CO with 5–10 ps of the exciting pulse, based on bleaching at 420 nm (absorbance mode; see Reynolds *et al.*, 1981). Although the two compounds dissociated at the same rate, the quantum yield for oxymyoglobin was much lower than carboxymyoglobin's, showing that in the former compound, most photon energy was dissipated in some other mechanism.

The functioning of visual pigments may be followed using picosecond methods, allowing a direct detection of postulated intermediates. For example, fast transient Raman spectra of 11-*cis*-rhodopsin showed that there is little change in the bonding configuration on excitation with visible light (Braiman and Mathies, 1982) and thus failed to confirm the favored mechanism involving *cis–trans* isomerism (Hilinski and Rentzepis, 1983).

8.7 Newer Methods Which Minimize Fluorescence Interference

8.7.1 Coherent Anti-Stokes Raman Spectroscopy (CARS) or Four-Wave Mixing Spectroscopy

In ordinary Raman (and resonance Raman) the inelastically scattered photon is shifted to a lower frequency by an amount equivalent to the frequency of the vibrational quantum absorbed by the molecule, as shown in Figure 8-12, i.e., a Stokes line is produced at a longer wavelength. If fluorescence is induced by the exciting source, it is also Stokes-shifted and may lead to obscuration of the weak Raman line, as shown in Figure

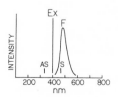

Figure 8-12. Avoidance of fluorescence interference by means of anti-Stokes detection of Raman transitions. In the example shown here, the laser excitation wavelength (Ex) at ~400 nm is absorbed by the substance in question or an interfering impurity, leading to a fluorescence band (F) at longer wavelengths. The latter obscures the weak Stokes Raman line (S), also located at a longer wavelength. In contrast, the anti-Stokes Raman line (AS) is found in the uncluttered region of the spectrum.

8-12. In fact, some degree of sample fluorescence is the rule rather than the exception in biological materials, and most spectra show an undulating or sloping baseline. From Figure 8-12 it is evident that an anti-Stokes Raman scattering, i.e., one in which the molecule loses rather than absorbs the vibrational photon, will lie in a quiescent region of the spectrum, free from fluorescence.

It is in fact possible to obtain the anti-Stokes Raman spectrum (Terhune, 1963; Minck et al., 1963) using the four-photon process of Figure 8-13, which makes use of the unique quality of laser radiation. In Figure 8-13(A), the coherent mixing involves two photons of the irradiating source (v_P), a photon corresponding to the Stokes frequency (v_S), and a photon at the anti-Stokes frequency (v_{AS}). It should be noted that a nonlinear process of this type requires high-intensity (generally pulsed) sources, which may raise questions of possible damage to a biological sample.

In a practical spectrometer the coherent beams v_P and v_S cross at an acute angle $\theta \sim 1°$. Frequency v_S is tuned over a range (requiring several dye lasers), and when v_S corresponds to a Raman line, v_{AS} is scattered coherently at angle $\phi \sim 2\theta$ relative to the v_S beam path. The exact value of ϕ depends on the vector summation of Eq. (8-9), which is shown graphically in Figure 8-13(B).

$$v_{AS} = 2v_P - v_S \tag{8-9}$$

Anti-Stokes scattering thus appears as a laser-like beam emerging in a direction differing from either v_S or v_P. The essentials of the spectrometer are shown in Figure 8-14, where the pumping laser (v_P) has the twofold function of supplying dye excitation and the v_P beam. Lens L insures that the v_P and v_S beams cross in the sample cuvette.

It should be noted that the coherent anti-Stokes Raman spectrum resembles resonance Raman (see Section 8.2) in its potential for resonance enhancement (Nestor et al., 1976). By contrast with these, conventional

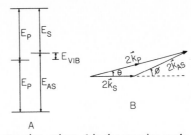

Figure 8-13. The four-photon mixing process of CARS. A. The combined energies of two pump laser photons (E_P) are decremented by the energy of a Stokes Raman photon (E_S) on interaction with the substance of interest. This creates a virtual state (with respect to ground) having the anti-Stokes Raman photon energy (E_{AS}). Marked enhancement of anti-Stokes Raman scattering is thus effected. B. The geometric properties of the system are seen in the four-photon wave vector construction, where θ is the crossing angle between the pump and Stokes resonant beams. Anti-Stokes scattering at angle ϕ results in a beam that is easily isolated from the other two.

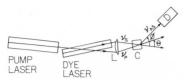

Figure 8-14. Components of a CARS spectrometer. The pump laser operates at a fixed frequency,
providing excitation for the dye laser and the pump
frequency reference beam. The dye laser is tunable, and a CARS spectrum is constructed by varying the dye laser (ν_S) frequency. (Actually, several
dye lasers are needed to cover a broad spectral range.) The lens (L) causes the two laser
beams to cross in the sample cuvette (C) and the anti-Stokes Raman scattering is routed
into a photodetector (D). CARS spectrometers use pulsed lasers.

Raman produces about one inelastic scattering event for $\sim 10^7$ incident
photons. Unlike resonance Raman, the coherent method is not applicable
if the sample absorbs strongly at ν_P or ν_S, i.e., the effect is quenched as
a result of laser nonlinearity. Another disadvantage is the necessity of
relatively higher sample concentrations. The major advantage of the method
lies in its potential for avoiding fluorescent emission bands; consequently,
CARS has been used to obtain resonance-enhanced spectra from flavo-
proteins (Dutta and Spiro, 1980). The flavins are difficult due to fluorescence.

Other coherence methods exist, e.g., the Raman-induced Kerr effect
(RIKES) (Heiman *et al.*, 1976) and inverse Raman spectroscopy (IRS)
(Owyoung, 1978).

8.7.2 Time-Resolved Raman Spectroscopy

A different method for minimizing fluorescence obscuration in Raman
spectra takes advantage of the "sluggishness" of fluorescence emission
relative to inelastic scattering events (Woodruff and Farquharson, 1978).
The laser is pulsed repetitively rather than operated continuously, and
the off-beam intensity is sampled at the times corresponding to a maximum
scattering/fluorescence ratio, as shown in Figure 8-15. Total fluorescence
intensity is the integrated area under the time-ordered curve. The detector
samples intensity data only during the laser pulse interval, and the selected

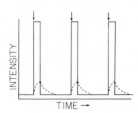

Figure 8-15. Time resolution fluorescence avoidance. Flou-
rescence interference is minimized by recording Raman scattering intensity near the leading edge (arrows) of the excitation
laser pulse. Fluorescence emission (dashed line) builds up
more slowly and is minimal at these points.

interpulse time (> pulse time) permits fluorescence to decay to an acceptably low level. The signal improvement by this method is limited to one or two orders of magnitude and may not even be realized in some cases (Yaney, 1972). Pulsed methods also provide an opportunity for kinetic studies in functioning pigment systems (Lyons et al., 1978).

References

Adar, F., and Erecinska, M., 1978. Resonance Raman spectra of whole mitochondria, *Biochemistry* 17:5484.

Barrow, G. M., 1962. *Introduction to Molecular Spectroscopy*, McGraw-Hill, New York.

Bertoluzza, A., Bonora, S., Fini, G., Morelli, M. A., and Simoni, R., 1983. Phospholipid–protein molecular interactions in relation to immunological processes, *J. Raman Spectrosc.* 14:393.

Bote, M. A. L., and Montero, S., 1984. Raman tensor tables. MX₃ molecules, *J. Raman Spectrosc.* 15:4.

Braiman, M., and Mathies, R., 1982. Resonance raman spectroscopy of bacteriorhodopsin's primary photoproduct: Evidence for a distorted 13-*cis* retinal chromophore, *Proc. Natl. Acad. Sci. U.S.A.* 79:403.

Brewer, R. G., and Mooradian, A., 1974. *Laser spectroscopy, Proceedings of an international conference held in Vail, Colorado, June 25–29, 1973*, Plenum Press, New York.

Brodersen, S., 1979. High-resolution rotation-vibration Raman spectroscopy, in *Topics in Current Physics: Raman Spectroscopy of Gases and Liquids*, Vol. 11 (A. Weber, ed.), Springer-Verlag, New York, pp. 7–69.

Bullock, P. A., and Myer, Y. P., 1978. Circular dichroism and resonance Raman studies of cytochrome *b* from *E. coli.*, *Biochemistry* 17:3084.

Butler, J. F., Nill, K. W., Mantz, A. W., and Eng, R. S., 1978. Application of tunable–diode–laser IR spectroscopy to chemical analysis, *ACS Symposium Series*, Vol. 85, *New Applications of Lasers to Chemistry* (G. M. Hieftje, ed.), American Chemical Society Publications, Washington, D.C., pp. 12–23.

Carey, P. R., 1978. Resonance Raman spectroscopy in biochemistry and biology, *Quart. Rev. Biophys.* 11:309.

Chinsky, L., Hubert-Habart, M., Laigle, A., and Turpin, P. Y., 1983. Carbonyl stretching vibrations of uracil studied by oxygen-18 isotopic substitutions with UV resonance Raman spectroscopy, *J. Raman Spectrosc.* 14:322.

Collman, J. P., Sorrell, T. N., and Hoffman, B. M., 1975. Models for cytochrome P-450, *J. Am. Chem. Soc.* 97:913.

Cotton, F. A., 1967. *Chemical Applications of Group Theory*, John Wiley & Sons, New York.

Desbois, A., Lutz, M., and Banerjee, R., 1978. Low-frequency vibrations in resonance Raman spectra of myoglobin: Iron–ligand and iron–nitrogen modes, *C.R. Acad. Sci., Ser. D.* 287:349.

Dhamelincourt, P., Wallart, F., Leclercq, M., Nguyen, A. T., and Landon, D. O., 1979. Laser Raman molecular microprobe (MOLE), *Anal. Chem.* 51:414A, 420A.

Duff, L. L., Appelman, E. H., Shriver, D. F., and Klotz, I. M., 1979. Steric disposition of oxygen in oxyhemoglobin as revealed by its resonance Raman spectrum, *Biochem. Biophys. Res. Commun.* 90:1098.

Dutta, P. K., and Spiro, T. G., 1980. Resonance coherent anti-Stokes Raman scattering spectra of oxidized and semiquinone forms of clostridium MP flavodoxin, *Biochemistry* 19:1590.

Eickman, N. C., Solomon, E. I., Larrabee, J. A., Spiro, T. G., and Lerch, K., 1978. Ultraviolet resonance Raman study of oxytyrosinase. Comparison with oxyhemocyanins, *J. Am. Chem. Soc.* 100:6529.

Freedman, T. B., Santillo, F. S., Zimba, C. G. Nafie, L. A., and Dabrowiak, J. C., 1983. Raman spectral studies of bleomycin A2 and related structural fragments: A probe for bleomycin–DNA interactions, *J. Raman Spectrosc.* 14:266.

Frushour, B. G., and Koenig, J. L., 1975. Raman spectroscopy of proteins, in *Advances in Infrared and Raman Spectroscopy* (R. J. H. Clark and R. E. Hester, eds.), Heyden and Son, New York.

Gaber, B. P., and Peticolas, W. L., 1977. On the quantitative interpretation of biomembrane structure by Raman spectroscopy, *Biochim. Biophys. Acta* 465:260.

Ghiretti, F. (ed.), 1968. *Physiology and Biochemistry of Hemocyanins,* Academic Press, New York.

Heiman, D., Hellwarth, R. W., Levenson, M. D., and Martin, G., 1976. Raman-induced Kerr effect, *Phys. Rev. Lett.* 36:189.

Hilinski, E. F., and Rentzepis, P. M., 1983. Biological applications of picosecond spectroscopy, *Nature (London)* 302:481.

Horak, M., and Horak, A. V., 1979. *Interpretation and Processing of Vibrational Spectra,* John Wiley & Sons, New York.

Hori, H., and Kitagawa, T., 1980. Iron–ligand stretching band in the resonance Raman spectra of ferrous iron porphyrin derivatives. Importance as a probe band for quaternary structure of hemoglobin, *J. Am. Chem. Soc.* 102:3608.

Horrocks, W. D., Rhee, M., Snyder, A. P., and Sudnick, D. R., 1980. Laser-induced metal ion luminescence: Interlanthanide ion energy transfer distance measurements in calcium binding proteins, parvalbumin and thermolysin, *J. Am. Chem. Soc.* 102:2650.

Kannan, K. K., Notstrand, B., Fridborg, L., Lorgren, S., Ohlsson, A., and Petef, M., 1975. Crystal structure of human erythrocyte carbonic anhydrase B. Three-dimensional structure at a nominal 2.2-ang. resolution, *Proc. Natl. Acad. Sci. U.S.A.* 72:51.

Katz, J. J., 1973. Chlorophyll interactions and light conversion in photosynthesis, *Naturwissenchaften* 60:32.

Keller, R. M., Wuethrich, K., and Debrunner, P. G., 1972. Proton magnetic resonance reveals high-spin iron(II) in ferrous cytochrome P450 from *Pseudomonas putida, Proc. Natl. Acad. Sci. U.S.A.* 69:2073.

Kitagawa, T., Ogoshi, H., Watanabe, E., and Yoshida, Z., 1975. Resonance Raman scattering from metallo-porphyrins. Metal and ligand dependence of the vibrational frequencies of octaethylporphyrins, *J. Phys. Chem.* 79:2629.

Kitagawa, T., Ozaki, Y., and Kyogoku, Y., 1978. Resonance Raman studies on the ligand–iron interactions in hemoproteins and metallo-porphyrins, *Adv. Biophys.* 11:153.

Kitagawa, T., Nishina, Y., Kyogoku, Y., Yamano, T., Ohishi, N., Takai-Suzuki, A., and Yagi, K., 1979. Resonance Raman spectra of carbon-13 and ntirogen-15-labeled riboflavin bound to egg-white flavoprotein, *Biochemistry* 18:1804.

Koch, S., Tang, S. C., Holm, R. H., Frankel, R. B., and Ibers, J. A., 1975. Ferric porphyrin thiolates. Possible relationship to cytochrome P-450 enzymes and the structure of (*p*-nitrobenzenethiolato) iron(III) proton porphyrin IX dimethyl ester, *J. Am. Chem. Soc.* 97:916.

Kurtz, D. M., Shriver, D. F., and Klotz, I. M., 1976. Resonance Raman spectroscopy with unsymmetrically isotopic ligands. Differentiation of possible structures of hemerythrin complexes, *J. Am. Chem. Soc.* 98:5033.

Kurtz, D. M., Shriver, D. F., and Klotz, I. M., 1977. Structural chemistry of hemerythrin, *Coord. Chem. Rev.* 24:145.

Larrabee, J. A., 1978. Resonance Raman studies of copper proteins, *Diss. Abstracts Int. B* 39:2773.

Lippert, J. L., Tyminski, D., and Desmeules, P. J., 1976. Determination of the secondary structure of proteins by laser Raman spectroscopy, *J. Am. Chem. Soc.* 98:7075.

Loehr, J. S., Freedman, T. B., and Loehr, T. M., 1974. Oxygen binding to hemocyanin. Resonance Raman spectroscopic study, *Biochem. Biophys. Res. Commun.* 56:510.

Long, T. V., and Loehr, T. M., 1970. The possible determination of iron coordination in nonheme iron proteins using laser-Raman spectroscopy. II. *Clostridium pasteurianium* rubredoxin in aqueous solution, *J. Am. Chem. Soc.* 93:1809.

Lontie, R., and Witters, R., 1973. Hemocyanin, in *Inorganic Biochemistry,* Vol. i (G. L. Eichhorn, ed.), Elsevier, Amsterdam, Chapter 12.

Lucia, T., and Garnier, A., 1979. Circular dichroism and resonance Raman spectra of the Cu(II)–Cu(I) complex of D-penicillamine. The CuS(cys) stretching mode in blue copper proteins, *Biochem. Biophys. Res. Commun.* 91:1273.

Luoma, G. A., and Marshall, A. G., 1978. Laser Raman evidence for new cloverleaf secondary structures for eukaryotic 5.8S RNA and prokaryotic 5S RNA, *Proc. Natl. Acad. Sci. U.S.A.* 75:4901.

Lutz, M., Kleo, J., Gilet, R., Henry, M., Plus, R., and Leicknam, J. R., 1976. Vibrational spectra of chlorophylls *a* and *b* labelled with magnesium-26 and nitrogen-15, in *Proc. Int. Conf. Stable Isotopes,* 2nd (1975), (E. R. Klein and P. D. Klein, eds.). NTIS, Springfield, Virginia.

Lutz, M., Brown, J. S., and Remy, R., 1979. Resonance Raman spectroscopy of chlorophyll–protein complexes, *Ciba Found. Symp.* 61(Chlorophyll Organ. Energy Transfer Photosynthesis):105.

Lyons, K. B., Friedman, J. M., and Fleury, P. A., 1978. Nanosecond transient Raman spectra of photolyzed carboxyhemoglobin, *Nature (London)* 275:565.

Marcus, M. A., 1978. Picosecond absorption and emission spectroscopy and kinetic and steady state resonance Raman spectroscopy of native isotopically labeled bacteriorhodopsin, *Diss. Abstracts Int. B* 39:1601.

Mathies, R., Freedman, T. B., and Stryer, L., 1977. Resonance Raman studies of the conformation of retinal in rhodopsin and isorhodopsin, *J. Mol. Biol.* 109:367.

Minck, R. W., Terhune, R. W., and Rado, W. G., 1963. Laser-stimulated Raman effect and resonant four-photon interactions in the gases H_2, D_2 and CH_4, *Appl. Phys. Lett.* 3:181.

Morris, M. D., and Wallan, D. J., 1979. Resonance Raman spectroscopy. Current applications and prospects, *Anal. Chem.* 51:182A.

Moses, V., Holm-Hansen, O., and Calvin, M., 1958. Response of chlorella to a deuterium environment, *Biochim. Biophys. Acta* 28:62.

Nestor, J. R., Spiro, T. G., and Klauminzer, G. K., 1976. Coherent anti-Stokes Raman scattering (CARS) spectra, with resonance enhancement of cytochrome-C and Vitamin B_{12} in dilute aqueous solution, *Proc. Natl. Acad. Sci. U.S.A.* 73:3329.

Nocentini, S., and Chinsky, L., 1983. *In vivo* studies of nucleic acid by ultraviolet resonance Raman spectroscopy on eukaryotic living cells, *J. Raman Spectrosc.* 14:9.

Omenotto, N., 1979. *Analytical Laser Spectroscopy,* John Wiley & Sons, New York.

Owyoung, A., 1978. Coherent Raman gain spectroscopy using CW laser sources, *IEEE J. Quantum Electron.* QE-141:192.

Ozaki, Y., Kitagawa, T., and Kyogoku, Y., 1978. Resonance Raman studies of hepatic microsomal cytochromes P-450: Evidence for strong pi basicity of the fifth ligand in the reduced and carbonyl complex forms, *Biochemistry* 17:5826.

Parker, F. S., 1983. *Applications of Infrared, Raman and Resonance Raman Spectroscopy in Biochemistry*, Plenum Press, New York.

Prescott, B., Steinmetz, W., and Thomas, G. J., Jr., 1984. Characterization of DNA structures by laser Raman spectroscopy, *Biopolymers* 23:235.

Reynolds, A. H., Rand, S. D., and Rentzepis, P. M., 1981. *Proc. Natl. Acad. Sci. U.S.A.* 78:2292.

Salares, V. R., Young, N. M., Bernstein, H. J., and Carey, P. R., 1978. Mechanisms of spectral shifts in lobster carotenoproteins. The resonance Raman spectra of ovoverdin and the crustacyanins, *Biochim. Biophys. Acta.* 576:176.

Salmeen, I., Rimai, L., and Babcock, G. T., 1978. Raman spectra of heme *a*, cytochrome oxidase–ligand complexes, and alkaline denatured oxidase, *Biochemistry* 17:800.

Siiman, O. and Carey, P. R., 1980. Resonance Raman spectra of some ferric and cupric thiolate complexes, *J. Inorg. Biochem.* 12:353.

Siiman, O., Young, N. M., and Carey, P. R., 1974. Resonance Raman studies of blue copper proteins, *J. Am. Chem. Soc.* 96:5583.

Solomon, E. I., Hare, J. W., and Gray, H. B., 1976. Spectroscopic studies and a structural model for blue copper centers in proteins, *Proc. Natl. Acad. Sci. U.S.A.* 73:1389.

Spiro, T. G., and Gaber, B. P., 1977. Laser Raman scattering as a probe of protein structure, *Annu. Rev. Biochem.* 46:553.

Spiro, T. G., and Strekas, T. C., 1974. Resonance Raman spectra of heme proteins. Effects of oxidation and spin state, *J. Am. Chem. Soc.* 96:338.

Steinfeld, J. I. (ed.), 1978. *Laser and Coherence Spectroscopy*, Plenum Press, New York.

Tang, S. P. W., Spiro, T. G., Antanaitas, C., Moss, T. H., Holm, R. H., Herskovitz, T., and Mortensen, L. E., 1975. Resonance Raman spectroscopic evidence for structural variation among bacterial ferredoxin, HIPIP, and $Fe_4S_4(SC_2PI)_4^{2-}$, *Biochem. Biophys. Res. Commun.* 62:1.

Terhune, R. W., 1963. Coherent anti-Stokes Raman spectroscopy, *Bull. Am. Phys. Soc.* 8:359.

Thomas, G. J., and Kyogoku, Y., 1977. In *Infrared and Raman Spectroscopy* (E. G. Brame and J. G. Grasselli, eds.), Part C (Biological Science), Marcel Dekker, New York, pp. 717–861.

Tsai, C. W., and Morris, M. D., 1975. Application of resonance Raman spectrometry to the determination of vitamin B_{12}, *Anal. Chim. Acta* 76:193.

Vergoten, G., Fleury, G., and Moschetto, Y., 1978. Low frequency vibrations of molecules with biological interest, in *Advances in Infrared and Raman Spectroscopy*, Vol. 4 (R. J. H. Clark and R. E. Hester, eds.), Heyden and Son, New York, pp. 195–269.

Woodruff, W. H. and Farquharson, S., 1978. Time-resolved resonance Raman spectroscopy (TR^3) and related vidicon Raman spectrography: Vibrational spectra in nanoseconds, in *New Applications of Lasers to Chemistry. ACS Symposium Series*, Vol. 85, (G. M. Hieftje, ed.), American Chemical Society Publications, Washington, D.C., pp. 216–236.

Yaney, P. P., 1972. Reduction of fluorescence background in Raman spectra by pulsed Raman technique, *J. Opt. Soc. Am.* 62:1297.

Circular Dichroism (CD) and Magnetic Circular Dichroism (MCD)

For reasons given in Chapter 7, ultraviolet and visible light absorption measurements offer very little information of a selective nature. Two other methods which use light in this wavelength range, optical rotatory dispersion (ORD) and circular dichroism (CD), provide evidence of a structural/stereochemical nature, although to gain such insight it is necessary to compare the results from ORD and CD with information from other methods. When the sample under investigation is subjected to an intense magnetic field (~ 50 kG) the phenomena observed in ORD and CD are extended, respectively, to magnetic optical rotatory dispersion (MORD) and magnetic circular dichroism (MCD). This chapter is concerned mostly with MCD. However, before attempting a discussion of MCD it will be necessary to first present a brief account of the basic characteristics of the related methods (Djerassi *et al.*, 1971; Stephens, 1974).

9.1 The Relationship between ORD, CD, MORD, and MCD

Plane-polarized light may be thought of as the resultant of left- and right-handed circularly polarized components, as shown in Figure 9-1. These components interact equally with matter as long as the medium through which they pass consists of symmetrical molecules. Unequal interaction usually occurs when the molecular field is asymmetric, forming the basis of the phenomena described below. Stated simply, ORD and MORD measure the wavelength dependence of refractive index differences between the left and right circularly polarized light while CD and MCD are due to a differential *absorption* of these two components.

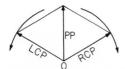

Figure 9-1. Plane-polarized light as a vector sum of left and right circularly polarized components. The line of propagation is from the eye through point 0. RCP and LCP are, respectively, the electric vectors of right and left circularly polarized waves, rotating about 0 in the directions indicated by the curved arrows.

The electric vector of plane-polarized light (PP) may be imagined as the vector sum of LCP and RCP, as shown here. When both vectors point straight up, PP = 2RCP = 2LCP, and when they are opposed, PP = 0.

9.1.1 Optical Rotatory Dispersion (ORD)

The refractive index, n, of a medium is the ratio of the velocity of light in a vacuum to its velocity in the medium. When the medium consists of a chiral substance, the left and right circularly polarized components of plane-polarized light interact differentially so that one experiences a lower velocity, i.e., there is a difference in refractive index. This has the effect of *progressively* rotating the plane of polarization as the ray passes through the substance. Fresnel's equation defines the rotation (in radians per unit length) as a function of the wavelength of the light, λ, and the refractive indices, n_L and n_R, experienced by the left and right circularly polarized components:

$$\phi = (\pi/\lambda)(n_L - n_R) \tag{9-1}$$

In actual ORD measurements the molecular rotation, $[\Phi]$, is recorded as a function of wavelength. Molecular rotation is determined from cell length, l (in decimeters), molar concentration, c, molecular weight, M_w, and the rotation of the plane of polarization, α (expressed in degrees):

$$[\Phi] = \alpha M_w/(100lc) \tag{9-2}$$

In this manner the ORD spectra of different substances may be compared on a molar basis, and $[\Phi]$ is thus comparable to the molar extinction coefficient of absorption spectroscopy. Under typical experimental conditions the value of α may amount to tens of degrees, although some cooperative phenomena based on molecular asymmetry, e.g., as in cholesteric liquid crystals, may lead to much larger rotatory effects.

Examples of ORD spectra are shown in Figure 9-2. Curve I is a *positive plain* ORD recording typical of substances which do not absorb light in the spectral region examined. Behavior of this type is predicted by the Drude equation (Djerassi, 1960), the first term of which is as follows:

$$[\Phi] \cong K/(\lambda^2 - \lambda_0^2) \tag{9-3}$$

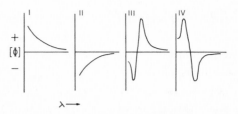

Figure 9-2. Plain and Cotton effect ORD curves. I. Positive plain; II. negative plain; III. positive Cotton effect; IV. negative Cotton effect. The ordinate is molecular rotation and the abscissa, wavelength.

The wavelength of the absorption maximum λ_0, for example, might lie in the vacuum ultraviolet. Curve II is a *negative plain* dispersion.

If the ORD spectrum contains an absorption band, then more complex effects may occur. First, the chromophore may possess a high degree of pseudosymmetry, and the ORD perturbations may be small. On the other hand, *Cotton effect* curves like III (positive) and IV (negative) in Figure 9-2 will result when the chromophore is intrinsically asymmetric or else inherently symmetric with a nearby asymmetric grouping. The sign of the Cotton effect depends on the chirality (left- or right-handedness) of the chromophore or the chiral effect induced by its immediate environment. More "distant" structural features have less effect. In the case of carbonyl chromophores the well-known Octant rule (Moffitt *et al.*, 1961) is a powerful tool for assigning absolute stereochemistry at the chromophore. ORD has been used to establish the structural features of biomolecules including those of protein substructures. However, the technique does not of itself possess element specificity unless it can be established by other means, for example, that the absorption is due to a particular entity such as a transition metal ion.

9.1.2 Circular Dichroism (CD)

Polarized light interacting with an optically active chromophore will undergo the rotatory effects described in the previous paragraph. In addition, the left- and right-handed circular components will experience a differential absorption as required by the following equation, where ϵ denotes molar extinction coefficients:

$$\Delta\epsilon = \epsilon_L - \epsilon_R \qquad (9\text{-}4)$$

The variable $\Delta\epsilon$ is known as the differential dichroic absorption. In a Cotton effect interaction the polarization axis is not only rotated but also, due to the unbalanced absorption, a degree of elliptical polarization is introduced. The molecular ellipticity, $[\theta]$, is analogous to the $[\Phi]$ variable discussed earlier:

$$[\theta] = 3300 \, \Delta\epsilon \qquad (9\text{-}5)$$

A typical CD spectrum is shown in Figure 9-3, and it presents only one extremum (in comparison with the two observed by ORD). Also, the sign of the CD maximum, which may be positive or negative, corresponds with the sign of the Cotton effect. The midpoint between extrema in an ORD spectrum closely matches the wavelength of the absorption maximum, whereas in CD the single extremum almost coincides with it. In general, the information contained in ORD and CD spectra is highly redundant, and one type of spectrum may be estimated from the other. For example, the amplitude, a, of the ORD curve, i.e., the vertical distance between extrema (Figure 9-2), may be calculated from the maximum value of $[\theta]$:

$$a = 0.0122 \, [\theta] \qquad (9\text{-}6)$$

9.1.3 Magnetic Optical Rotatory Dispersion (MORD) and Magnetic Circular Dichroism (MCD)

The methods of ORD and CD, discussed in the previous section, require that the substance under investigation be optically active. MORD and MCD are fundamentally different since the rotation and absorption effects are *induced* even if the substance under scrutiny is *not* optically active. The Drude equation applies as in ORD, and the variables $[\Phi]$ and $[\theta]$ have the same meaning; however, the differential influences on left and right circularly polarized light derive from the Faraday effect (Faraday, 1846), in which an intense magnetic field is applied to the sample with the lines of force parallel to the light beam. In this arrangement the signs of $[\Phi]$ and $[\theta]$ are reversed simply by reversing the magnetic field vector.

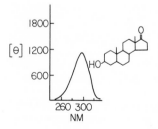

Figure 9-3. Circular dichroic spectrum of 3β-hydroxy-5α-androstan-17-one. The ordinate is molecular ellipticity and the abscissa, wavelength. Dichroic absorption by this steroid is associated with the ketone functional group. The enantiomer (mirror image) of this steroid will produce a *negative* CD curve, i.e., the spectrum only changes in the sign of molecular ellipticity.

Simplified schematics of MCD and MORD spectrometers are shown in Figure 9-4. It will be observed that MORD requires *two* tandem cells in the beam path each with its own magnet, and the fields must be *opposed*. The second cell contains solvent only, and the solvent's contribution to rotation created in the first cell is thus cancelled in the second, leaving only the effect of the solute under investigation. MCD is inherently simpler because a solvent may be chosen which does not absorb in the 200–800 nm range (e.g., water). Effects will then be due to magnetically induced dichroic absorption in the solute under investigation. Emphasis on MCD to date is partly due to economics since an apparatus which requires *two* superconducting magnets costs much more initially and consumes more liquid helium.

The element selectivity of MCD parallels that of CD in being relatively low unless an absorption band in question has been definitely linked to a particular atom or grouping of atoms, e.g., a metal ion. Quite often the pigmentation of an enzyme is due to a transition metal ion, and nearly as often the same atomic unit is intimately involved in the catalysis mechanism. Both CD and MCD are valuable adjuncts to other physical approaches since they provide stereochemical information about coordination sites. In addition, MCD is able to probe paramagnetic properties. Thus it would appear that these methods are most applicable in studies of the transition elements. One of the advantages of MCD is its potential for extreme sensitivity, e.g., some porphyrin and chlorin derivatives are detectable in the 10-ng/ml range (Djerassi *et al.*, 1971).

9.2 Sample Considerations

The dichroic methods compare with nmr, esr, and resonance Raman in their ability to accommodate aqueous samples. In principle it should be possible to conduct *in vivo* measurements, provided that the species contains mostly one pigment. However, in practice the metalloproteins

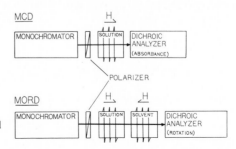

Figure 9-4. Basic features of MCD and MORD spectrometers.

and metalloenzymes which will yield element-selective information are usually minor components. Thus, most biological applications of CD and MCD are oriented toward *in vitro* studies of purified biopolymers and substrate-sized molecules in an aqueous solution. Solid-state samples are much more complex owing to the effects of crystal lattice anisotropy (Djerassi *et al.*, 1971).

9.3 Effects Observed in MCD Spectra

Some of the features of MCD spectra depend on the presence or absence of degenerate energy levels in either the ground or the excited state of the chromophore. By applying a magnetic field the degeneracy is removed, creating a pair of transitions which interact differentially with left and right circularly polarized light. Also, the Faraday effect causes a similar differential dichroic interaction in nondegenerate transitions.

The theoretical basis of MCD was formed in 1932 (Serber, 1932) when it was shown that the magnetically induced molecular ellipticity, $[\theta]_m$ (units: degree deciliter decimeter^{-1} mole^{-1} Gauss^{-1}), may be described by an expression of three terms, denoted A, B, and C, and two functions f_1 and f_2 of the absorption maximum, ν_0, the frequency variable, ν, and the absorption bandwith Γ:

$$[\theta]_m = -21.3458\{f_1(\nu_0,\nu,\Gamma)A + f_2(\nu_0,\nu,\Gamma)[B + C/kT]\} \qquad (9\text{-}7)$$

In a real substance the equation may be simplified if, for example, the A-term is zero. Figure 9-5 shows how the variables A, B, and C are related

Figure 9-5. Energy levels and magnetic splitting leading to A-, B-, and C-terms in MCD spectra. Insets below each energy level diagram refer only to cases in which a strong magnetic field has been applied, and these show how differential absorption of left (L) and right (R) circularly polarized components leads to the observed spectral effects. The applied field removes degeneracy in both A- and C-term spectra. In C-term spectra the a'

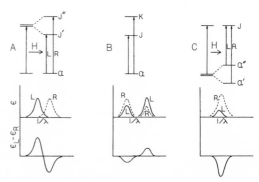

and a'' levels are not widely separated, and their relative populations will depend on temperature; i.e., a'' will become less populated as the temperature is lowered, and the C-term amplitude (negative) will *increase*.

to chromophore energy levels and observed MCD absorptions. In the cases considered below, the substance is assumed to be optically inactive.

9.3.1 A-Term Spectra

The A-term is restricted to molecules with at least a threefold symmetry axis, which results in degeneracy in the excited state. As shown in Figure 9-5, the applied field removes the degeneracy to create two new levels, j' and j'', arranged symmetrically above and below the original level j. The associated transitions $a \rightarrow j'$ and $a \rightarrow j''$ are observed in the absorption spectrum as two bands on either side of the original ν_0. Furthermore, these absorptions are specific for left and right circularly polarized light, resulting in the characteristic S-shaped differential dichroic absorption curve shown in Figure 9-5. Thus, an MCD absorption which resembles the Cotton effect curves of ORD is referred to as an "A-term."

Symmetry requirements would appear to minimize the likelihood of observing A-terms in the typically chiral environment of biomolecules. However, in some circumstances the chromophore may possess a rather high degree of pseudosymmetry, and A-terms may occur in the MCD spectra of biomolecules.

9.3.2 B-Term Spectra

A more likely situation in the biomolecular environment is the chromophore with a single ground state and two or more excited states. B-Terms result when the applied magnetic field causes a mixing of the ground and two or more excited states, which in turn creates *partial* preferences for left and right circularly polarized light, as shown in Figure 9-5. The differential dichroic absorption curve shows one positive and one negative MCD absorption, each located at the original absorption band position.

When the separation between levels is small, the MCD intensity *increases* (i.e., magnetic mixing is more pronounced), and the spectrum begins to resemble an A-term. Also, if the transition moments of the two bands are parallel, MCD absorption will be zero. The absorption is maximum when the two moments are orthogonal. B-Terms are common in MCD spectra.

9.3.3 C-Term Spectra

It will be noted in Eq. (9-7) that the C-term is the coefficient of the factor $1/kT$, where k is the Boltzmann constant and T the absolute temperature. The C-term is thus associated with a temperature-dependent phenomenon. A model for such behavior is shown in the right example of Figure 9-5. In this case the zero-field chromophore possesses a degenerate ground state (e.g., as a consequence of symmetry). An applied field removes the degeneracy, creating two new levels which are populated to an extent determined by the energy difference between a' and a'' and the average thermal energy, kT.

The transitions $a' \rightarrow j$ and $a'' \rightarrow j$ have opposite preferences toward right and left circularly polarized light. However, since level a'' is less populated, the $a'' \rightarrow j$ absorption will be weaker, resulting in the differential dichroic absorption curve shown below. It should be noted that C-terms not only produce positive or negative maxima of $[\theta]_m$ but also frequency shifts. They are easily identified through temperature variation measurements.

Since ground state degeneracy occurs in paramagnetic ions, C-term spectra are of particular interest in MCD studies of metalloproteins and metalloenzymes. Evidence gained through MCD is often complementary to that provided by esr and static magnetic susceptibility measurements.

9.3.4 MCD Spectra of Optically Active Chromophores

The above examples assumed that the substance under examination was not optically active. Optically active chromophores present a more complex picture. One approach is to obtain the difference spectrum between the CD and MCD responses, which helps to delineate between intrinsic and extrinsic asymmetry effects.

9.4 Biochemical Applications

The temperature-dependent C-terms occurring in MCD spectra provide a means of distinguishing between diamagnetic and paramagnetic entities (Stephens, 1974), e.g., as in the case of the iron–sulfur clusters found in ferredoxins (Stephens et al., 1978a,b). Ironically, there have been relatively few temperature-dependent MCD studies of transition metal-containing enzymes and proteins, although it seems reasonable to expect that the number will increase as instrumentation becomes more widely available. The ability to measure the absorption features associated with unpaired electrons allows one to distinguish between the d–d transitions occurring within the metal ion (or ions) and other types of light absorption originating in the organic portion of the biomolecule. Thus, MCD has a degree of selectivity toward paramagnetic transition metal ions. In comparison, the more conventional forms of optical spectroscopy (light absorption and CD) produce spectra which are generally more difficult to interpret.

The splitting induced by the applied field is low. Thus, temperature dependent C-terms are best studied at low temperatures, generally below 77 K. Also, C-terms are not necessarily observed in paramagnetic complexes; for example, the MCD spectrum of Cu(II)-pheophorbide-a was found to be temperature independent between 23 and 63 K (Schreiner *et al.*, 1978). In this case, the d-orbital containing the lone unpaired electron ($d_{x^2-y^2}$) apparently does not interact significantly with the pheophorbide π-electron orbitals involved in light absorption.

Although many applications of MCD have not used the potential for characterizing paramagnetic ions, these "other" methods are of interest in biological studies and will be included in the following discussion. MCD must be viewed as an auxiliary method to be used in conjunction with CD and conventional optical spectroscopy. The reader should examine several older works which were significant steps in the development of MCD for biological studies (Weser *et al.*, 1971; Ulmer *et al.*, 1973; Holmquist and Vallee, 1973; Mori *et al.*, 1975; Stephens *et al.*, 1978a,b; and Holmquist and Vallee, 1978.

9.4.1 Sample Preparations

Measurements in the visible portions of the spectrum are relatively straightforward. These may be extended further into the near infrared by replacing H_2O with D_2O. Also, it is preferable to prepare the protein samples as thin films on an infrared-transmitting substrate in order to strictly limit the amount of possible O—H and O—D vibrational absorption in the spectrum (Solomon *et al.*, 1980).

9.4.2 MCD and Heme Structures

A general review of MCD and hemoproteins has been presented by Hatano and Nozawa (1978). MCD is useful in the characterization of heme groups in metalloproteins and is able to distinguish between the oxidation and spin states of the central ion atom (Shimizu *et al.*, 1975). Absorptions at wavelengths near 700 nm ($\sim 14,300$ cm^{-1}) are ascribed to d-electron transitions, and the utility of MCD in clarifying the origin of such absorptions is exemplified in a study of a cytochrome P-450 conducted by Nozawa *et al.* (1978).

Absorption, CD, and MCD spectra of reduced P-450 are shown in Figure 9-6, which encompasses the red and near-infrared regions of the spectrum. On varying the temperature it was found that the feature marked by the arrow (690 nm) was temperature dependent; all others were temperature independent. Thus, the absorption at 690 nm has a C-term component and is associated with unpaired iron d-electrons while the other MCD features arise from A- and B-terms. The temperature dependence of the 690-nm absorption is shown at the right in Figure 9-6. Oxidized P-450 (not shown) exhibited a similar C-term at 760 nm.

The visible and near-ultraviolet regions were also revealing of metal-associated changes. The spin state of the iron changed from high to low on going from high temperature to low temperature (Nozawa et al., 1978).

In an earlier study of cytochrome P-450 it was noted that the MCD bands could be detected in complex biological preparations, which permitted a correlation with the MCD features of isolated P-450 (Dolinger et al., 1974). This contrasts with absorption measurements, which are compromised by light scattering in biological suspension. It was also noted that MCD was able to detect changes in the heme moiety due to interaction with a substrate. The P-450 enzyme used in the study was isolated from Pseudomonas putida, which can be grown on camphor as the sole carbon source and is presumably able to initiate the metabolism of camphor via a P-450 hydroxylation.

Camphor binds with the P-450 enzyme, producing marked changes in the latter's MCD spectrum. In particular, the iron is high-spin in the camphor/enzyme complex, and the MCD spectrum is dominated by B- and C-terms. The camphor-free complex is low-spin, and its MCD spectrum consists mostly of A-terms. This agrees with the anticipated strong mixing between iron d-orbitals and the porphyrin π-orbitals in the high-

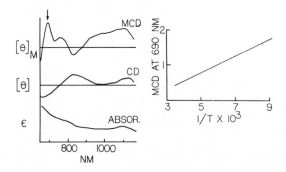

Figure 9-6. MCD, CD, and absorption spectra of *Pseudomonas putida* cytochrome P-450 in the near-infrared region. The MCD feature at 690 nm (arrow) is a C-term, as indicated by its temperature dependence in the plot of $[\theta]_m$ versus $1/T$ shown at the right. Adapted from Nozawa et al., 1978.

spin iron complex (Dolinger *et al.*, 1974). Quantum mixing is expected to be more attenuated in the low-spin form. Changes in axial ligation affect the MCD spectrum (Vickery *et al.*, 1976).

The MCD properties of cytochrome a_3 are consistent with a high-spin ferrihemoprotein in which the iron is antiferromagnetically coupled to one of the two Cu^{2+} ions present in the enzyme (Palmer *et al.*, 1976). This interpretation is consistent with the results of Mössbauer, esr, and static magnetic susceptibility measurements. MCD is complementary to the latter methods in probing paramagnetic ions. It should be noted that nmr and ESCA are also sensitive to paramagnetic atoms (see Chapters 2 and 7). A combination of several of these methods is the preferred strategy in approaching the problems presented by metalloproteins and metalloenzymes.

In other applications, MCD may be used, much as conventional absorption spectroscopy, in the quantitation of specific substances, often with a high degree of selectivity. For example, oxyhemoglobin can be detected in preparations of aquomethemoglobin at the 0.1% contamination level (Linder *et al.*, 1978).

9.4.3 Iron–Sulfur Cluster Proteins

In MCD studies of the four-iron clusters of ferredoxins and high-potential iron proteins (Stephens *et al.*, 1978a,b) the spectral region between 33,000 cm^{-1} (UV) and 5000 cm^{-1} (near IR) was found to contain distinct MCD features, as shown in Figure 9-7, where redox changes are observed to produce marked differences. These may be used to distinguish between the three cluster oxidation states (e.g., C^{1-}, C^{2-}, and C^{3-}). Moreover, the MCD spectrum is found to be insensitive to the protein itself, reporting only the status of the cluster moiety. Comparisons of the MCD spectra of four-iron and two-iron ferredoxins indicated that the method is also able to distinguish between these types.

One- and two-iron clusters have also been examined at low temperatures (Rivoal *et al.*, 1977; Thomson *et al.*, 1977), and similar correlations

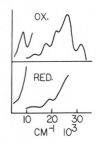

Figure 9-7. MCD spectra of oxidized and reduced *Clostridium pasteurianum* ferredoxin. These forms are easily distinguished by means of MCD. The left-hand portions of these spectra are shown at high gain to reveal spectroscopic differences. Adapted from Stephens *et al.*, 1978.

with cluster identity were observed. The latter experiments produced temperature-dependent spectra, thus making full use of the MCD method.

MCD is similar to absorption spectroscopy in being insensitive to the protein associated with the chromophoric group, but is like CD in producing a more structured spectrum (i.e., one with more features and thus more information). The ferredoxins are good subjects for evaluating MCD effects since the cluster may be excised and examined separately (Holm and Ibers, 1977).

9.4.4 *d*-Electron Transitions in the Blue Copper Proteins

The *d*-electron energy levels of an isolated transition metal ion are degenerate (i.e., have the same energy) but are split, as shown in Figure 9-8, when subjected to various configurations of coordinating ligands. It will be noted that degeneracy is still present in some coordination geometries (e.g, octahedral and tetrahedral). Also, given the same ligand strength, some symmetries produce a more marked splitting of the levels. For example, an octahedral field will have about twice the effect of a tetrahedral field (Figure 9-8). The energy differences between the *d*-levels are functions of the strength of the coordinating ligands when geometry is held to a particular type; otherwise, the relative energies of the *d*-orbitals are as shown in Figure 9-8. For example, d_{xz} and d_{yz} always are the lowest (degenerate) levels in a square planar complex.

The observed colors of transition metals ions and features in the CD spectra and MCD spectra of samples containing such ions are sometimes due to the absorption of a photon promoting an electron from an occupied *d*-level to an unfilled upper *d*-level. The frequency of the photon is determined by the energy difference between the levels, Δ, where $\Delta = h\nu$. Thus, by measuring the wavelengths of the observed maxima in optical

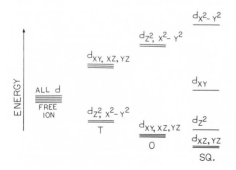

Figure 9-8. Transition metal *d*-orbital splitting and degeneracy in different coordination geometries. All five *d*-orbitals are degenerate in the free ion. Relative energy levels and *d*-orbital identities are shown for tetrahedral (T), octahedral (O), and square planar (SQ) symmetries.

spectra it is, in principle, possible to determine the d-level spacings, from which ligand strength and coordination geometry may be inferred (Figure 9-8). Actually, it is seldom easy to do this in biological studies, where the ligands themselves may be unknown and the d-level spectral features may often be obscured by the absorptions associated both with ligand–metal charge transfer and the electron energy levels of organic structures present in the protein. MCD may be used in conjunction with CD and absorption spectroscopy to aid in the identification of d–d transitions. Examples are found in studies of the blue copper proteins.

Until recently the blue copper proteins have resisted efforts to obtain a clear picture of their copper coordination sites. Colman et $al.$ (1978) obtained an X-ray crystallographic density map of poplar plastocyanin at 2.7-angstrom resolution, establishing a distorted tetrahedral geometry about the Cu(II) ion. The ligands consist of two sulfur and two nitrogen donors, i.e., N,N,S,S coordination. A similar conclusion was reached earlier by Solomon et $al.$ (1976), based on MCD and other spectroscopic evidence. A structural model of the site is shown in Figure 9-9, which also includes a representation of the σ, π, and d-electron energy levels at the right.

In combined MCD, CD, and absorption spectrum studies of plasto-cyanin, stellacyanin, and azurin (Solomon et $al.$, 1980), it was found that the CD and MCD features attributable to d–d transitions, i.e., those between the top two energy levels of Figure 9-9, were located in the near-infrared region of the spectrum with $\bar{\nu} < 11{,}500 \text{ cm}^{-1}$ (the infrared/visible boundary is at about $12{,}500 \text{ cm}^{-1}$) and thus reflect a weak ligand field. By contrast, the d–d transitions of tetragonal CuN_6^{2+} complexes are found well into the visible region of the spectrum at $\bar{\nu} > 13{,}400 \text{ cm}^{-1}$ while square planar Cu(II) species are generally observed with $\bar{\nu} > 15{,}400 \text{ cm}^{-1}$. Five-coordinate Cu(II) complexes are also blue-shifted. The weak coordination environment observed for these proteins is in good agreement

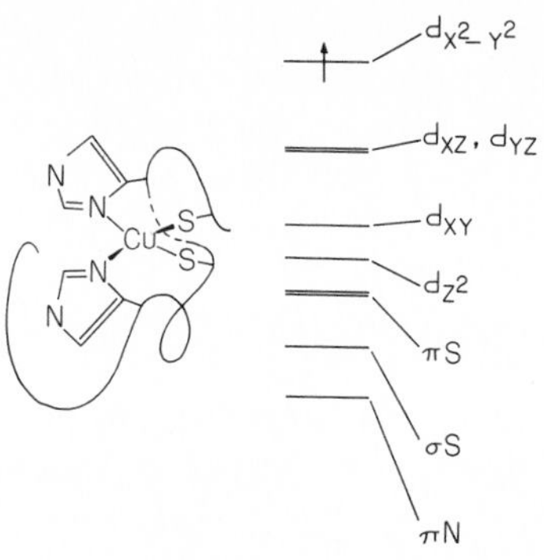

Figure 9-9. A distorted tetrahedral model proposed for the blue copper proteins, and the energy levels associated with it. The $d_{x^2-y^2}$ orbital contains an unpaired electron; all others have paired electrons (not shown). Adapted from Solomon et $al.$, 1980.

with ligand field calculations for a N,N,S,S tetrahedral structure distorted 6° toward square planar geometry (Solomon *et al.*, 1976, 1980).

9.4.5 Extrinsic Probes and MCD Effects

Applications of MCD vary from the relatively simple to those which attempt to account for detailed spectral features in terms of structure and the chromophore orbital symmetries, degeneracies, etc. In a straightforward application, Richardson and Behnke (1978) used the MCD spectrum of Pr^{3+} to probe interactions of Pr^{3+} with concanavalin A. Aqueous solutions of Pr^{3+} produced MCD spectra which did not become altered on addition of lysozyme, a protein with which Pr^{3+} appears to have no specific interaction. However, concanavalin A binds Pr^{3+}, and its addition to an aqueous solution of Pr^{3+} leads to an altered MCD spectrum. The MCD spectrum of the Pr^{3+}–concanavalin A complex is not changed by added Ca^{2+}, indicating that Pr^{3+} and Ca^{2+} are not competing for the same binding site, in agreement with the X-ray crystallographic structure (Becker *et al.*, 1975). An approach of this type is useful in establishing whether or not one is substituting a La^{3+} ion for Ca^{2+} in extrinsic probe studies. In many cases the lanthanide ion does interact at the calcium binding site, but this condition must be verified before valid conclusions may be drawn.

In the same study, circular dichroic spectra of Co^{2+}–concanavalin A were altered when Gd^{3+} was added, consistent with competition for the same binding site (but different from the Pr^{3+} site).

9.4.6 MCD Studies of Systems Which Do Not Contain Paramagnetic Metal Ions

The removal of orbital degeneracy by an applied magnetic field, e.g., as in MCD, often reveals characteristics which are not evident in the absorption or CD spectra. An example is found in the antibiotic netropsin, which forms a complex with DNA. Solutions of netropsin alone do not possess natural CD, while the CD spectrum of DNA [Figure 9-10(A)] is strongly dependent on polynucleotide secondary structure, e.g., the presence or absence of the helix.

On mixing DNA with netropsin, the CD spectrum is observed to change drastically, as shown in Figure 9-10(B). It is tempting to conclude that such changes simply reflect alterations in the helical structure of the double-stranded DNA on binding with netropsin. However, one might also imagine that the secondary structure of DNA remains largely unaltered and that the new CD spectral features originate in netropsin, being

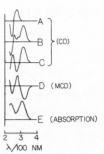

Figure 9-10. Absorption, CD, and MCD spectra of DNA and DNA/netropsin complexes. Adapted from Sutherland *et al.*, 1978.

induced (e.g., through a removal of degeneracy) in the bound state (Sutherland *et al.*, 1978).

Following this assumption, Figure 9-10(C) shows the difference spectrum derived by subtracting Figure 9-10(A) from Figure 9-10(B), i.e., the presumed component of the CD spectrum [in Figure 9-10(B)] originating in netropsin. The difference spectrum, Figure 9-10(C), may be compared with the MCD spectrum of netropsin, shown in Figure 9-10(D), where the inflections and peaks are seen to differ only in sign (which could be rectified by reversing the applied MCD magnetic field). Thus, there is strong evidence that the chirality of DNA removes degeneracy in the bound form of netropsin, resulting in a dichroic spectrum closely com-

Table 9-1. Tryptophan Content and MCD Absorption[a,b]

Protein	λ_{max} (± 0.2 nm)	No. of Trp residues found by MCD	Literature value
Ribonuclease	0		0
Lysozyme	292.8	5.96, 5.88, 6.20, 6.02	6
Pepsin	292.6	4.92, 4.90, 4.91, 5.08, 5.07, 4.95	5
Chymotrypsin	293.2	8.05, 8.16, 8.10, 8.04	8
Chymotrypsinogen	293.3	8.21, 8.60, 8.16, 8.80	8
Thermolysin	292.5	2.91, 3.30, 3.30, 3.22, 3.21	3
Deoxyribonuclease	293.2	2.90, 2.99, 2.82, 2.98, 2.82	3
Carbonic anhydrase	292.5	6.91, 7.52, 7.55, 6.42, 7.92	7
Carboxypeptidase A	292.8	7.50, 7.40, 7.84, 7.94	7
Staphylococcus nuclease	293.0	1.12	1
Neutral protease	292.0	4.20	4
Ovidin	293.0	16.5	16
Bovine serum albumin	293.3	1.82, 1.73	2
Aldolase	292.5	11.6	12

[a]Measured in 0.05 M Tris (pH 7.5)–0.1 M NaCl.
[b]Adapted from Holmquist and Vallee, 1973, and Barth *et al.*, 1972.

parable to that produced by the applied field of MCD, which also removes degeneracy.

As a further point, it should be noted that the MCD spectrum of netropsin does not match the envelope of the simple absorption spectrum [Figure 9-10(E)]. As a general rule, absorption arising in one chromophoric component will result in an MCD spectrum with the same shape as the absorption spectrum. This is not the case for netropsin, showing conclusively that there are two (or more) chromophoric components under the absorption band.

MCD may also be applied in the strict sense as a qualitative analytical tool, e.g., the spectra of chlorins *a* and *b* are distinctively different (Schreiner *et al.*, 1978), forming the basis for a selective quantitative analysis. The tryptophan content of proteins may be determined quantitatively from the intensity of a characteristic and selective absorption band near 293 nm (Holmquist and Vallee, 1973; Barth *et al.*, 1972). A comparison of chemically determined and MCD-based counts of tryptophan is presented in Table 9-1. Ivanetich *et al.* (1984) have described a fast method for measuring porphyrins in urine using MCD.

References

Barth, G., Bunnenberg, E., and Djerassi, C., 1972. Magnetic circular dichroism studies. XIX. Determination of the tyrosine: tryptophan ratios in proteins, *Anal. Biochem.* 48:471.

Becker, J. W., Reeke, G. N., Wang, J. L., Cunningham, B. A., and Edelman, G. M., 1975. The covalent and three-dimensional structure of concanavalin-A, III. Structure of the monomer and its interactions with metals and saccharides, *J. Biol. Chem.* 250:1513.

Colman, P. M., Freeman, H. C., Guss, J. M., Murata, M., Norris, V. A., Ramshaw, J. A. M., and Yenkatappa, M. P., 1978. X-ray crystal structure of plastocyanin at 2.7 angstrom resolution, *Nature (London)* 272:319.

Djerassi, C., 1960. *Optical Rotatory Dispersion: Applications to Organic Chemistry,* McGraw-Hill, New York.

Djerassi, C., Bunnenberg, E., and Elder, D. L., 1971. Organic chemical applications of magnetic circular dichroism, *Pure Appl. Chem.* 25:57.

Dolinger, P. M., Kielczewski, M., Trudel, J. R., Barth, G., Linder, R. E., Bunnenberg, E., and Djerassi, C., 1974. Magnetic circular dichroism studies XXV. A preliminary investigation of microsomal cytochromes, *Proc. Natl. Acad. Sci. U.S.A.* 71:399.

Faraday, M., 1846, *Phil. Trans. Roy. Soc. London* 3:1.

Hatano, M., and Nozawa, T., 1978. Magnetic circular dichroism approach to hemoprotein analysis, *Adv. Biophys.* 11:95.

Holm, R. H., and Ibers, J. A., 1977. Synthetic analogs of the active sites of iron–sulfur proteins, in *Iron–Sulfur Proteins,* Vol. 3 (W. Lovenberg, ed.), Academic Press, New York, pp. 205–281.

Holmquist, B., and Vallee, B. L., 1973. Tryptophan quantitation by magnetic circular dichroism in native and modified proteins, *Biochemistry* 12:4409.

Holmquist, B., and Vallee, B. L., 1978. Magnetic circular dichroism, *Methods in Enzymology*, 49:149.

Ivanetich, K. M., Movsowitz, C., and Moore, M. R., 1984. Rapid semiquantitative measurement of total porphyrins in urine and feces by magnetic circular dichroism, *Clin. Chem.* 30:391.

Linder, R. E., Records, R., Barth, G., Bunnenberg, E., Djerassi, C., Hedlung, B. E., Rosenberg, A., Benson, E. S., Seamans, L., and Moscowitz, A., 1978. Magnetic circular dichroism studies. Part LIV. Partial reduction of aquomethemoglobin on a Sephadex G-25 column as detected by magnetic circular dichroism spectroscopy and revised extinction coefficients for aquomethemoglobin, *Anal. Biochem.* 90:474.

Moffitt, W., Woodward, R. B., Moscowitz, A., Klyne, W., and Djerassi, C., 1961. Structure and the optical rotatory dispersion of saturated ketones, *J. Am. Chem. Soc.* 83:4013.

Mori, W., Yamauchi, O., Nakao, Y., and Nakahara, A., 1975. Spectroscopic studies on the active site of *Sepioteuthis lessoniana* hemocyanin, *Biochem. Biophys. Res. Commun.* 66:725.

Nozawa, T., Shimizu, T., Hatano, M., Shimada, H., Iizuka, T., and Ishimura, Y., 1978. Magnetic circular dichroism of *Pseudomonas putida* cytochrome P-450 in near infrared region, *Biochim. Biophys. Acta* 534:285.

Palmer, G., Babcock, G. T., and Vickery, L. E., 1976. A model for cytochrome oxidase, *Proc. Natl. Acad. Sci. U.S.A.* 73:2206.

Richardson, C. E., and Behnke, W. D., 1978. Physical studies of lanthanide binding to concanavalin-A, *Biochim. Biophys. Acta* 534:267.

Rivoal, J. C., Briat, B., Cammack, R., Hall, D. O., Rao, K. K., Douglas, I. M., and Thomson, A. J., 1977. The low temperature magnetic circular dichroism spectra of iron–sulfur proteins. I. Oxidized rubredoxins, *Biochim. Biophys. Acta* 493:122.

Schreiner, A. F., Gunter, J. D., Hamm, D. J., Jones, I. D., and White, R. C., 1978. Magnetic CD spectra of chlorophylls, chlorophyllides, and zinc(II) and copper(II) pheophytins and pheophorbides, *Inorg. Chim. Acta* 26:151.

Serber, R., 1932. *Phys. Rev.* 41:489.

Shimizu, T., Nozawa, T., Hatano, M., Imai, Y., and Sato, R., 1975. Magnetic circular dichroism studies of hepatic microsomal cytochrome P-450, *Biochemistry* 14:4172.

Solomon, E. I., Hare, J. W., and Gray, H. B., 1976. Spectroscopic studies and a structural model for blue copper centers in proteins, *Proc. Natl. Acad. Sci. U.S.A.* 73:1389.

Solomon, E. I., Hare, J. W., Dooley, D. M., Dawson, J. H., Stephens, P. J., and Gray, H. B., 1980. Spectroscopic studies of stellacyanin, plastocyanin and azurin. Electronic structure of the blue copper sites, *J. Am. Chem. Soc.* 102:168.

Stephens, P. J., 1974. Magnetic circular dichroism, *Annu. Rev. Phys. Chem.* 25:201.

Stephens, P. J., Thomson, A. J., Dunn, J. B. R., Keiderling, T. A., Rawlings, J., Rao, K. K., and Hall, D. O., 1978a. Circular dichroism and magnetic circular dichroism of iron–sulfur proteins, *Biochemistry* 17:4770.

Stephens, P. J., Thomson, A. J., Keiderling, T. A., Rawlings, J., Rao, K. K., and Hall, D. O., 1978b. Cluster characterization in iron–sulfur proteins by magnetic circular dichroism, *Proc. Natl. Acad. Sci. U.S.A.* 75:5273.

Sutherland, J. C., Duval, J. F., and Griffin, K. P., 1978. Magnetic circular dichroism of netropsin and natural circular dichroism of the netropsin–DNA complex, *Biochemistry* 17:5088.

Thomson, A. J., Commack, R., Hall, D. O., Rao, K. K., Brait, B., Rivoal, J. C., and

Badoz, J., 1977. The low temperature magnetic circular dichroism spectra of iron–sulfur proteins; II. Two iron ferredoxins, *Biochim. Biophys. Acta* 493:132.

Ulmer, D. D., Holmquist, B., and Vallee, B. L., 1973. Magnetic circular dichroism of nonheme iron proteins, *Biochem. Biophys. Res. Commun.* 51:1054.

Vickery, L., Nozawa, T., and Sauer, K., 1976. Magnetic circular dichroism studies of myoglobin complexes—Correlations with heme spin state and axial ligation, *J. Am. Chem. Soc.* 98:343.

Weser, U., Bunnenberg, E., Commack, R., Djerassi, C., Flohe, L., Thomas, G., and Voelter, W., 1971. A study on purified bovine erythrocuprein, *Biochim. Biophys. Acta* 243:203.

Kinetic Methods

10

10.1 Introduction

In the early fall of 1965 at Arden House, Harriman, N.Y., a score of people were talking about a mysterious serum copper protein, ceruloplasmin. They were participants in a symposium on copper in biological systems (Peisach *et al.,* 1966). All knew that ceruloplasmin is a copper protein circulating in the bloodstream, having seven or eight copper atoms per protein molecule, with an enzyme activity similar to laccase, but weaker. One said the enzyme activity of ceruloplasmin was just accidental. Others were still trying to find the role of ceruloplasmin (or its contained copper) in a seemingly endless search for physiological activity. Broman (1967) discussed the possible physiological function of ceruloplasmin as a "moot question." A variety of observations were considered before he suggested that the function of ceruloplasmin is to transfer copper to cytochrome oxidase. Later, catecholamines were proposed as physiological substrates (Walaas, 1967). Then ceruloplasmin was found to be an iron oxidase in serum, and a possible physiological role was identified (Osaki, 1966). Ceruloplasmin is now called ferroxidase (EC 1.16.3.1: Osaki, 1966; Osaki and Walaas, 1967). This is an example of classical steady state kinetic methods revealing one role of an element.

Although the scope of this chapter is large, it is not our intention to cover the full range of kinetic methods, and we will not consider examples for all of the biological elements. Rather, we will attempt to illustrate the importance of these techniques by citing a few good examples. It will be helpful to classify these methods according to the kind of biological phenomena under investigation. These areas are: (1) kinetics of an element in whole body metabolism such as ferrokinetics or the mobilization of iron by ferroxidase *in vivo* and *in vitro* (Gross, 1964; Ragan *et al.,* 1969;

Osaki *et al.,* 1970); (2) enzyme kinetics (Neurath, 1970), which often involves metallic ions located in the active centers, and both metallic and non-metallic elements as modifiers; (3) kinetics of biological pigments such as hemoglobin (Fe), hemocyanin (Cu), plastocyanin (Cu), cytochromes (Cu and Fe), and chlorophylls (containing Mg) as they interact with other molecules; and (4) the effect of an element itself on a biological function, examples being the influence of sodium or potassium ions on membrane transport. The methods used to study such systems are further classified according to the physical means used to establish and measure time-ordered processes associated with a particular biological function.

10.2 Methods

10.2.1 Flow Techniques

In most kinetic measurements, one begins by establishing a state of thermodynamic nonequilibrium. For example, if two solutions containing reactive compounds are mixed, an ensuing chemical reaction is started, and the system proceeds toward equilibrium. The initial displacement from equilibrium is usually large. Of the various techniques considered here, the stopped-flow procedure (Kustin, 1969) is perhaps the one used most frequently, and if the mixing compartment and cuvette volumes are kept small, thorough mixing takes place in about 10^{-3} to 10^{-4} s (Figure 10-1). Obviously, events which happen in a shorter time frame cannot be characterized by this method.

Transient chemical intermediates are often detected spectrophoto-metrically at ultraviolet and visible wavelengths. Other spectroscopic instruments capable of accommodating liquid samples are equally applicable, and the possibilities are numerous, e.g., esr, nmr, infrared, etc.

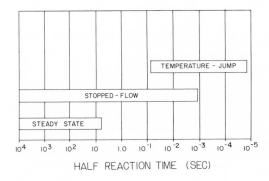

Figure 10-1. Approximate relationship between various kinetic methods and half-reaction time.

Some detection methods do not respond to rapidly changing signals. For example, one might monitor a specific chemical environment by following the increase or decrease of a particular nmr line, but even if a strong signal may be acquired in a single pulse, the decay time of the FID signal (which may be seconds; see Section 2.4.4.4) becomes a limiting factor (but see also Section 10.2.4 for an entirely different nmr approach).

In ultraviolet and visible spectrometry, the situation is reversed, and rate of mixing becomes the limiting factor. The reacting entities do not necessarily have to absorb in the optical spectrum. For example, in stopped-flow studies of the rate of release of lanthanide ions from the calcium-binding protein, parvalbumin, a large excess of the dye xylenol orange was mixed with a solution of the parvalbumin/lanthanide ion complex. Xylenol orange rapidly chelates with the released lanthanide ion and experiences a spectral change in the process. This has allowed a characterization of the kinetics of release of lanthanide ions from parvalbumin's multiple binding sites (Corson *et al.*, 1983). In other cases, interaction may be detected by changes in an absorption band belonging to one of the reactants, as in kinetic measurements of the rate of intercalation of the antitumor antibiotic echinomycin with DNA (Fox and Waring, 1984). As a general rule, methods capable of following the more fleeting chemical transients are of greater power and sophistication because they offer some hope of observing the elementary processes of a complex reaction mechanism.

10.2.2 Flash Photolysis and Related Techniques

Intense but brief pulses of laser light may be used to create substantial deviations from equilibrium. This approach is especially suitable for characterizing photochemical reactions, and if the laser emits ultraviolet light, free radicals are usually produced (visible light may have the same effect). Laser flash photolysis/esr spectrometer combinations which operate at ultraviolet and visible wavelengths are elegant tools for studying biological pigments, but these have not been much exploited in spite of their long-standing availability (Atkins *et al.*, 1970). If flash photolytic chemical transients are detected by ultraviolet or visible spectrophotometric instruments, one may observe events in the picosecond time frame. The possible modes of detection include absorption, fluorescence, and Raman spectroscopies (see Section 8.6.2.8).

A related approach is pulsed radiolysis. This requires an electron accelerator which can generate intense submicrosecond pulses of high-energy (MeV range) electrons. In a typical arrangement, the electron beam passes from the accelerator's evacuated beam tube, through a thin mylar window, into the sample cuvette, where the pulse of energy is deposited. A cursory examination might lead the reader to think that the method is restricted to problems of radiation biology (involving ionizing radiation), but innovations are possible. For example, one may create the superoxide anion in a time frame unattainable by the more conventional flow techniques, and the method is thus suitable for studies of the superoxide dismutase metalloenzymes (Lengfelder and Elstner, 1979; Lengfelder, 1980).

10.2.3 Chemical Relaxation Methods

Rapid equilibrium perturbations are possible since equilibrium constants are functions of the intensive thermodynamic variables (for example, temperature or pressure). This condition is generally true, provided ΔH, ΔV, ΔS, or any other state variable is non-zero. A sudden change in the intensive variable forces the system toward a new equilibrium. The rate of attainment of equlibrium is in turn a function of the most favorable energy barriers and entropies presented to the species involved, resulting in a particular mechanism of competing or sequential chemical reactions (or a combination of both). Obviously, very fast perturbations allow us to detect and characterize rapid reactions so that we may (at least in principle) explore the whole range of chemical interest (Schelly and Eyring, 1971; Hammes, 1968), which involves time scales as small as 10^{-10} s (Marcandalli et al., 1984).

The basis of the so-called temperature jump method developed by Eigen (see Eigen and Kruse, 1963, and references cited therein) is:

$$\frac{\partial \ln K_P}{\partial T} = \frac{\Delta H^0}{RT^2} \tag{10-1}$$

where ΔH^0 is the standard enthalpy of reaction. A very rapid temperature increment may be attained by dissipating the charge of a large capacitor in a small-volume sample cuvette. The sample must contain dissolved electrolyte for conductivity, i.e., the time constant ($t = RC$) of the electric circuit must be small, and since the capacitance (C) is large, cell resistance (R) must be as small as possible. Alternatively, the sample may be heated by means of a powerful infrared pulse from an iodine laser (Marcandalli et al., 1984). The latter approach circumvents electrostatic field artifacts and permits studies at low ionic strength.

Other variables may be changed suddenly. For pressure jump, the basis is:

$$\frac{\partial \ln K_P}{\partial P} = -\frac{\Delta V^0}{RT} \qquad (10\text{-}2)$$

where ΔV^0 is the standard volume change of reaction. Similarly, a strong electric field may be applied to the sample. The appropriate relationship is:

$$\frac{\partial \ln K_P}{\partial E} = \frac{\Delta D^0}{RT} \qquad (10\text{-}3)$$

where ΔD^0 is the standard difference of partial molal polarization.

The three methods just described make use of step perturbations, but measurements may also depend on sinusoidal variations of the intensive properties. For example, the specimen may be subjected to strong ultrasonic irradiation, which produces quasi-adiabatic variations in pressure and temperature (at least initially, before there is heating on the sample). The observed spectroscopic property will vary sinusoidally, but with a phase lag which depends on the rate processes involved.

All of these methods involve relatively small perturbations. The advantage of small equilibrium perturbations is mathematical simplification (compared with other kinetic methods). The rate equations reduce to linear, first-order differential equations of the following general form:

$$-\frac{d\Delta c_i}{dt} = \sum_{j=1}^{n} a_{ij}\, \Delta c_j \qquad (10\text{-}4)$$

where the a_{ij} variables are functions of rate constants and equilibrium concentrations, and the Δc_j variables are concentration deviations. The solution to Eq. (10-4) is:

$$\{c(t)\} = \left[\sum_{j} e^{-\lambda_j t}\, f(\{A\}) \right] \{C(0)\} \qquad (10\text{-}5)$$

where $\{c\}$ and $\{A\}$ are matrices of concentration deviations, t is time, and values for a_{ij} and λ_j are obtained by solving the determinant:

$$\begin{vmatrix} a_{11}-\lambda_1 & a_{12} & \cdots & a_{1n} \\ a_{21} & a_{22}-\lambda_2 & \cdots & a_{2n} \\ \cdots & \cdots & \cdots & \cdots \\ a_{n1} & \cdots & \cdots & a_{nn}-\lambda_j \end{vmatrix} = 0 \qquad (10\text{-}6)$$

Each eigenvalue, λ_j, is a reciprocal relaxation time, and an $n \times n$ determinant yields n relaxation times for n, independent concentration variables, i.e., a relaxation spectrum. Thus, the following reaction:

$$E + S \rightleftharpoons A \rightleftharpoons B \rightleftharpoons C \rightleftharpoons D \qquad (10\text{-}7)$$

yields four characteristic relaxation times. For a sequential mechanism such as the above, the relaxation times are coupled, i.e., functions of all equilibrium concentrations and rate constants. However, in practice they are actually uncoupled or at least partially uncoupled when there are significant rate differences in sequential steps—apparently a common situation in enzymic catalysis—and a further mathematical simplification is introduced.

10.2.4 Spectroscopic Line Broadening and Lineshape Effects Related to Species Lifetime, the Uncertainty Principle, and Exchange Equilibria

The Heisenberg Uncertainty Principle relates spectroscopic linewidth to species lifetime, and the relationship was presented in Eq. (5-10). From this simple equation it is seen that as the species lifetime (Δt) becomes shorter, spectral lines become broader due to the attendant increase in ΔE. The reader should note that the phenomenon is a general one, applying to all forms of electromagnetic spectroscopy. Other sections of this volume should be consulted for a more detailed treatment (see Sections 2.3.2 and 5.2.5). The reader is warned that several other phenomena may influence linewidth.

Rate of chemical exchange may affect lineshapes in nmr spectra. For example, if protons exchange slowly between two spectroscopically distinct environments, one observes two spectral lines separated by a difference in chemical shift, $\Delta \nu$. At higher temperatures, the rate of equilibration is increased, and if the equilibration is rapid enough, a single line averaging the two environments replaces the original pair. In the intermediate case (a broadened line), the rate of equilibration is given by $\sim \Delta \nu$. Rates obtained by nmr lineshape analysis do not require the establishment of a state of nonequilibrium. Also, the method is applicable to any nmr-active isotope. Jeener *et al.* (1979) have described a two-dimensional nmr method which displays an exchange spectrum. Their 2D presentation is a direct visualization of the kinetic matrix of the reacting system.

10.2.5 Other Kinetic Methods

The importance of various classical methods should not be overlooked, e.g., simple progress curves with the substrate/enzyme ratio large (zero-order kinetics) and studies of the rate of distribution of a radioisotope between various chemical species. In steady state kinetic methods the relationship between reaction rates and changes in the environment of a particular element is usually indirect. Nevertheless, depletion and repletion studies of a particular element are often based on a kinetic procedure to detect and measure the amount of active enzyme present. The amount of the enzyme is determined from the slope of a zero-order progress curve, and if the latter is found to be dependent on the element in question, one infers that the element is a structural or catalytic component of the enzyme (Frieden and Osaki, 1970). Confirmation that the protein contains the element (by chemical analysis, atomic absorbance) should accompany the kinetic evidence since it is entirely possible that an effect may be indirect, especially if the study is being carried out *in vivo*. The use of an inhibitor or activator of the reaction is often helpful, and examples will be cited in the applications section of this chapter.

Many different elements (often metals) serve as the cofactors of enzymes. Some are tightly bound; others are easily dissociable. The examples of Table 10-1 are mostly metalloenzymes which have metals tightly bound to the protein. There are many more enzymes which require divalent ions such as Mg(II) and Ca(II) for their activities. The role of an element in an *isolated* biological system, e.g., a purified enzyme or perhaps a crude fraction containing an enzyme system, may also be detected by the type of depletion and repletion experiments just described. However, there is need for caution when dealing with an *in vitro* experiment. It is possible that the particular element may not be removed without irreversibly denaturing the enzyme. Also, such procedures, if they work, only indicate that a certain element is necessary for a biological function. They do not show us *how* the element functions.

10.2.6 Choice of Detection Methods

The choice of a method in a kinetic study depends on several factors: (1) the rate of the reaction; (2) the electromagnetic properties of the element under study; (3) the optical absorption spectrum of the species which undergoes change during the reaction; (4) whether the species does or does not show fluorescence; and (5) the sensitivity of the equilibrium point to temperature and other intensive variables. Figure 10-1 presents a comparison of three major methodologies, and considerations of reaction half-time will obviously favor one of these approaches. The spectroscopic detection method selected (e.g., nmr, esr, fluorescence, etc.) should take advantage of the concept of element selectivity presented in the introduction to this volume. If the necessary instrumentation is available, most reacting systems are more likely to yield to a combination of methods.

The outcome of a kinetic study should be a better picture (hopefully!)

Table 10-1. Enzymes Using Various Elements as Cofactors[a]

Name of enzyme	Element	Role of element, if known
Amylase	Ca, Cl	Active center, activator of subunit to subunit binding interactions.
Pyruvate carboxylase	Mn, Mg	Active center, activator
Ferredoxin	Fe	Intermolecular electron transport
Metapyrocatechase	Fe	Active center
Pyrocatechase	Fe	Active center
Protocatechase	Fe	Active center
NADH dehydrogenase	Fe	Active center
Succinate dehydrogenase	Fe	Active center
Dihydroorotate dehydrogenase	Fe	Active center
Aldehyde oxidase	Fe, Mo	Active center
Xanthine oxidase	Fe, Mo	Active center
Monoamine oxidase	Cu	Active center
Amine oxidase	Cu	Active center
Diamine oxidase	Cu	Active center
D-Galactose oxidase	Cu	Active center
Uricase	Cu	Active center
Tyrosinase	Cu	Active center
Dopamine β-hydroxylase	Cu	Active center
Laccase	Cu	Active center, intramolecular electron transport
Ascorbate oxidase	Cu	Active center, intramolecular electron transport
Ferroxidase	Cu	Active center, intramolecular electron transport
Cytochrome oxidase	Cu, Fe (Hemo)	Active center
Carbonic anhydrase	Zn	Active center
Carboxypeptidase A,B	Zn	Active center
Alcohol dehydrogenase	Zn	Active center
Glutamic dehydrogenase	Zn	Active center
D-Glyceraldehyde-3-phosphate dehydrogenase	Zn	Active center
Lactic dehydrogenase	Zn	Active center
Malic dehydrogenase	Zn	Active center
Alkaline phosphatase	Zn	Active center, activator
Nitrate reductase	Mo	Active center
Hexokinase	Mg	Activator
Pyruvate phosphokinase	K	Activator
Plasma membrane AtPase	Na, K, Mg	Activator, regulator

[a]Adopted from Neurath, 1970.

of the relationship between molecular structure, as in the active site of an enzyme, and the reaction mechanisms and conformational changes involved in the reaction under scrutiny. Such studies are therefore directly concerned with the true nature of biological functions. Even so, a kinetic investigation can only support, not prove, a hypothesis about a reaction mechanism. If there were only two possibilities to be considered, then the data might strongly favor one of them, but biological systems tend to present a much more complex picture.

10.3 Applications

10.3.1 Enzyme Kinetics

Keeping the limitations in mind, we will first consider cases no more sophisticated than simple inhibition or activation experiments. For example, steady state kinetic studies of the hydrolysis of β-glycerophosphate with pig kidney alkaline phosphatase (EC 3.1.3.1) at various pH values, substrate concentrations, and Mg^{2+} concentrations (Ahlers, 1975) suggested that Mg^{2+} is the essential activator of alkaline phosphatase since there is no residual activity in the absence of the ion. Also, the binding of the substrate occurred independently of the binding of the Mg^{2+}. From an extensive analysis of the kinetic data, Ahlers suggested that magnesium ions act as allosteric effectors (Brunel and Cathala, 1973), i.e., Mg^{2+} induces a conformational change at the active site, which results in an optimal structure around the catalytic center (Zn^{2+}). The kinetic results also indicated the presence of an unknown negative charge in the vicinity of the active site, where Mg^{2+} could bind. Ahlers hypothesized that a hydroxy group coordinated to Zn^{2+} liberates the alcoholic residue by nucleophilic attack on phosphate.

A similar regulatory action of Mg^{2+} in a different enzyme, creatine phosphokinase (EC 2.7.3.2), was suggested from kinetic studies of the enzyme and a computer analysis of the results. In the presence of all substrates and products, the ratio of the rates of forward and reverse reactions is effectively regulated by the concentration of magnesium ion. At low magnesium ion concentration, creatine phosphate is synthesized, while at high Mg^{2+} concentration ATP synthesis is promoted. Mathematical analysis of the kinetic model suggested that the observed regulatory effect of Mg^{2+} on the overall reaction rate could be elucidated by: (1) a sigmoidal variation in the MgADP concentration resulting from a competition between ATP and ADP for Mg^{2+}, and (2) a high affinity of the enzyme toward MgADP.

10.3.2 Blue Copper Enzymes

Studies of the role of copper in "blue copper enzymes" are of fundamental interest but inherently complex. Examples of the "blue copper enzymes" are ascorbate oxidase (EC 1.10.3.3, L-ascorbate: O_2 oxidoreductase), ferroxidase (EC 1.16.3.1, iron(II): O_2 oxidoreductase), and laccase (EC 1.10.3.2, benzenediol: O_2 oxidoreductase). The main interest in kinetic studies of these enzymes is that they have at least three different types of copper content: Type 1 copper is "blue copper" with a strong optical absorption at 610 nm and an epr spectrum with an unusually small hyperfine coupling constant. Type 2 copper has little visible absorption but presents an epr signal comparable to that of inorganic Cu^{2+} (Malmstrom et $al.$, 1970). Type 3 copper has an absorption at 330–340 nm but no epr signal, and it participates in an oxidative mechanism (DeLey and Osaki, 1975). A kinetic study of ascorbate oxidase by Strothkamp and Dawson (1977) suggested that azide and fluoride, as well as the substrate, ascorbate, bind to type 2 copper, whereas in a study of laccase Andreasson et $al.$ (1973) found, using spectrophotometric and epr techniques, that a type 2 copper bound by H_2O_2 resembled the intermediate of laccase reoxidation. They proposed that laccase might be reoxidized by consecutive two-electron transfer steps through type 2 copper. The reduction of ferroxidase (Osaki and Walaas, 1967) and its reoxidation (Carrico et $al.$, Manabee et $al.$, 1973; Osaki and Walaas, 1967) were extensively studied using stopped-flow spectrophotometry to observe a fast reaction and ordinary (slow) spectrophotometry to record controlled slow reactions involving chelators and activators (DeLey and Osaki, 1975). The studies indicated that there are two different type 1 copper atoms in ferroxidase, one auto-oxidizable via an intramolecular electron transport system, and the other needing a trace amount of iron or copper to be reoxidized.

A schematic diagram of the three different types of copper in ferroxidase is shown in Figure 10-2, illustrating three different roles of copper in one enzyme molecule. The type 1 copper accepts electrons from the substrate and passes them to the type 2 copper, which in turn transfers them to the type 3 copper cluster, where the reduction of an oxygen molecule to water occurs. One of the type 1 coppers cannot be reoxidized unless a metal ion (Fe or Cu) is in a position to facilitate the electron transport.

These findings also suggest a mechanism for controlling iron metabolism, namely that of substrate activation of ferroxidase. A sudden influx of iron into the bloodstream will be met with activation of ferroxidase by iron, an increase in the rate of transferrin formation, and an effective

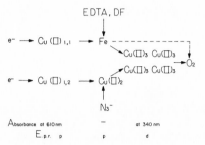

Figure 10-2. Schematic representation of the proposed intramolecular electron transport system in ferroxidase. The assumption is made that the two different type 1 copper atoms belong to the same enzyme molecule. DF, desferral; p, paramagnetic; d, diamagnetic. Adapted from DeLey and Osaki, 1975.

depletion of the undesirable free iron. This substrate (Fe^{2+}) activation has been clearly demonstrated (Osaki, 1966). The kinetics of the catalysis of Fe^{2+} by ferroxidase yielded two K_m values: $K_{m_1} = 0.5$ μM and $K_{m_2} = 50$ μM. It is likely that the larger K_m value was actually an activation constant (involving the substrate, Fe^{2+}).

10.3.3 In Vivo Kinetic Studies of the Role of Elements

Study of the metabolic process of iron turnover in an animal body, including the human body, is called ferrokinetics. A radioactive form of iron, such as ^{59}Fe, may be used as a tracer of these chemical events. An extensive description of ferrokinetics has been published (Gross, 1964). It is now well known that severe hypocupremia causes an anemia very similar to iron deficiency, with the exception that the reticuloendothelial system retains a large amount of stored iron (the latter being absent in true iron deficiency anemia). Studies of these processes are longstanding. After an experimental animal had been made copper deficient (and thus seemingly hypoferric) Lee et al. (1968) were able to show the mobilization of stored iron into the animal's plasma after an injection of a relatively large amount of inorganic copper (100 $\mu g/kg$ body weight).

Osaki (1966) postulated a biological role of ceruloplasmin in which ceruloplasmin promoted the rate of iron saturation of transferrin and stimulated iron utilization. Thus, the name ferroxidase (EC 1.16.3.1) was proposed based on the enzymic activity of this copper protein in serum. Ragan et al. (1969) tested this hypothesis using copper-deficient swine. They chose a homologous ferroxidase with a copper content low enough that it would not cause immediate iron mobilization if injected as inorganic copper. The results, shown in Table 10-2, were dramatic. The increase in plasma iron could be caused by two possibilities: (1) by an increase of influx of iron into the plasma, or (2) by a decrease of iron outflow from the plasma. In ferrokinetic experiments using radioactive iron and com-

Table 10-2. Increase in Plasma Iron after Injection of Homologous
Ferroxidase or Copper Sulfate[a]

Time (min)	Increase in plasma iron (mg/100 ml)		Paired t-test
	Ferroxidase-injected pigs[b]	CuSO_4-injected pigs	
5	18	3	0.01
15	26	3	0.005
30	84	2	0.005
60	136	14	0.001
240	225	20	—

[a]Adapted from Ragan et al., 1969.
[b]Four animals in each group.

puter simulation methods, the best match between observed and com-
puter-simulated data was obtained by holding the plasma iron outflow
unchanged and increasing influx initially 5-fold (Ragan et al., 1969).

The mobilization of iron by ferroxidase was then demonstrated in
perfused liver (Osaki et al., 1970). The latter results established that one
role of copper in the biological system is, in the form of ferroxidase, in
regulating iron inflow from storage or the intestine. How ferroxidase acts
in this vital control of iron mobilization requires further investigation.
Such experiments should be conducted in a controlled atmosphere (low
oxygen and high bicarbonate), at physiological pH, and they should be
carried out using the capillary system of fresh liver (e.g., by perfusion).
It is possible that an intact biological membrane plays a cooperative role
in iron mobilization by ferroxidase.

10.3.4 Substitution Experiments

When the situation permits, we can replace elements in an enzyme
which has no light-absorbing activity. Thus, it is possible to introduce
elements which provide optical or paramagnetic absorptions. In some
cases this may be done with retention of activity, and a good example is
found in alkaline phosphatase. Since zinc-containing alkaline phosphatase
has no visible absorption, Gottesman et al. (1969) prepared from the
apoenzyme of E. coli and cobaltous sulfate a cobalt-containing alkaline
phosphatase which had many kinetic similarities to the native zinc-
containing enzyme, but which also possessed visible absorption. Any
change in cobalt ion at the substituted enzyme, such as substrate binding,

could then be observed spectrophotometrically. These investigators found that the spectral changes associated with the cobalt ions were in good agreement with the kinetically determined catalytic rate (at or below saturation). It was also concluded that the interaction involved in forming the catalytically optimal conformation of the active site must be very complex. Although cobalt-substituted alkaline phosphatase is not the native zinc enzyme, the study of the substituted enzyme gave valuable information about the role of the catalytic center (and presumably zinc).

Carboxypeptidase is yet another enzyme amenable to substitution. Native carboxypeptidase (EC 3.4.17) contains zinc as the active center, but this enzyme is also functional with various metal ions acting in place of zinc. The functional ions include Mn^{2+}, Co^{2+}, and Cd^{2+}. As discussed previously, it is essential to make the active center optically visible in order to study interactions of the metal with its environment or with the substrate. It is particularly interesting if one can detect events at the active site during catalysis. In 1970 Latt *et al.* described experiments using a "surveyor substrate," and cobalt-substituted carboxypeptidase was employed in these studies. This procedure could be applied to carboxypeptidase since the identity of an essential component of the active site was known. The fluorescent substrate (*N*-dansylated surveyor) measured distances within the active center. The cobalt-substituted enzyme was chosen because the native zinc enzyme cannot quench the fluorescences originating from the dansyl group.

Two energy donor–acceptor pairs were chosen for the observation. One pair measured the distance from a known, fixed active center (cobalt), which served as the acceptor of the energy donated by a dansyl group at the *N*-terminus of the substrate, while the other pair indicated the formation of the enzyme/substrate complex through energy transfer from the tryptophanyl group (fluorescent) of the enzyme to the dansyl group (absorbing). Thus the dansyl group played a dual role. The concept is depicted in Figure 10-3. Based on stopped-flow measurements of fluorescence intensities and quantum yield, it was possible to calculate the distance between the dansyl group and the cobalt ion. Table 10-3 presents the distances between the dansyl group (on substrates) and the catalytic center of cobalt carboxypeptidase.

The reader will note that X-ray analysis of an enzyme's structure of a static picture matches well with the results of the surveyor substrate method described above which provides a picture of the microcosmic condition of the enzyme during the dynamic stage of catalysis. The experiment also clearly demonstrated the location of the metal (cobalt) in the active center region during the catalysis.

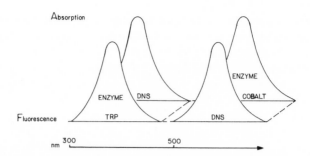

Figure 10-3. Schematic representation of the spectral overlap relationships between tryptophan, the dansyl group, and cobalt carboxypeptidase, the three moieties making up the
two donor–acceptor pairs utilized in the application of the surveyor substrate method to
the study of carboxypeptidase. Adopted from Latt *et al.*, 1970.

10.3.5 Paramagnetism

When an enzyme contains a paramagnetic species such as manganese
or copper, it is possible to use electron paramagnetic resonance (epr or
esr) to observe the binding of the metal to the enzyme moiety and to
measure the metal–substrate interaction by means of the proton relaxation
rate of water. p-Methylaspartase (EC 4.3.1.2), which requires Mn^{2+} and
K^+ for its activity, was studied by the above described method (Field
and Bright, 1970). It was concluded that Mn^{2+} acts as a bridge between
the substrate and the enzyme. Esr methods for probing metalloenzymes
are treated in Chapter 5.

10.3.6 Inhibitors

The role of copper in dopamine β-hydroxylase (EC 1.14.17.1) was
studied using 2-mercaptoethylguanidine (Dilberto *et al.*, 1973). It was

Table 10-3. Dansyl–Cobalt Distances[a]

Substrate	Indicated by surveyor substrate	Predicted from X-ray crystallography
DNS–Gly–Phe	7 Å	7 Å
DNS–Gly–Gly–L-Phe	10 Å	10 Å
DNS–Gly–Gly–Gly–L-Phe	12 Å	13 Å

[a]Adapted from Latt *et al.*, 1970.

shown kinetically that *two* 2-mercaptoethylguanidine molecules inhibited the enzyme, indicating that two copper ions must be available to the inhibitor at the active site. The involvement of copper could be concluded from the study, but it is not necessary to limit the number of copper ions to two, e.g., there might be another inaccessible copper ion in the enzyme with which the inhibitor could not interact.

10.3.7 Temperature Jump Characterizations

The binding of Ca^{2+} and acetylcholine was studied by Newmann and Chang (1976) using a temperature jump perturbation based on the properties of Eq. (10-1). Since neither acetylcholine nor Ca^{2+} have optical absorptions, murexide was used as an indicator of the Ca^{2+} ion (the murexide–Ca^{2+} complex absorbs at 480 nm). The chemical relaxation spectrum (see Section 10.2.3) obtained for the Ca–AcCh receptor system reflected multiple interactions between Ca^{2+} and the receptor. At least three relaxation times were observed. Two slower relaxation times showed that slower intramolecular reactions, such as conformational changes, might be involved in the overall Ca–AcCh receptor reaction. The dependence of acetylcholine binding constants on Ca^{2+} concentration also suggests competition between acetylcholine and Ca^{2+} for the AcCh receptor molecule.

10.3.8 Circular Dichroism

There are several tRNA species which can undergo reversible changes from their native conformation to a relatively stable but biologically inactive, denatured conformation (Lindahl *et al.*, 1967). This interconversion of tRNA could be involved in biological control mechanisms. Hawkins *et al.* (1977) studied one such tRNA (a tRNA^{leu} species) from baker's yeast.

The change in circular dichroism of tRNA^{leu} was used to follow the renaturing process. When Mg^{2+} was added, the first step in the conversion of the denatured structure to the native structure happened very rapidly, followed by a second slower, temperature-dependent step which restored the tRNA to its biologically active state. It appears likely, based on the studies, that Mg^{2+} plays an important role in obtaining and maintaining the native structure of tRNA^{leu}.

10.3.9 Chelators

There are numerous enzymes which require divalent metal ions, such as Mg^{2+}, Ca^{2+}, Mn^{2+}, Mo^{2+}, Co^{2+}, Zn^{2+}, Cd^{2+}, Cu^{2+}, etc., for their activities. Those enzymes are usually inhibited by the addition of chelators (such as EDTA) to the reaction mixture. (Note that there are a few exceptions, such as blue copper enzymes, which contain tightly bound metal ions that are not removed by chelating agents.) After a dialysis of the enzyme against an EDTA solution (alternatively, a Chelex-100 resin slurry) and subsequent removal of the EDTA–metal complex, the addition of various metal ions to the inactive apoenzyme solution can help to identify the possible species of the metal ion responsible for enzyme activity (activity is determined from the slope of a zero-order kinetic plot). Leucine aminopeptidase (EC 3.4.11.1) from *Aspergillus,* for example, is inhibited to varying degrees by several chelating agents (Sugiura *et al.,* 1978), as shown in Table 10-4. The native aminopeptidase was dialyzed against 2 mM EDTA in 10 mM phosphate buffer at pH 7.5, and this treatment resulted in a complete loss of enzymic activity. The effect of various metal ions on the apoenzyme solution could then be tested. Table 10-5 shows the recovery of activity on the addition of 11 different salts. The K_m values and the activation energies of the native enzyme, the zinc-reactivated enzyme, and the cobalt-reactivated enzyme were then compared (Table 10-6). From metal analysis, the leucine aminopeptidase of *Aspergillus* contains 1 g-atom of zinc per mole of the enzyme. The above results strongly suggested that this fungal leucine aminopeptidase is a zinc enzyme and that the metal plays a key role in the enzymic catalysis.

Table 10-4. Effect of Various
Chelating Agents in Inhibiting
Leucine Aminopeptidase[a]

Inhibitor	PI[b]
EDTA	5.7
O-Phenanthroline	4.2
2,6-Pyridinedicarboxylic acid	3.9
8-Hydroxyquinoline	3.2
α-Dipyridyl	2.4
Iminodiacetic acid	2.2

[a]Adapted from Sugiura *et al.,* 1978.
[b]50% inhibition of the enzyme activities.

Table 10-5. Effect of Various Metals on
the Reactivation of EDTA-Treated
Aminopeptidase[a]

Metal ion (0.1mM)	Relative activity (%)
None	0
CoCl$_2$	84
NiCl$_2$	0
MnCl$_2$	7
MgCl$_2$	0
CaCl$_2$	3
CdCl$_2$	0
CuCl$_2$	0
HgCl$_2$	0
ZnCl$_2$	100[b]
FeCl$_3$	0
NaCl	0

[a]Adapted from Sugiura *et al.*, 1978.
[b]Relative to the Zn^{2+} activated enzyme.

The reader will note that such studies may also reveal nonphysiological substitutions which nevertheless lead to catalytic activity (see Section 10.3.4). As a result of possible subsitution alternatives, enzyme kinetic studies must be supplemented with a very careful and thorough elemental analysis of the native enzyme, using atomic absorbance or sensitive methods of the type described in Chapter 12. An ambiguity may remain if the enzyme is activated by any one of several loosely held ions. Conclusions will then be guided by a knowledge of the ions present in the enzyme's compartment, or perhaps by the relative efficacy of the activating ions. On the latter point, it is not necessarily true that the best catalytic ion is the one involved under *in vivo* conditions.

Table 10-6. Comparison of K_m and Activation
Energy of Native and Zinc- and Cobalt-Reactivated
Enzymes[a]

	K_m (mM)	E_{act}[b] (10^3 cal/mol)
Native	0.25	9.2
Zinc reactivated	0.25	9.8
Cobalt reactivated	0.064	15.6

[a]Adapted from Sugiura *et al.*, 1978.
[b]Estimated from Arrhenius plots.

References

Ahlers, J., 1975. The mechanism of hydrolysis of β-glycerophosphate by kidney alkaline phosphatase, *Biochem. J.* 149:535.

Andreasson, L. E., Branden, R., Malmstrom, B. G., and Vangard, T., 1973. An intermediate in the reaction of reduced laccase with oxygen, *FEBS Lett.* 32:187.

Atkins, P. W., McLauchlan, K. A., and Simpson, A. F., 1970. A flash-correlated one microsecond response electron spin resonance spectrometer for flash photolysis studies, *J. Phys. E* 3:547.

Broman, L., 1967. The function of ceruloplasmin—A moot question in *Molecular Basis of Some Aspects of Mental Activity*, Vol. 2 (O. Walaas, ed.), Academic Press, London.

Brunel, C., and Cathala, G., 1973. Activation and inhibition processes of alkaline phosphatase from ions (magnesium (2^+) and zinc (2^+) ions), *Biochim. Biophys. Acta* 309:104.

Carrico, R. J., Malmstrom, B. G., and Vangard, T., 1971. A study of the reduction and oxidation of human ceruloplasmin, evidence that a diamagnetic chromophore in the enzyme participates in the oxidase mechanism, *Eur. J. Biochem.* 22:127.

Corson, D. C., Williams, T. C., and Sykes, B. D., 1983. Calcium binding proteins: Optical stopped-flow and proton nuclear magnetic resonance studies of the binding of the lanthanide series of metal ions to parvalbumin, *Biochemistry* 22:5882.

Deley, M., and Osaki, S., 1975. Intramolecular electron transport in human ferroxidase, *Biochem. J.* 151:561.

Diliberto, E. J., DiStefano, V., and Smith, J. C., 1973. Mechanism and kinetics of the inhibition of dopamine β-hydroxylase by 2–mercaptoethylguanidine, *Biochem. Pharmacol.* 22:2961.

Eigen, M., and Kruse, W., 1963. *Z. Naturforsch.* 186:857, and references cited therein.

Fields, G. A., and Bright, H. J., 1970. Magnetic resonance and kinetic studies of the activation of β–methylaspartase by manganese, *Biochemistry* 9:3801.

Fox, K. R., and Waring, M. J., 1984. Stopped-flow kinetic studies on the interaction between echinomycin and DNA, *Biochemistry* 23:2627.

Frieden, E., and Osaki, S., 1970. Ceruloplasmin: A possible missing link between copper and iron, in *Effects of Metals on Cells, Subcellular Elements and Macromolecules* (J. Maniloff, J. R. Coleman, and M. W. Miller, eds.), Charles C. Thomas Publisher, Springfield, Illinois.

Gottesman, M., Simpson, R. T., and Vallee, B. L., 1969. Kinetic properties of cobalt alkaline phosphatase, *Biochemistry* 8:3776.

Gross, F. (ed), 1964. *Iron Metabolism*. Springer-Verlag, Berlin.

Hammes, G. G., 1968. Relaxation spectrometry of biological systems, *Adv. Protein Chem.* 23:1.

Hawkins, E. R., Chang, S. H., and Mattice, W., 1977. Kinetics of the renaturation of yeast tRNA[leu], *Biopolymers* 16:1557.

Jeener, J., Meier, B. H., Bachmann, P., and Ernst, R. R., 1979. Investigation of exchange processes by two-dimensional nmr spectroscopy, *J. Chem. Phys.* 71:4546.

Kustin, K. (ed), 1969. Fast reactions, in *Methods in Enzymology*, Vol. XVI, Academic Press, New York.

Latt, S. A., Auld, D. S., and Vallee, B. L., 1970. Surveyor substrates: Energy transfer gauges of active center topography during catalysis, *Proc. Nat. Acad. Sci. U.S.A.* 67:1383.

Lee, G. R., Nacht, R. S., Lukens, J. N., and Cartwright, G. E., 1968. Iron metabolism in copper deficient swine, *J. Clin. Invest.* 47:2058.

Lengfelder, E., 1980, A coaxial electron beam attenuator for a Febetron 705, *Radiat. Phys. Chem.* 16:405.

Lengfelder, E., and Elstner, E. F., 1979. Cyanide insensitive iron superoxide dismutase in *Euglena gracilis,* comparison of the reliabilities of different test systems for superoxide dismutases, *Z. Naturforsch.* 34:374.

Lindahl, T., Adams, A., and Fresco, J. R., 1967. Isolation of "renaturable" transfer ribonucleic acids, *J. Biol. Chem.* 242:3129.

Malmstrom, B. G., Reinhammer, B. and Vangard, T., 1970. The state of copper in stellacyanin and laccase from the laquer tree *Rhus verniaifera, Biochem. Biophys. Acta* 205:48.

Manabe, T., Hatano, H., and Hiromi, K., 1973. Kinetic studies on the aerobic oxidation of reduced human ceruloplasmin, *J. Biochem. (Tokyo)* 73:1169.

Marcandalli, B., Winzek, C., and Holzwarth, J. F., 1984. A laser temperature jump investigation of the interaction between proflavine and calf-thymus deoxyribonucleic acid at low and high ionic strength avoiding electric field effects, *Ber. Bunsenges. Phys. Chem.* 88:368.

Neurath, H. (ed.), 1970. *The Proteins, Composition, Structure, and Function, Vol. V, Metalloproteins,* Academic Press, New York.

Newmann, E., and Chang, H. W., 1976. Dynamic properties of isolated acetylcholine recepter protein: Kinetics of the binding of acetylcholine and Ca ions, *Biophys. J.* 73:3994.

Osaki, S., 1966. Kinetic studies of ferrous ion oxidation with crystalline human ferroxidase (ceruloplasmin), *J. Biol. Chem.* 241:5053.

Osaki, S., and Walaas, O., 1967. Kinetic studies of ferrous ion oxidation with crystalline human ferroxidase. II. Rate constants at various steps and formation of a possible enzyme substrate complex, *J. Biol. Chem.* 242:2653.

Osaki, S., Johnson, D. A., and Frieden, E., 1970. The mobilization of iron from the perfused mammalian liver by a serum copper enzyme, Ferroxidase I, *J. Biol. Chem.* 24:3018.

Peisach, J., Aisen, P., and Blumberg, W. E. (eds.), 1966. *The Biochemistry of Copper,* Academic Press, New York.

Ragan, H. A., Nacht, S., Lee, G. R., Bishop, C. R., and Cartwright, G. E., 1969. Effect of ceruloplasmin on plasma iron in copper deficient swine, *Am. J. Physiol.* 217:1320.

Schelly, Z. A., and Eyring, E. M., 1971. Step perturbation relaxation techniques, *J. Chem. Ed.* 48:A639.

Strothkamp, R. E., and Dawson, C. R., 1977. A spectroscopic and kinetic investigation of anion binding to ascorbate oxidase, *Biochemistry* 16:1926.

Sugiura, M., Ishikawa, M., and Sakaki,M., 1978. Some properties of leucine aminopeptidase from *Aspergillus japonica* as a metalloenzyme, *Chem. Pharm. Bull.* 26:3101.

Walaas, E., 1967. *Catecholamines as substrates for ceruloplasmin,* in *Molecular Basis of Some Aspects of Mental Activity,* Vol. 2 (O. Walaas, ed.), Academic Press, London.

Bioinorganic Topochemistry: Microprobe Methods of Analysis

11

11.1 Electron Probe Microanalysis

11.1.1 Introduction

Direct elemental ultramicroanalysis of biological samples is now possible, based on X-ray emission stimulated by a highly focused electron beam. This is an elegant extension of the X-ray spectrometric identification and quantitation method for specific elements. The method is made possible by combining either an energy- or a wavelength-dispersive X-ray spectrometer with a scanning electron microscope. The hybrid is appropriately described as a tool for inorganic topochemistry, since the beam may impinge upon a resolved structural feature while the X-ray emission spectrum is recorded. This procedure is especially valuable in that all of the elements heavier than beryllium may be specifically determined. Figure 11-1 presents a schematic representation on a typical electron microprobe instrument.

The collisions of high-velocity electrons with atoms cause the inner electrons of the atoms to vacate their normal orbitals. The return of a higher-energy electron to the inner shell is accompanied by emission of energy in the form of X-rays, and the characteristic radiation, especially that resulting from transitions to the lowest atomic level, is related to the atomic number of the atom. Thus, element-specific X-ray lines are emitted when an element is excited by the electron beam. Elements are determined by analysis of the wavelengths of the X-ray photons emitted. Quantitation may follow by measurement of the intensities or profile areas of the characteristic X-ray lines, which correspond to energy transitions of a particular element (Cowley, 1982; Hren *et al.*, 1979).

Analysis of the emission is accomplished either in a wavelength-

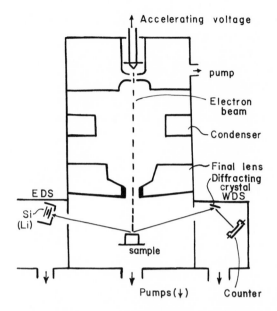

Figure 11-1. Schematic diagram of an electron microprobe. Adapted from Reed, 1975, and from Geller, 1977.

dispersive spectrometer (WDS) or an energy-dispersive spectrometer (EDS). The former is the older method and depends on a natural diffractive element—a crystal with atomic planes spaced regularly. The Bragg law is the basis of spectroscopic dispersion, i.e., $n\lambda = 2d \sin \theta$. Here d is the spacing between crystal planes, n is an integer (1,2,3. . .), λ is the X-ray wavelength, and θ is the diffractive angle. Properly positioned detection slits or a scanning goniometer may be used to sort specific wavelengths into an X-ray counter. Thus, the unique X-ray emissions of an element may be detected and quantified. This design resembles a conventional optical (UV-visible or infrared) spectrometer, where the crystal grating is analogous to an ordinary ruled optical grating.

In the energy-dispersive method of analysis, X-rays are received by a solid-state detector based on silicon, usually of the lithium-drifted type. This semiconducting detector absorbs X-rays, and the deposited energy raises electrons to a conduction band. An applied high voltage sweeps these electrons into a counting circuit, where the X-ray photon energy from the excited sample is proportional to the number of electrons collected; i.e., the voltage deflection or pulse height is proportional to photon energy. A spectrum is generated after amplification of these pulses and further electronic sorting in a pulse height analyzer (PHA). The PHA is an analog-to-digital device in which narrow ranges of pulse heights are sorted into counting channels. In this type of spectrometer an X-ray line's intensity and its lineshape are determined by the accumulated counts in a narrow group of counting channels. The spectrum is channel number (energy) versus pulses counted (intensity).

Advantages and features of the EDS and WDS methods may be compared. The two methods are similar in accuracy and precision. WDS can be tenfold more sensitive than EDS, depending on the sample geometry and thickness (Geller, 1977). The WDS method may give better resolution, and it can detect boron and atoms of higher atomic number. The EDS method has advantages in: (a) faster analysis times, (b) simultaneous data measurement for all energies, and (c) a high collection efficiency. However, the EDS approach is not suitable for atomic numbers smaller than ten (Heywood, 1979).

Quantitation of the data may be obtained from the ratio of unknown and standard line intensities (or profile areas), e.g., as $C_x = C_{st}(I_x/I_{st})$, where C_x and I_x are, respectively, the concentration and line intensity of the unknown sample; C_{st} and I_{st} are likewise the concentration and intensity of a standard.

Selectivity for microscopic sample features is based on the fact that a volume of only several cubic microns may be analyzed by the electron beam at one time. Methods for the treatment of biological samples have been reviewed, e.g., see Lechene and Warner (1977), Robison (1973), and Echlin and Saubermann (1977). Three distinct types of electron microprobe analyses are possible: point analysis, elemental line scans, and elemental mapping (Heywood, 1979). Point analysis can pinpoint selected areas of a sample, and can compare the composition of each area to the others. This type of analysis is illustrated in Figure 11-2 where diatoms and algae are subjected to the probe. Elemental line scans give the distribution of the element of interest along a given line chosen in the sample. Elemental mapping shows the location of the element of interest within the two-dimensional view of the sample (see Figure 11-3).

Further discussions of the theory of electron probe microanalysis may be found in articles or books by Hall et al. (1964), Andersen (1967), Eramus (1978), Heinrich (1968), and Hall (1968).

11.1.2 Applications

Nearly two decades ago, Hall (1968) pointed out that the electron microprobe method should be making a greater contribution in biology because as little as 10^{-15} g of elements of biological interest could be

A

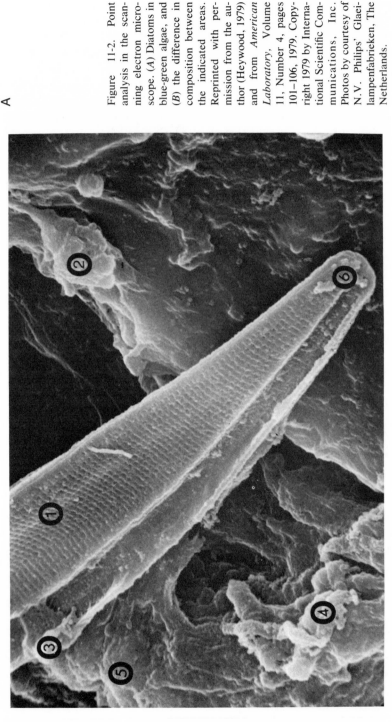

Figure 11-2. Point analysis in the scanning electron microscope. (A) Diatoms in blue-green algae, and (B) the difference in composition between the indicated areas. Reprinted with permission from the author (Heywood, 1979) and from *American Laboratory*, Volume 11, Number 4, pages 101–106, 1979. Copyright 1979 by International Scientific Communications, Inc. Photos by courtesy of N.V. Philips' Glaeilampenfabrieken, The Netherlands.

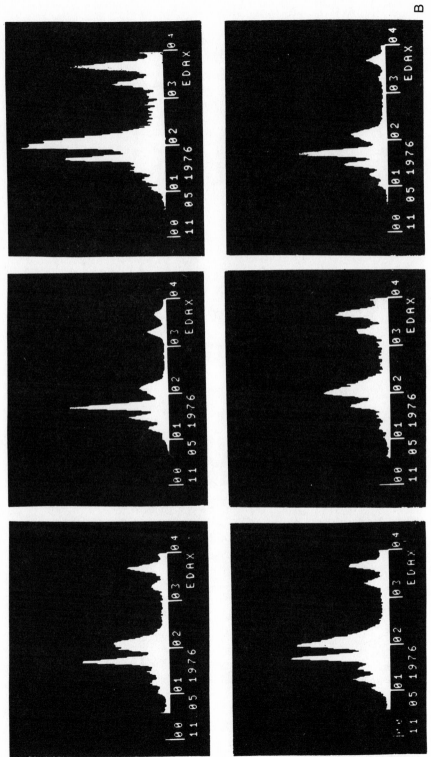

B

A

B

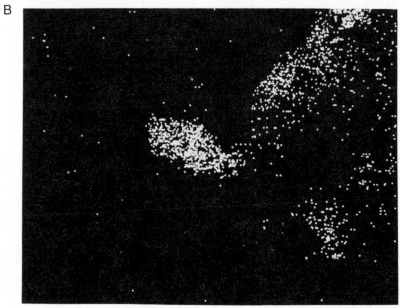

Figure 11-3. Elemental mapping in the scanning electron microscope. Specimen is a silver catalyst: (*A*) Secondary electron image of a bridge between two grains (1450×), (*B*) calcium distribution over the same area, and (*C*) silver distribution over the same area. Reprinted with permission as for Figure 11-2.

detected within a spatial resolution of one micrometer. The following examples show that biological applications of the electron microprobe method have indeed been rewarding.

One of the earliest applications of this physical method to biology and medicine is found in the work of Mellors and Carroll (1961), who demonstrated the uptake of chromium and iron into tissue in the vicinity of a surgical metallic implantation (hip joint prosthesis). Other investigations soon followed, e.g., in the study of calcium–phosphate relationships in bone (Mellors *et al.*, 1964) and in measurements of the distribution of calcium and sulfur in transverse sections of normal and sclerotic human arteries (Hall *et al.*, 1964).

Tousimis and Adler (1963) analyzed tissue sections of cornea obtained at autopsy from patients who had Wilson's disease (hepatolenticular degeneration). The electron microprobe method showed that copper was in the highest concentration on the endothelial side of Descemet's membrane, while almost no copper was found in the anterior half of this membrane or in the membrane of the center of the cornea. These results agree with those of other workers using completely different methods of analysis for copper in the same biological tissue.

Scanning electron microscopy (SEM) coupled with an energy-dispersive spectrometer allowed for nondestructive, pinpoint detections of sodium and heavier elements in tropical woody plants (Taniguchi *et al.,* 1982). Two new types of mineral deposits in plants were found, each containing titanium.

Specific ectodermal cells of the chick chorioallantoic membrane contain high concentrations of calcium, according to correlated electron probe and electron microscope analyses (Coleman *et al.,* 1970). These may be specialized cells involved in the trans-epithelial transport of calcium, and represent "compartments" of sequestered calcium.

The X-ray spectrum of a normal centriole in guinea pig renal tubule was obtained with lithium fluoride, ammonium dihydrogen phosphate, and gypsum crystals using a scanning (WDS) spectrometer (Shafer and Chandler, 1970). The results indicated that chlorine, potassium, silicon, and sulfur are elements native to the centriole. The finding of silicon was said to be unexpected but exciting. However, *any* evidence of silicon should be regarded carefully in view of its frequent presence as a contaminant.

The plaque deposits of Alzheimer's dementia have been subjected to a biochemical investigation using the electron microprobe (Austin, 1978). Alzheimer senile plaques and their inner cores were isolated from pathological brain tissue. Although the element aluminum had been expected—based on experimental administration of aluminum to rabbits, resulting in Alzheimer-like neurofibrillar material—no aluminum was found in the diseased human tissue. Instead, silicon proved to be the prominent element, even after correction by controls. Colorimetric assays confirmed the finding of silicon, but any role of silicon in such disease remains speculative. It was considered that silicon in localized high concentration represents an epiphenomenon rather than any causative phenomenon (Austin, 1978).

Although the above results appear to be valid, special care needs to be taken against contamination of biological tissues by silicon (Smith, 1979), and the investigator must be aware of potential sources of artifactual silicon. For example, oils or greases used during specimen preparation can give rise to spurious silicon. The vacuum system itself can introduce a false positive for silicon. Contamination should especially be suspected if the silicon peak increases with time (see Figure 11-4). A common source of silicon contamination is from the vacuum grease used with freeze-drying equipment.

Another consideration is the lowest possible intensity of the Si

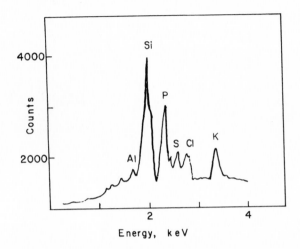

Figure 11-4. X-ray spectrum of mouse hepatocyte nuclei, showing a strong contamination of Si which increased with time. Adapted from Smith (1979).

X-ray peak, which cannot be less than the peak of Si internal fluorescence from the dead layer of the detector (in other words, a background lacking silicon is essentially impossible with silicon detectors). These problems are important since silicon is now thought to be a biological trace element, and its distribution in tissue is an open question.

Kierszenbaum *et al.* (1971) studied seminiferous tubules with the electron microprobe. Pyroantimonate-precipitable cations (Mg^{2+}, Ca^{2+}, Na^+) were found to be localized in the walls of the seminiferous tubules, in two concentric layers corresponding to the inner and outer layers of the tubular wall. It is possible that this localization corresponds closely to that of acid mucoproteins.

It has been suggested that the diagnosis of some thyroid malignancies may be aided by a determination of the location of iodine in tissue, based on electron microprobe analysis (Banfield *et al.*, 1971), e.g., diagnosis would depend on the presence or absence of a characteristic pathological distribution.

Marshall and Forrest (1977) used freeze-substituted animal tissues as test specimens to determine the effects of different accelerating voltages on X-ray microanalysis in a transmission electron microscope. Contrary to reports by other workers, an increase in the accelerating voltage gave a substantial increase in sensitivity (Table 11-1). This discrepancy may be explained by different section thicknesses and geometries. Marshall and Forrest (1977) suggest that the instrument operator should establish

Table 11-1. P/b Ratio[a] from a Section
(0.5 μm Thick) of Freeze-substituted
Malpighian Tubules of the Cricket[b]

Element	P/b at the indicated accelerating voltages[c]		
	60 kV	80 kV	100 kV
Ca	1.0	2.0	2.4
K	1.0	2.0	2.3
Cl	1.0	1.9	2.1
P	1.0	2.0	2.3
Mg	1.0	1.4	2.7

[a]P/b = peak-to-background ratio.
[b]Adapted from Marshall and Forrest, 1977.
[c]Values for 60 kV are normalized to 1.0.

empirical parameters giving the optimum peak-to-background ratio and counting rate for the microscope in use.

Cantley *et al.* (1977) found vanadium in ATP prepared from muscle [the vanadate ion is a potent inhibitor of (Na,K)-ATPase]. This identification was performed by electron probe microanalysis and was confirmed by microwave-induced emission spectroscopy (Kawaguchi and Vallee, 1975) and also by electron paramagnetic resonance spectroscopy. An interesting discussion involved the point that $H_2VO_4^-$ (at natural pH) is structurally similar to the phosphate ion. The orthovanadate ion is tetrahedral with V—O bond lengths of 1.66 Å, while in tetrahedral orthophosphate the P—O bonds are 1.55 Å. It was suggested that orthovanadate might bind to a phosphate site on the (Na,K)-ATPase enzyme.

Nerve terminals have been studied with respect to the intracellular location of sodium, potassium, phosphate, and chloride ions (Silbergeld and Costa, 1979). As expected, the cytoplasm was relatively enriched in Na and Cl, whereas mitochondria were enriched in Ca and P contained in noncrystalline electron-dense accumulations. Also, *in vitro* treatments of synaptosomes and isolated mitochondria gave the predicted results for calcium uptake and retention, as assayed with the electron microprobe. This line of work serves as a reminder that an electron microprobe coupled with scanning transmission electron microscopy can avoid the problems of translocation and homogenization of minerals in tissues. In related work, Galvan *et al.* (1984) measured intracellular elements in rat sympathetic neurons, using energy-dispersive electron microprobe analysis. Values obtained for the concentrations of Na, K, and Cl were 5, 196, and

32 mM, respectively. Nuclear and cytoplasmic values were not significantly different from each other.

In another case involving calcium, an electron microscope/EDS spectrometer system was used to measure calcium uptake by sarcoplasmic reticulum vesicles in a comparison of normal and batrachotoxin-denervated muscle tissue (Wan *et al.*, 1982). The data indicated that batrachotoxin acts by blocking both axonal transport and impulse activity, resulting in a chemical denervation of the muscle.

Energy dispersive X-ray microanalysis was employed in an investigation of the physiological distribution of the elements in skin, including different layers of epidermis (Forslind, 1984). The electrolytes Na, K, Ca, and Mg were of interest in this study of contact reactions.

Computer programs have been developed for the integration of on-line data collection and data reduction, using the features of magnetic disk storage devices: random access, program linkage, high speed, and data storage (Hamilton *et al.*, 1977). Electron microprobes (and the related instruments described in the following section) are able to generate volumes of data at a rapid rate, necessitating some kind of data processing capability, e.g., a dedicated computer (Gavrilovic and Brooks, 1982).

11.2 Ion, Laser, and Proton Microprobe Analysis of Elements

These methods, like the electron microprobe, also provide for a sophisticated and sensitive analysis of elements of biological interest. Ion microprobes combine the properties and functions of an ion emission microscope and a mass spectrometer. All of the elements of the periodic table may be detected, sometimes at the parts per billion level of concentration (Karasek, 1970). A high-energy beam of ions is focused and bombarded onto the surface of the sample being analyzed. Secondary ions are produced from each element actually present on the sample surface, and these secondary ions are stripped from the surface. This is clearly a sample-destructive method. The secondary ions are collected and interpreted by a computerized mass analyzer. Data which may be obtained include chemical composition; composition and thickness of thin films; the nature of embedded inclusions; isotope ratios; and the chemical structure of inorganic and some organic compounds (Gavrilovic and Majewski, 1977). For some investigations, electron microprobe analysis could be coupled with data from the ion microprobe mass analyzer; the high precision and versatility of the former method would complement the extreme sensitivity of the latter.

Laser microprobe optical emission spectroscopy has been discussed in terms of its use with biological samples (Tretyl *et al.*, 1972). It was pointed out that for biological material, the spectral emission of an element in a laser-generated plasma is strongly dependent on the matrix in which the element is located. Thus detection limits for an unknown must be based on standards with a composition similar to the unknown.

The laser microprobe detection limits for elements such as Li, Mg, Ca, Cu, Zn, Hg, and Pb are given in Table 11-2, and are compared to detection limits of various elements determined by some other physical techniques. Laser microprobe samples may be internally calibrated (Verbueken *et al.*, 1984). Chelex-100 ion-chelating resin beads containing elements such as Pt, Al, Na, K, and Ca are co-embedded with the biological sample, which is then cut to the desired thickness. The microscopic sample thus contains easily identified (circular) zones of known composition in close proximity to the object of interest.

Proton microprobe analysis is another sensitive tool for the determination of trace elements. The electron microprobe approach is less sensitive because of its limiting factor, i.e., a relatively high bremsstrahlung background (the X-ray continuum from decelerating electrons). A proton beam results in a markedly reduced background and, therefore, a better signal-to-noise ratio. Proton microprobe analysis begins with the acceleration of a charged-particle beam of protons, which is then focused accurately on the subject to be analyzed. This causes the emission of characteristic X-rays of the elements present in the sample (Valkovic, 1980)—the same principle as involved in electron microprobe analysis.

A proton microprobe analyzer was used to test a hypothesis that a toxin—nickel carbonyl—was involved in the deaths of victims of Legionnaire's disease (Chen *et al.*, 1977). It did appear that higher levels of

Table 11-2. Comparison of Sensitivities of
Elemental Analytical Methods[a]

Method	Detection limits	
	in grams	in ppm[b]
Ion microprobe[c]	10^{-17}–10^{-19}	10^{0}–10^{-2}
Electron microprobe	10^{-15}–10^{-17}	10^{3}–10^{1}
Laser microprobe	10^{-12}–10^{-15}	10^{2}–10^{-1}
Neutron activation	10^{-10}–10^{-13}	10^{-3}–10^{-6}
Atomic absorption	10^{-9}–10^{-12}	10^{-1}–10^{-5}

[a]Adapted from Tretyl *et al.*, 1972.
[b]Parts per million.
[c]A proton microprobe is also reported to be more sensitive than the electron microprobe (Bosch *et al.*, 1978).

nickel were found in the lungs of the victims (compared to controls). However, difficulties in ruling out nickel contamination made the test inconclusive. This illustrates a basic problem in the use of *any* very sensitive method—the avoidance of minute but significant contamination in biological samples (see Section 5.1). It may be pointed out that the prokarocyte actually responsible for Legionnaire's disease has since been characterized (Chandler *et al.*, 1977; Katz and Nash, 1978).

Proton microprobe analysis has been used for the chemical analysis of lunar trace elements in samples gathered by Apollo 17 astronauts (Bosch *et al.*, 1978). Two minerals—ilmenite ($FeTiO_3$) and baddeleyite (ZrO_2)—had not been detected by electron microprobe analysis but were found with proton microprobe analysis, illustrating the sensitivity differences discussed above. Bosch *et al.* (1978) also added a note that the proton microprobe method should be applicable to biological objects with a size of approximately 10 μm; sea urchin eggs (40 μm in diameter) gave the expected Ca/S ratios when they were subjected to the proton microprobe method. Shock-freezing of biological samples is a necessary preparative step; this prevents the migration of water-soluble ions such as K^+, Na^+, and Cl^-.

Russell *et al.* (1981) used the proton microprobe method to detect trace elements in environmental and biological specimens. They found that proton energies of 2–3 MeV yielded the lowest X-ray continuum background and therefore the optimum detection limits. At lower energies, the K-shell ionization cross section falls off rapidly; at higher energies, both the electron bremsstrahlung background (due to secondary processes) and the nuclear γ-ray background increase. The cited reference reports useful data on blank values for backing foils of various types (plastics, carbon, etc.).

A scanning proton microprobe analyzer has been used for the medical diagnosis of primary biliary cirrhosis (Watt *et al.*, 1984). This disease is characterized by progressive destruction of the small intrahepatic bile ducts, cholestasis, and high levels of copper within the liver. The microprobe constructed maps of 7-μm sections of diseased liver. Copper was found to be distributed in deposits less than 5 μm in size at specific locations in the liver. Also, a 1:1 atomic equivalence existed between Cu and S. The nature of the sulfur (e.g., as part of a protein–Cu structure or as an inorganic salt) was not established.

X-ray fluorescence and neutron activation analysis also detect parts per million concentrations of an element [see Chapter 12 for examples of neutron activation analysis; the PIXE X-ray method for trace element analysis is exemplified in Budnar and Starc (1982)]. However, these bulk methods do not analyze selected small regions of a microscopic specimen.

Proton and electron microprobes are relatively nondestructive, while the ion probe and laser mass spectrometer destroy the subject of interest.

A variety of microprobe methods have been discussed in this chapter, and some other physical methods of analysis have also been mentioned. One should bear in mind that the absolute detection limit would not be the sole determining factor in selecting a method for analysis, as different problems or sample types may be best studied with one or another of the various physical techniques available. A great deal of interest centers on how the various biological elements are distributed within cells or tissues, and the microprobe methods appear to be well-suited for answering such questions. The reader will benefit by examining issue C2 of *J. Phys. Colloq.*, 1984, which is a collection of papers on microprobe methods.

References

Andersen, C. A., 1967. An introduction to the electron probe microanalyzer and its application to biochemistry, in *Methods of Biochemical Analysis*, Vol. 15 (D. Glick, ed.), Wiley-Interscience, New York, pp. 147–270.

Austin, J. H., 1978. Silicon levels in human tissues, *Biochemistry of Silicon and Related Problems* (G. Bendz and I. Lindquist, eds.), Plenum Press, New York, pp. 255–268.

Banfield, W. G., Grimley, P. M., Hammond, W. G., Taylor, C. M., deFlorio, B., and Tousimis, A. J., 1971. Electron probe analysis for iodine in human thyroid and parathyroid glands, normal and neoplastic, *J. Natl. Cancer Inst.* 46:269.

Bosch, F., Goresy, A. E., Martin, B., Bogdan, P., Nobiling, R., Schwalm, D., and Taxel, K., 1978. The proton microprobe: A powerful tool for nondestructive trace element analysis, *Science* 199:765.

Budnar, M., and Starc, V., 1982. Determination of trace elements in human urine, *Period. Biol.* 84:119.

Cantley, L. C., Josephson, L., Warner, R., Yanagisawa, M., Lechene, C., and Guidotti, G., 1977. Vanadate is a potent (Na,K)-ATPase inhibitor found in ATP derived from muscle, *J. Biol. Chem.* 252:7421.

Chandler, F. W., Hicklin, M. D., and Blackman, J. A., 1977. Demonstration of the agent of Legionnaire's disease in tissue, *N. Engl. J. Med.* 297:1218.

Chen, J. R., Francisco, R. B., and Miller, T. E., 1977. Legionnaire's disease: Nickel levels, *Science* 196:906.

Coleman, J. R., DeWitt, S. M., Batt, P., and Terepka, A. R., 1970. Electron probe analysis of calcium distribution during active transport in chick chorioallantoic membrane, *Exp. Cell Res.* 63:216.

Cowley, J. M., 1982. Electron microscopy, *Anal. Chem.* 54:83R.

Echlin, P., and Saubermann, A. J., 1977. Preparation of biological specimens for X-ray microanalysis, in *Scanning Electron Microscopy*, Vol. 1, IIT Research Institute, Chicago, pp. 621–634.

Erasmus, D. (ed.), 1978. *Electron Probe Microanalysis in Biology*, Chapman and Hall, London.

Forslind, B., 1984. Clinical applications of scanning electron microscopy and X-ray microanalysis in dermatology, in *Scanning Electron Microscopy* (O. Johari, ed.), Scanning Electron Microscopy, Inc., Chicago, pp. 183–206.

Galvan, M., Doerge, A., Beck, F., and Rick, R., 1984. Intracellular electrolyte concentrations in rat sympathetic neurons measured with an electron microprobe, *Pfluegers Arch.* 400:274.

Gavrilovic, J., and Brooks, D. A., 1982. Problems associated with computerized analysis of a large number of small particles, *Microbeam Anal.* 17:495.

Gavrilovic, J., and Majewski, E., 1977. Use of ion and electron microprobes for full characterization of particulate matter, *Am. Lab.* 9:19.

Geller, J. E., 1977. A comparison of minimum detection limits using energy and wavelength dispersive spectrometers, *Scanning Electron Microsc.* 10:281.

Hall, T., 1968. Some aspects of the microprobe analysis of biological specimens, in *Quantitative Electron Probe Microanalysis* (K. F. J. Heinrich, ed.), National Bureau of Standards Special Publication 298, NBS, Washington, D.C., pp. 269–299.

Hall, T. A., Hale, A. J., and Switsur, V. R., 1964. Some applications of microprobe analysis in biology and medicine, in *The Electron Microprobe* (T. D. McKinley, K. F. J. Heinrich, and D. B. Wittry, eds.), John Wiley and Sons, New York, pp. 805–833.

Hamilton, W. J., Hinthorne, J. R., Ray, L. A., and Whatley, T. A., 1977. Automated electron microprobe analysis: A system for the ARL-SEMQ computer based on mass storage and speed capabilities of the flexible magnetic disk, *Proc. Ann. Conf. Microbeam Anal. Soc.* 12:52A.

Heinrich, K. F. J. (ed.), 1968. Quantitative Electron Probe Microanalysis, National Bureau of Standards Special Publication 298, NBS, Washington, D.C., pp. 1–299.

Heywood, J. A., 1979. Elemental analysis in the scanning electron microscope, *Am. Lab.* 11:101.

Hren, J. J., Goldstein, J. I., and Joy, D. C. (eds.), 1979. *Introduction to Analytical Electron Microscopy,* Plenum Press, New York.

Karasek, F. W., 1970. The ion microanalyzer, *Research/Develop.* 21:32.

Katz, S. M., and Nash, P. N., 1978. Legionnaire's disease: Structural characteristics of the organism, *Science* 199:896.

Kawaguchi, H., and Vallee, B. L., 1975. Microwave excitation emission spectrometry; determination of picogram quantities of metals in metalloenzymes, *Anal. Chem.* 47:1029.

Kierszenbaum, A. L., Libanati, C. M., and Tandler, C. J., 1971. The distribution of inorganic cations in mouse testis: Electron microscope and microprobe analysis, *J. Cell Biol.* 48:314.

Lechene, C. P., and Warner, R. R., 1977. Ultramicroanalysis: X-ray spectrometry by electron probe excitation, *Annu. Rev. Biochem. Biophys.* 6:57.

Marshall, A. T., and Forrest, Q. G., 1977. X-ray microanalysis in the transmission electron microscope at high accelerating voltages, *Micron* 8:135.

Mellors, R. C., and Carroll, K. G., 1961. A new method for local chemical analysis of human tissue, *Nature* 192:1090.

Mellors, R. C., Carroll, K. G., and Solberg, T., 1964. Quantitative analysis of Ca/P molar ratios in bone tissue with the electron microprobe, in *The Electron Microprobe* (T. D. McKinley, K. F. J. Heinrich, and D. B. Wittry, eds.), John Wiley and Sons, New York, pp. 834–840.

Robison, W. L., 1973. Application of the electron microprobe to the analysis of biological material in *Microprobe Analysis* (C. A. Andersen, ed.), Wiley-Interscience, New York, pp. 271–321.

Russell, S. B., Schulte, C. W., Faiq, S., and Campbell, J. L., 1981. Specimen backings for proton-induced X-ray emission analysis, *Anal. Chem.* 53:571.

Shafer, P. W., and Chandler, J. A., 1970. Electron probe X-ray microanalysis of a normal centriole, *Science* 170:1204.

Silbergeld, E. K., and Costa, J. L., 1979. Synaptosomal Ca metabolism studies by electron microprobe analysis, *Exp. Neurol.* 63:277.

Smith, N. K. R., 1979. A review of sources of spurious silicon peaks in electron microprobe X-ray spectra of biological specimens, *Anal. Biochem.* 94:100.

Taniguchi, T., Harada, H., and Nakato, K., 1982. Mineral deposits in some tropical woody plants, *Ann. Botany* 50:559.

Tousimis, A. J., and Adler, I., 1963. Electron probe X-ray microanalyzer study of copper within Descemet's membrane of Wilson's disease, *J. Histochem. Cytochem.* 11:40.

Tretyl, W. J., Orenberg, J. B., Marich, K. W., Saffir, A. J., and Glick, D., 1972. Detection limits in analysis of metals in biological materials by laser microprobe optical emission spectrometry, *Anal. Chem.* 44:1903.

Valkovic, V., 1980. *Analysis of Biological Material for Trace Elements Using X-ray Spectroscopy,* CRC Press, Boca Raton, Florida.

Verbueken, A. H., Van Grieken, R. E., Paulus, G. J., and De Bruijn, W. C., 1984. Embedded ion exchange beads as standards for laser microprobe mass analysis of biological specimens, *Anal. Chem.* 56:1362.

Wan, K. K., Boegman, R. J., and Barnett, R. I., 1982. Biochemical and morphological characteristics of calcium uptake by denervated skeletal muscle, *Exp. Neurol.* 78:205.

Watt, F., Grime, G. W., Takacs, J., and Vaux, D. J. T., 1984. Oxford scanning proton microprobe; A medical diagnostic application, *Gov. Rep. Announce. Index (U.S.)* 84:59.

Neutron Activation Analysis 12

12.1 Introduction

The discovery of the neutron by Chadwick in 1932 is a relatively recent event in the history of science (see Heller, 1976). Soon after that discovery, neturons had been used to activate elements, and Bowen (1956) showed that neutron activation could be a highly sensitive method of analysis.

Neutron activation analysis is a method for the determination of trace elements, and it is based on thermal neutron-induced synthesis of radio-active isotopes (radionuclides or radioisotopes) from stable nuclides. These products are then measured by γ-ray spectroscopy (or a liquid scintillation counter in some cases). The method has been made possible by the availability of intense neutron sources (e.g., reactors) working in conjunction with efficient photon and particle counters. As used here, the term "efficient" means ability to detect (i.e., count) most of the nuclear decompositions taking place within a sample. An activated isotope emits a measurable and characteristic radiation, and the half-lives of the isotopes and their gamma or particle radiation characteristics allow for identification and quantitation.

More specifically, the fission process in nuclear reactors gives off neutrons; those which have been slowed down by graphite or other material to an energy of about 0.025 eV are termed thermal neutrons (Olson et al., 1973). Thus, the sample is exposed to a thermal neutron flux of about 10^{11} to 10^{13} neutrons cm^{-2} s^{-1} and for a period on the order of several half-lives of the isotope in question (if the isotope is short-lived). Nuclei of the bombarded nuclides are activated and become radioactive by their capture of thermal neutrons, mostly via the (n,γ) reaction. The mass of the nuclide increases by 1, and γ-ray emission results from nuclear stabilization. Gamma rays emitted from the activated sample possess certain energies which depend on the kind of nuclear products present and thus on the kind of elements which were present beforehand. As stated above, this is a selective method for elemental identification and quantitation.

After activation, the sample is removed from the reactor and counted. A specific isotope produced from any given element will have an amount of activity defined by the equation:

$$A = N\sigma f(1 - e^{-\lambda t}) \qquad (12\text{-}1)$$

where A is the activity in disintegrations s^{-1}; N is the number of atoms of the specific isotope activated; σ is interaction probability in cm^2; f is the neutron flux in neutrons $cm^{-2}\ s^{-1}$; λ is the decay constant of the radionuclide formed, in s^{-1}; and t is the irradiation time, in s (see DeSoete *et al.*, 1972).

The contained sample weight of the element in question is easily determined by the following indirect but precise procedure: A standard of a known amount of the pure element(s) in question is irradiated along with the unknown for the same time interval and under the same condition of neutron flux (i.e., with the sample containers in close proximity). Then the radioactivity of each sample is likewise measured under the same conditions. A simple direct proportionality expression may then be used for a determination of the amount of the element in the unknown:

$$W_u = (W_s \times C_u)/C_s \qquad (12\text{-}2)$$

where W_u and W_s are the weights of the unknown and standard, respectively; C_u and C_s are the net counting rates (counts per second) for the unknown and the standard.

As with many modern methods of analysis of trace elements, stringent controls must be in effect in order to prevent extraneous contamination of the sample before neutron irradiation. Other factors to consider include chemical removal of abundant elements from the sample, such as sodium, chlorine (as chloride), and phosphorus (usually as phosphate ion), which are easily activated and may obscure the element in question. The term *radiochemical analysis* implies chemical isolation of the element in question (either before or after activation), whereas an *instrument analysis* depends on well-resolved gamma spectra. Both approaches have disadvantages, and they are often used in combination. Spectral obscuration of the type described above may compromise a direct instrument approach, and it is not always a simple matter to achieve complete quantitation in the isolation of a trace quantity of an element.

12.2 Applications and Examples of Neutron Activation

A number of recent reviews have been written on neutron activation analysis as it relates to a very wide range of endeavors, from forensic chemistry, to soil science, to biochemistry. Leddicotte (1971) has given a thorough article on the activation analysis of biological trace elements. Additional articles, which include material on the procedures and applications of the method, have been written by Kay *et al.* (1973), Kramer and Wahl (1968), and Hoste *et al.* (1971). A useful book consisting entirely of tables has been compiled by Erdtmann (1976); necessary irradiation times, half-lives of the products, and decays per second per microgram of the activated elements are given in the latter useful reference. There are two recent reviews of biological and clinical trace element research involving neutron activation analysis (Braetter, 1983; Heydorn, 1984). These give additional details on sampling procedures and treat the topic of trace elements in human disease.

Manganese may be determined by neutron activation *after* the isolation of this element from blood (Papavasiliou and Cotzias, 1961). Manganese is oxidized to MnO_4^- using H_2O_2, then precipitated with tetraphenylarsonium ion. The precipitate is collected on filter paper, activated, and counted. Figure 12-1 illustrates how the data appear with and without the purification step. Clearly, the fractionation of manganese is necessary

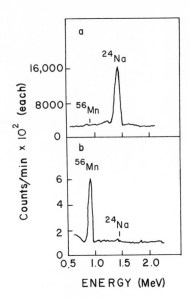

Figure 12-1. The γ-ray spectrum of neutron-activated whole blood (100 μl): (*a*) without purification of manganese; (*b*) after purification of manganese by oxidation and precipitation (see the text). Adapted from Papavasiliou and Cotzias, 1961.

for its detection: otherwise, intense activity from the much larger amount of sodium present overshadows the weaker ^{56}Mn activity.

Blood may also be analyzed using a combination of controlled potential electrolysis and neutron activation analysis (Jorstad *et al.*, 1981). The serum fraction is freeze-dried, neutron-irradiated, and electrolyzed. Radioactive species which deposit on a mercury cathode (and which are determined simultaneously by instrumet resolution) are derived from the elements Ag, Au, Cd, Co, Fe, Hg, Sb, and Zn. Other elements remaining in solution (Br, Ca, Cs, Na, Rb, and Se) may be determined separately.

Cerebrospinal fluid has been analyzed for its manganese and copper content (Kanabrocki *et al.*, 1964a). Eighty-eight normal human adults gave a value of 0.22 ± 0.15 μg % for Mn and 19.6 ± 8.1 μg % for Cu. These were the values for the nondialyzable elements; therefore, it may be presumed that these amounts were bound to protein or other macromolecules. The same workers presented further data on cerebrospinal fluid, using neutron activation analysis (Kanabrocki *et al.*, 1964b). Spectra obtained for copper, manganese (and sodium) are shown in Figure 12-2.

Iodoamino acids and thyroid hormones in urine have been detected and quantitated at the nanogram level, using a combination of high-performance liquid chromatography (HPLC) and neutron activation (Firouzbakht *et al.*, 1981). Sensitivity and specificity for ^{128}I activity is an advantage in studies of iodine.

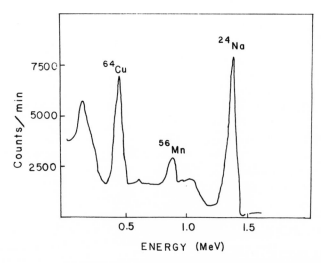

Figure 12-2. The γ-ray spectrum of cerebrospinal fluid, about one hour after a 30-min irradiation. The manganese and copper are presumably bound to macromolecules, as the sample had been dialyzed before analysis. Adapted from Kanabrocki *et al.*, 1964b.

A neutron activation analysis for the determination of gold and platinum in cancerous tissues of patients treated with cisplatin has been described (Tjioe et al., 1984). Tissues from healthy individuals were used as controls. The radiochemical separation is based on the selective removal of radioactive Au and Pt as small metallic nuggets after the reactions ^{197}Au (n,γ) ^{198}Au and ^{198}Pt (n,γ) ^{199}Pt $\rightarrow$ ^{199}Au.

In a thorough study on the structure of nitrate reductase from E. coli, neutron activation analysis yielded a value of 3.2 ± 0.5 mol Mo per mol of enzyme (MacGregor et al., 1974). These data correlated with an overall model of the enzyme having four large subunits, four small subunits, and four atoms of molybdenum per molecule of protein (nitrate reductase).

Molybdenum may be solvent extracted from seawater using pyrrolidone dithiocarbamate and diethyldithiocarbamate at pH 1.4 as the chelating agents and chloroform as the solvent (Mok and Wai, 1984). The recovery of molybdenum was quantitative. Uranium was not extracted at this pH value, eliminating interference from the ^{235}U $(n,\text{fission})$ ^{99}Mo reaction. The extraction process also precludes interference from sodium and bromine. This method should be adaptable to biological materials.

Iron, zinc, selenium, rubidium, and cesium were determined simultaneously in either serum or packed blood cells (Versieck et al., 1977). Precautions against activating any contaminations (which might be introduced in the samples before neutron activation) were also described in this report. Blood samples were drawn with a plastic cannula trocar and collected in high-purity quartz tubes. Careful cleaning procedures and dust-free conditions are essential requirements in an activation analysis, and the cited case is a good example of carefully controlled measurements.

The same laboratory also reported on the determination of chromium and cobalt in human serum (Versieck et al., 1978). After irradiation, cobalt could be measured directly via the 1173.1-keV or 1332.4-keV gamma peak of ^{60}Co. Chromium could not be determined directly because ^{32}P interferes with the 320.0-keV gamma peak of ^{51}Cr. Therefore, chromium was heated with perchloric acid and nitric acid and then distilled as CrO_2Cl_2. The recovery of chromium was quantitative while cobalt and phosphate remained in the residue.

The work of Versieck et al. (1978) also illustrates another principle involved in neutron activation analysis. The fast neutron flux of the reactor caused fast neutron-induced reactions of iron present in the samples, also leading to ^{51}Cr, i.e., by ^{54}Fe (n,α) ^{51}Cr. However, interference with the ^{50}Cr (n,γ) ^{51}Cr process was not significant, as the fast neutron activation cross section is negligible compared to the thermal cross section (see Table 12-1). The possibility of activation from a different element must not be overlooked.

Table 12-1. Nuclear Data for the Activation of Chromium, Cobalt,
and Iron in Serum[a]

Nuclear reaction	Cross sections, thermal or fast (barns)	Half-life ($t_{1/2}$)	Photopeaks	
			keV	Intensity (%)
^{50}Cr (n,γ) ^{51}Cr	16	27.8 days	320.0	9.0
^{59}Co (n,γ) ^{60}Co	37	5.25 years	1173.1	99.9
^{54}Fe (n,α) ^{51}Cr	0.00074	27.8 days	320.0	9.0

[a]Adapted from Versieck et al., 1978.

Environmental material (orchard leaves, "bovine liver," and sub-bituminous coal) were simultaneously analyzed for As, Sb, Cd, Cr, Cu, and Se (Gallorini et al., 1978). For purification of the samples, an inorganic ion exchanger of hydrated manganese dioxide was used for the retention and subsequent determination of As, Cr, Sb, and Se; solvent extraction with diethyldithiocarbamate compounds was used for Cd and Cu. The neutron activation method proved to be applicable to markedly different types of samples.

Another type of ion exchange resin—Srafion NMRR—has been shown to be effective in radiochemical separations. Nadkarni and Morrison (1978) used this purification procedure to separate molybdenum after neutron activation of steels, and in geological, biological, and environmental samples and standard reference materials.

Selenium and other elements have been determined in platelets using neutron activation analysis (Kasperek et al., 1979). The selenium content of platelets (in ng/g) was found to be at least twice that of literature values for erythrocytes and spleen, heart, and pancreas tissues, and comparable to the amount in liver. The enzyme glutathione peroxidase contains and requires selenium for catalysis. When there is a selenium deficiency, the activity of this enzyme is lowered. Kasperek postulates that glutathione peroxidase activity would be critical for the life of the platelet, and decreased selenium would lead to the aggregation of platelets, causing increased thrombi. In other words, the known influence of selenium on cardiovascular disease may be due to increased disruption of platelets when total amount of selenium is too low.

A method for the preparation of platelets for trace element analysis by neutron activation has been presented. The technique includes a protocol for avoiding contamination (Iyengar et al., 1979; Kiem et al., 1979).

Selenium in other types of samples—such as lung, eye lens, enzymes,

ACTIVATION ANALYSIS SENSITIVITIES

Sensitivities are expressed as the micrograms of the element that must be in the sample to be detected and determined by the Activation Analysis Service. THE SENSITIVITIES OF MANY ELEMENTS CAN BE INCREASED UP TO 100-FOLD. Sensitivities are for interference-free conditions. * For more information, contact Lawrence E. Kovar.

* Interference-free implies that only the element of interest will become radio-active in the sample. If interferences exist, radiochemistry may be necessary for optimum sensitivity.

1	2	3	4	5	6	7	8	9	10	11	12	13	14	15	16	17	18
1 H NA																	2 He NA
3 Li NA	4 Be NA											5 B NA	6 C NA	7 N NA	8 O NA	9 F 0.4	10 Ne 2.
11 Na 0.004	12 Mg 0.5											13 Al 0.004	14 Si 1.fs	15 P NA	16 S NA	17 Cl 0.05	18 Ar 0.002
19 K 0.2	20 Ca 4.	21 Sc 0.001	22 Ti 0.1	23 V 0.002	24 Cr 0.3	25 Mn 0.0001	26 Fe 2.fs	27 Co 0.01	28 Ni 0.7	29 Cu 0.002	30 Zn 0.1	31 Ga 0.002	32 Ge 0.1	33 As 0.005	34 Se 0.01	35 Br 0.003	36 Kr 0.01
37 Rb 0.02	38 Sr 0.005	39 Y 0.4	40 Zr 0.8	41 Nb 3.	42 Mo 0.1	43 Tc NA	44 Ru 0.04	45 Rh 0.005	46 Pd 0.03	47 Ag 0.004	48 Cd 0.005	49 In 0.00006	50 Sn 0.03	51 Sb 0.007	52 Te 0.03	53 I 0.002	54 Xe 0.1
55 Cs 0.001	56 Ba 0.02	57 La 0.005	72 Hf 0.0006	73 Ta 0.1	74 W 0.004	75 Re 0.0008	76 Os 1.	77 Ir 0.0003	78 Pt 0.1	79 Au 0.0005	80 Hg 0.003	81 Tl NA	82 Pb NA	83 Bi NA	84 Po NA	85 At NA	86 Rn NA
87 Fr NA	88 Ra NA	89 Ac NA	104 (Rf) NA	105 (Ha) NA	106												

58 Ce 0.2	59 Pr 0.03	60 Nd 0.03	61 Pm NA	62 Sm 0.001	63 Eu 0.0001	64 Gd 0.007	65 Tb 0.03	66 Dy 0.00003	67 Ho 0.003	68 Er 0.002	69 Tm 0.2	70 Yb 0.02	71 Lu 0.0003
90 Th 0.2	91 Pa NA	92 U 0.003	93 Np NA	94 Pu NA	95 Am NA	96 Cm NA	97 Bk NA	98 Cf NA	99 Es NA	100 Fm NA	101 Md NA	102 No NA	103 (Lr) NA

fs – FAST NEUTRONS, FISSION SPECTRUM
p – REACTOR PULSE
NA – ANALYSIS NOT NORMALLY PERFORMED

KEY
ATOMIC NUMBER 33 As ← SYMBOL
0.005 ← SENSITIVITY IN MICROGRAMS (INTERFERENCE FREE)

COUNTING BY GAMMA–RAY SPECTROMETRY, UNLESS OTHERWISE STATED

Figure 12-3. Periodic table showing neutron activation analysis sensitivities. Courtesy of General Activation Analysis, Inc.

and tobacco, either with or without destruction—has also been determined by neutron activation analysis (Olson *et al.*, 1973; Polkowska-Motrenko *et al.*, 1982).

From the foregoing, it should be evident that neutron activation analysis is an alternative to other element-specific detection and quantitation methods, such as atomic absorbance. It is clear from an inspection of Figure 12-3 that the method is especially appropriate for biological elements such as V, Cu, and especially Mn. Other examples of neutron activation analysis include: (1) determining trace metals such as Ag, Cd, Cu, and Mn after coprecipitation with lead phosphate (Holzbecher and Ryan, 1982); (2) determining Hg in environmental and biological samples after preconcentration with lead diethyldithiocarbamate (Lo *et al.*, 1982); (3) recording as many as seventeen trace elements in plants (Asubiojo *et al.*, 1982); and (4) determining levels of Ba, Sr, V, Cu, Zn, U, and Th in marine biological specimens and sediment samples prior to oil-drilling activities (DeLancey, 1982).

These methods may have been overlooked by many investigators simply as a result of not being much emphasized in the educational process (possibly also from a mood which rejects almost anything nuclear). Neutron activation analysis has been introduced as an instruction experiment at the college level. Pickering (1972) described an experiment which may be used as an introduction to the method and as a starting point for teaching radiochemical techniques in the freshman year. A wire containing sufficient ^{55}Mn is exposed to a neutron flux to produce ^{56}Mn, which has a half-life of 2.55 h. The experiment is considered to be relatively safe. Rengan (1978) presented "an elegant neutron activation analysis" for undergraduates; he suggests activating an alloy containing Al, Mn, Ti, and V.

References

Asubiojo, O., Guinn, V., and Okumaga, A., 1982. Multielement analysis of Nigerian chewing sticks by instrumental neutron activation analysis, *J. Radioanal. Chem.* 74:149.

Bowen, H. J. N., 1956. Neutron activation, *J. Nucl. Energy* 3:18.

Braetter, P., 1983. On the application of neutron activation analysis in the life sciences, *Radiochim. Acta* 34:85.

DeLancey, K., 1982. Radiochemical neutron activation analysis studies of biological samples in connection with marine production of petroleum and instrumental neutron activation analysis of sedimentary rocks related to uranium exploration, *Diss. Abstracts Int. B.* 43:417.

DeSoete, D., Gijbels, R., and Hoste, J., 1972. Neutron activation analysis, *Chemical Analysis Monograms* 34:140.

Erdtmann, G., 1976. *Neutron Activation Tables*, Weinheimverlag Chemie, New York.

Firouzbakht, M. L., Garmestani, S. K., Rack, E. P., and Blotcky, A. J., 1981. Determination of iodoamino acids and thyroid hormones in a urine matrix by neutron activation analysis, *Anal. Chem.* 53:1746.

Gallorini, M., Greenberg, R. R., and Gills, T. E., 1978. Simultaneous determination of arsenic, antimony, cadmium, chromium, copper, and selenium in environmental material by radiochemical neutron activation analysis, *Anal. Chem.* 50:1479.

Heller, R. A., 1976. Before neutrons, *J. Chem. Ed.* 53:714.

Heydorn, K., 1984. *Neutron Activation Analysis for Clinical Trace Element Research*, Vols. 1 and 2, CRC Press, Boca Raton, Florida.

Holzbecher, J., and Ryan, D. E., 1982. Determination of trace metals by neutron activation after coprecipitation with lead phosphate, *J. Radioanal. Chem.* 74:25.

Hoste, J., OpDeBeeck, J., Gijbels, R., Adams, F., VanDen Winkel, P., and DeSoete, D., 1971. *Instrumental and Radio Chemical Activation Analysis*, CRC Press, Cleveland, Ohio.

Iyengar, G. G., Borberg, H., Kasperek, K., Kiem, J., Siegers, M., Feinendegen, E., and Gross, R., 1979, Elemental composition of platelets. Part I. Sampling and sample preparation of platelets for trace-element analysis, *Clin. Chem.* 25:699.

Jorstad, K., Salbu, B., and Pappas, A. C., 1981. Multielement analysis of human blood serum by neutron activation and controlled potential electrolysis, *Anal. Chem.* 53:1398.

Kanabrocki, E. L., Case, L. F., Miller, E. B., Kaplan, E., and Oester, Y. T., 1964a. A study of human cerebrospinal fluid: copper and manganese, *J. Nucl. Med.* 5:643.

Kanabrocki, E. L., Fields, T., Decker, C. F., Case, L. F., Miller, E. B., Kaplan, E., and Oester, Y. T., 1964b. Neutron activation studies of biological fluids: manganese and copper, *Int. J. Appl. Radiat. Isotop.* 15:175.

Kasperek, K., Iyengar, G. V., Kiem, J., Borberg, H., and Feinendegen, L. E., 1979. Elemental composition of platelets. Part III. Determination of Ag, Au, Cd, Co, Cr, Cs, Mo, Rb, Sb, and Se in normal human platelets by neutron activation analysis, *Clin. Chem.* 25:711.

Kay, M. A., McKown, D. M., Gray, D. H., Eichor, M. E., and Vogt, J. R., 1973. Neutron activation analysis in environmental chemistry, *Am. Lab.* 5:39.

Kiem, J., Borgerg, H., Lyengar, B. V., Kasperek, K., Siegers, M., Feinendegen, L. E., and Gross, R., 1979. Elemental composition of platelets. Part II. Water content of normal human platelets and measurements of their concentrations of Cu, Fe, K, and Zn by neutron activation analysis, *Clin. Chem.* 25:705.

Kramer, H. H., and Wahl, W. H., 1968. Neutron activation analysis, in *Principles of Nuclear Medicine* (H. N. Wagner, ed.), W. B. Saunders Co., Philadelphia, pp. 811–832.

Leddicotte, G. W., 1971. Activation analysis of the biological trace elements, in *Methods of Biochemical Analysis*, Vol. 19 (D. Glick, ed.), Interscience Publishers, New York, pp. 345–434.

Lo, J. M., Wei, J. C., Yang, M. H., and Yeh, S. J., 1982. Preconcentration of mercury with lead diethyldithiocarbamate for neutron activation analysis of biological and environmental samples, *J. Radioanal. Chem.* 72:571.

MacGregor, C. H., Schnaitman, C. A., Normansell, D. E., and Hodgins, M. G., 1974. Purification and properties of nitrate reductase from *E. coli* K12, *J. Biol. Chem.* 249:5321.

Mok, W. M., and Wai, C. W., 1984. Preconcentration with dithiocarbamate extraction for determination of molybdenum in seawater by neutron activation analysis, *Anal. Chem.* 56:27.

Nadkarni, R. A., and Morrison, G. H., 1978. Determination of molybdenum by neutron activation and Srafion NMRR ion exchange resin separation, *Anal. Chem.* 50:294.

Olson, O. E., Palmer, I. S., and Whitehead, E. I., 1973. Determination of selenium in biological materials, in *Methods of Biochemical Analysis,* Vol. 21 (D. Glick, ed.), Interscience Publishers, New York, pp. 58–66.

Papavasiliou, P. S., and Cotzias, G. C., 1961. Neutron activation analysis: the determination of manganese, *J. Biol. Chem.* 236:2365.

Pickering, M., 1972. A freshman experiment in neutron activation analysis, *J. Chem. Ed.* 49:430.

Polkowska-Motrenko, H., Dermelj, M., Byrne, A. R., Fajgelij, A., Stegnar, P., and Kosta, L., 1982. Radiochemical neutron activation analysis of selenium using carbamate extraction, *Radiochem. Radioanal. Lett.* 53:319.

Rengan, K., 1978. An elegant neutron activation analysis; an undergraduate experiment, *J. Chem. Ed.* 55:203.

Tjioe, P. S., Volkers, K. J., Kroon, J. J., and De Goeij, J. J. M., 1984. Determination of gold and platinum traces in biological materials as a part of a multielement radiochemical activation analysis system, *Int. J. Environ. Anal. Chem.* 17:13.

Versieck, J., Hoste, J., Barbier, F., Michels, H., and De Rudder, J., 1977. Simultaneous determination of iron, zinc, selenium, rubidium, and cesium in serum and packed blood cells by neutron activation analysis, *Clin. Chem.* 23:1301.

Versieck, J., Hoste, J., Barbier, F., Steyaert, H., De Rudder, J., and Michels, H., 1978. Determination of chromium and cobalt in human serum by neutron activation analysis, *Clin. Chem.* 24:303.

Index

DATE DUE